PROGRESS IN OPTICS

VOLUME XXXVI

EDITORIAL ADVISORY BOARD

G. S. AGARWAL,	*Ahmedabad, India*
T. ASAKURA,	*Sapporo, Japan*
C. COHEN-TANNOUDJI,	*Paris, France*
V. L. GINZBURG,	*Moscow, Russia*
F. GORI,	*Rome, Italy*
A. KUJAWSKI,	*Warsaw, Poland*
J. PEŘINA,	*Olomouc, Czech Republic*
R. M. SILLITTO,	*Edinburgh, Scotland*
H. WALTHER,	*Garching, Germany*

PROGRESS IN OPTICS

VOLUME XXXVI

EDITED BY

E. WOLF

University of Rochester, N.Y., U.S.A.

Contributors

L.A. APRESYAN, M. BERTERO, M. BERTOLOTTI, I. BIALYNICKI-BIRULA,
V. CHUMASH, I. COJOCARU, C. DE MOL, E. FAZIO,
P. HARIHARAN, Yu.A. KRAVTSOV, F. MICHELOTTI, B.C. SANDERS

1996

ELSEVIER
AMSTERDAM · LAUSANNE · NEW YORK · OXFORD · SHANNON · TOKYO

ELSEVIER SCIENCE B.V.
SARA BURGERHARTSTRAAT 25
P.O. BOX 211
1000 AE AMSTERDAM
THE NETHERLANDS

Library of Congress Catalog Card Number: 61-19297
ISBN Volume XXXVI: 0 444 82530 4

© 1996 ELSEVIER SCIENCE B.V. All rights reserved.

No part of this publication may be reproduced, stored in a retrieval system, or transmitted, in any form or by any means, electronic, mechanical, photocopying, recording or otherwise, without the prior written permission of the publisher, Elsevier Science B.V., Rights & Permissions Department, P.O. Box 521, 1000 AM Amsterdam, The Netherlands.

Special regulations for readers in the USA: This publication has been registered with the Copyright Clearance Center Inc. (CCC), 222 Rosewood Drive, Danvers, MA 01923. Information can be obtained from the CCC about conditions under which photocopies of parts of this publication may be made in the USA.

All other copyright questions, including photocopying outside of the USA, should be referred to the Publisher, unless otherwise specified.

No responsibility is assumed by the publisher for any injury and/or damage to persons or property as a matter of products liability, negligence or otherwise, or from any use or operation of any methods, products, instructions or ideas contained in the material herein.

PRINTED ON ACID-FREE PAPER

PRINTED IN THE NETHERLANDS

PREFACE

This volume presents five review articles that cover a broad range of topics which are likely to be of interest to many scientists concerned with optics and related subjects.

The first article, by V. Chumash, I. Cojocaru, F. Fazio, F. Michelotti and M. Bertolotti, deals with nonlinear optical properties of chalcogenide glasses. These materials have many interesting structural properties some of which are useful for applications to integrated active optical devices. This article presents a review of experimental measurements of nonlinear absorption coefficients and nonlinear refractive indices of such materials. A review of various models formulated to explain their properties is also included.

The second article, by P. Hariharan and B.C. Sanders, presents a review of quantum effects in optical interferometry. After a brief introduction concerning sources of nonclassical light, second- and fourth-order interference, the geometric phase, two-photon interferometry, complementarity and quantum limits are discussed. Experiments are also reviewed involving the generation of pairs of photons in entangled states, which are used to investigate some puzzling features of quantum mechanics, including tests of Bell's inequalities, quantum erasers and single-photon tunneling.

The article which follows, by M. Bertero and C. De Mol, reviews researches on super-resolution, i.e. the possibility of overcoming the classical diffraction limit of about half a wavelength. The problem is shown to be essentially equivalent to extrapolating the spatial frequency spectrum of the object beyond the spectral band of the optical system. It is demonstrated that in the presence of noise significant super-resolution can be achieved when the linear dimensions of the object are comparable with the resolution limit of the system. Some practical applications are also considered, particularly in the field of confocal scanning microscopy and in connection with inverse diffraction from far-field and near-field data.

The next article, by Yu.A. Kravtsov and L.A. Apresyan, is concerned with the theory of radiative energy transfer. The traditional theory is phenomenological, based largely on the intuitive concept of geometrical rays. More recently many attempts have been made to provide the theory with a sounder foundation. This article reviews such researches, which use the more modern techniques of wave theory, coherence theory and statistical physics. The article includes examples which demonstrate that the equation of radiative transfer may sometimes take diffraction into account, and discusses a number of effects which have been discovered relatively recently and which have a bearing on this subject, such as enhanced backscattering and the phenomenon of weak localization. Some nonlinear transport problems are also discussed.

The concluding article by I. Bialynicki-Birula deals with the somewhat elusive but potentially useful concept of the photon wave function. A review is presented of the century-old history of this subject. It is shown that the photon wave function bridges the gap between classical electromagnetic theory and quantum electrodynamics and it has a number of uses.

It is a pleasure to note that all the articles in this volume have been contributed by leading experts in the various fields.

Emil Wolf

Department of Physics and Astronomy
University of Rochester
Rochester, New York 14627, USA

October 1996

CONTENTS

I. NONLINEAR PROPAGATION OF STRONG LASER PULSES IN CHALCOGENIDE GLASS FILMS
by V. CHUMASH AND I. COJOCARU (CHISINAU, MOLDOVA),
E. FAZIO, F. MICHELOTTI AND M. BERTOLOTTI (ROME, ITALY)

§ 1. INTRODUCTION	3
§ 2. NONLINEAR TRANSMISSION OF CW LASER RADIATION THROUGH THIN ChG FILMS	4
§ 3. NONLINEAR ABSORPTION OF LASER PULSES IN ChG	11
§ 4. OPTICAL HYSTERESIS AND NONLINEAR ABSORPTION OF LASER PULSES IN ChG	19
§ 5. MODEL OF NONLINEAR LIGHT ABSORPTION INTO A HIGHLY EXCITED NONCRYSTALLINE SEMICONDUCTOR: PHENOMENOLOGICAL APPROACH	27
5.1. No participation of local acoustic phonons ($\gamma = 0$)	30
5.1.1. $n_a \approx n_a^0$; $n = n_0 \ll 1$	30
5.1.2. $n > n_0$	30
5.2. Participation of local acoustic phonons ($\gamma \neq 0$)	31
§ 6. REFRACTIVE INDEX CHANGES OF CHALCOGENIDE GLASSES	35
§ 7. CONCLUSIONS	43
ACKNOWLEDGEMENT	44
REFERENCES	44

II. QUANTUM PHENOMENA IN OPTICAL INTERFEROMETRY
by P. HARIHARAN (LINDFIELD, AUSTRALIA) AND B.C. SANDERS (NORTH RYDE, AUSTRALIA)

§ 1. INTRODUCTION	51
1.1. Quantum effects	51
1.2. Complementarity	51
1.3. Second-order coherence	52
1.4. Nonclassical states of light	52
1.5. Fourth-order coherence	53
1.6. Entangled states	53
1.7. Beam-splitting and tunneling	54
1.8. Quantum limits	54
§ 2. OPTICAL SOURCES	54
2.1. Quantum description of radiation	55
2.2. Independent sources	58

2.3. Two-atom sources	58
2.4. Sources of nonclassical light	59
2.5. Single-photon states	59
2.5.1. Atomic cascade	59
2.5.2. Parametric down-conversion	60
2.6. The beam splitter	60
2.7. Squeezed states of light	62
§ 3. SECOND-ORDER INTERFERENCE	65
3.1. Interference at the "single-photon" level	65
3.2. Interference with single-photon states	66
3.3. Interference with independent sources	67
3.4. Interference in the time domain	70
3.5. Superposition states	72
§ 4. THE GEOMETRIC PHASE	73
4.1. The geometric phase in optics	73
4.2. Observations at the single-photon level	74
4.3. Observations with single-photon states	76
§ 5. FOURTH-ORDER INTERFERENCE	78
5.1. Nonclassical fourth-order interference	79
5.2. Interference in separated interferometers	86
5.3. The geometric phase	89
5.4. Tests of quantum theory	90
§ 6. TWO-PHOTON INTERFEROMETRY	91
6.1. Entangled states and Bell's inequality	92
6.2. Interferometric tests of Bell's inequality	94
6.3. Other tests of local realism	99
6.4. Two-photon interference	100
6.5. Two-photon tests of Bell's inequality	106
§ 7. COMPLEMENTARITY	108
7.1. Quantum-nondemolition measurements	108
7.2. Delayed-choice experiments	109
7.3. The quantum eraser	110
7.4. Single-photon tunneling	112
7.4.1. Tunneling time	113
7.4.2. Dispersion cancellation	113
7.4.3. Measurements of tunneling time	114
7.5. Interaction-free measurements	115
§ 8. QUANTUM LIMITS TO INTERFEROMETRY	117
8.1. Number–phase uncertainty relation	117
8.2. The standard quantum limit	119
8.3. Interferometry below the SQL	120
8.4. Interferometers using active elements	121
§ 9. CONCLUSIONS	122
REFERENCES	123

III. SUPER-RESOLUTION BY DATA INVERSION
by M. Bertero (Genova, Italy) and C. De Mol (Brussels, Belgium)

§ 1. Introduction 131
§ 2. Resolution Limits and Bandwidth 134
    2.1. The Rayleigh resolution limit 134
    2.2. Band of an optical system and sampling theorems 135
    2.3. Deblurring and super-resolution 139
§ 3. Linear Inversion Methods and Filtering 142
    3.1. The overall impulse response 142
    3.2. Optimal filtering for convolution equations 144
    3.3. Filtered singular-system expansions for compact operators 149
§ 4. Out-of-Band Extrapolation 154
§ 5. Confocal Microscopy 162
§ 6. Inverse Diffraction and Near-Field Imaging 168
Acknowledgements 172
Appendix A 173
Appendix B 174
References 176

IV. RADIATIVE TRANSFER: NEW ASPECTS OF THE OLD THEORY
by Yu.A. Kravtsov and L.A. Apresyan (Moscow, Russian Federation)

§ 1. Introduction 181
    1.1. Classical radiative transfer theory 181
    1.2. RTE and basic equations of statistical-wave theory 183
    1.3. Diffraction content of radiative transport equation. Nonclassical radiometry . . . 186
    1.4. Purpose and content of the review 188
§ 2. Radiative Transfer in Free Space 188
    2.1. Radiometry and coherence 188
        2.1.1. Radiometric description of radiation in free space 189
        2.1.2. Radiance and the correlation function of the wave field 190
        2.1.3. Radiance as the spectrum of quasi-uniform fluctuations 191
        2.1.4. Wigner function as a local spectral density: advantages and disadvantages 193
        2.1.5. Quasi-uniform fields and their spectra 194
        2.1.6. Local quasi-uniform coherence function in the geometrical optics
            approximation 196
    2.2. Generalized radiance of plane sources 198
        2.2.1. Definition of generalized radiance 198
        2.2.2. Generalized radiance of plane sources and nonclassical radiometry . . . 199
    2.3. Wolf's red and blue shifts of spectral lines 199
§ 3. Phenomenological and Statistical-Wave Derivations of the Radiation Transfer
    Equation 200
    3.1. Radiation transfer equations in scattering media 200

		3.1.1. Phenomenological derivation	200
		3.1.2. Heuristic applicability conditions for RTE	201
	3.2.	Dyson and Bethe–Salpeter equations	203
		3.2.1. Discrete and continuous models of scattering media	203
		3.2.2. The Dyson equation and effective parameters of random media	204
§ 4.	New Aspects of the RTE	207
	4.1.	Transfer equations and the coherent field	207
	4.2.	Diffraction radiometry. The parabolic equation method	208
	4.3.	Correlation scattering cross sections	209
	4.4.	Limiting resolving power in radiometric measurements	211
§ 5.	New Application Fields of the RTE	212
	5.1.	Enhanced backscattering (weak localization)	212
	5.2.	Nonlinear RTE and strong localization	217
	5.3.	RTE with fluctuating parameters. The effect of translucence	220
	5.4.	Warming and cooling effects in scattering media	226
	5.5.	Thermal radio emission of rough surfaces	228
§ 6.	New Effects in Statistical Radiative Transfer	229
	6.1.	Scalar memory effect	229
	6.2.	Time reversed memory effect	232
	6.3.	Long correlation effects for intensities	232
	6.4.	Polarization effects	234
	6.5.	Correlation effects of imaging in diffuse scattering media	235
§ 7.	Conclusion	237
References	237

V. PHOTON WAVE FUNCTION
by I. Bialynicki-Birula (Warsaw, Poland and Rochester, NY, USA)

§ 1.	Introduction	248
	1.1.	Coordinate vs. momentum representation	249
	1.2.	Phase representation	250
	1.3.	Landau–Peierls wave function	251
	1.4.	Riemann–Silberstein wave function	252
§ 2.	Wave Equation for Photons	254
	2.1.	Wave equation for photons in free space	254
	2.2.	Wave equation for photons in a medium	256
	2.3.	Analogy with the Dirac equation	259
§ 3.	Photon Wave Function in Coordinate Representation	259
	3.1.	Photons have no antiparticles	260
	3.2.	Transformation properties of the photon wave function in coordinate representation	261
	3.3.	Photon Hamiltonian	262
§ 4.	Photon Wave Function in Momentum Representation	263
	4.1.	Photon wave function as a Fourier integral	264
	4.2.	Interpretation of Fourier coefficients	265

	4.3. Transformation properties of the photon wave function in momentum representation 267
§ 5.	PROBABILISTIC INTERPRETATION 267
	5.1. Scalar product 268
	5.2. Expectation values of physical quantities 269
	5.3. Connection with Landau–Peierls wave function 273
§ 6.	EIGENVALUE PROBLEMS FOR THE PHOTON WAVE FUNCTION 273
	6.1. Eigenvalue problems for momentum and angular momentum 274
	6.2. Eigenvalue problem for the moment of energy 274
	6.3. Photon propagation along an optical fiber as a quantum-mechanical bound state problem 276
§ 7.	RELATIVISTIC INVARIANCE OF PHOTON WAVE MECHANICS 278
§ 8.	LOCALIZABILITY OF PHOTONS 279
§ 9.	PHASE-SPACE DESCRIPTION OF A PHOTON 281
§ 10.	HYDRODYNAMIC FORMULATION 283
§ 11.	PHOTON WAVE FUNCTION IN NON-CARTESIAN COORDINATE SYSTEMS AND IN CURVED SPACE 285
§ 12.	PHOTON WAVE FUNCTION AS A SPINOR FIELD 286
§ 13.	PHOTON WAVE FUNCTIONS AND MODE EXPANSION OF THE ELECTROMAGNETIC FIELD . . 288
§ 14.	SUMMARY 290
ACKNOWLEDGMENTS 291
REFERENCES 292

AUTHOR INDEX 295
SUBJECT INDEX 305
CONTENTS OF PREVIOUS VOLUMES 309
CUMULATIVE INDEX 317

E. WOLF, PROGRESS IN OPTICS XXXVI
© 1996 ELSEVIER SCIENCE B.V.
ALL RIGHTS RESERVED

I

NONLINEAR PROPAGATION OF STRONG LASER PULSES IN CHALCOGENIDE GLASS FILMS

BY

V. Chumash and I. Cojocaru

Center of Optoelectronics, I.A.P., Academy of Sciences of Moldova, Academiei str. 1, Chisinau, 277028, Moldova

AND

E. Fazio, F. Michelotti and M. Bertolotti

Università degli Studi di Roma "La Sapienza", Dipartimento di Energetica, GNEQP of CNR and INFM, Via A. Scarpa 16, 00161 Roma, Italy

CONTENTS

	PAGE
§ 1. INTRODUCTION	3
§ 2. NONLINEAR TRANSMISSION OF CW LASER RADIATION THROUGH THIN ChG FILMS	4
§ 3. NONLINEAR ABSORPTION OF LASER PULSES IN ChG	11
§ 4. OPTICAL HYSTERESIS AND NONLINEAR ABSORPTION OF LASER PULSES IN ChG	19
§ 5. MODEL OF NONLINEAR LIGHT ABSORPTION INTO A HIGHLY EXCITED NONCRYSTALLINE SEMICONDUCTOR: PHENOMENOLOGICAL APPROACH	27
§ 6. REFRACTIVE INDEX CHANGES OF CHALCOGENIDE GLASSES	35
§ 7. CONCLUSIONS	43
ACKNOWLEDGEMENT	44
REFERENCES	44

§ 1. Introduction

Nonlinear optical effects in semiconductors are studied less frequently in the amorphous than in the crystalline state. The chalcogenide glass semiconductor (ChG) is a material that is highly resistive to crystallization. Irradiation of light could cause a change in its optical properties that is not related to the crystalline–amorphous type of transformation.

It is well known that the optical properties of ChG near the equilibrium state are primarily determined by the spectrum of the localized states in the band gap, and by their carrier concentration. Most investigations do not address the question as to how the process of carrier excitation by light takes place, however, and how these carriers relax later, especially from extended into localized states. An understanding of the relaxation processes is a basic problem in amorphous-state physics, since they represent the first step for the localization process, the basic property of ChG. Furthermore, the knowledge of the kinetics peculiarities of the electron–hole pair relaxation and localization permits the testing of different theoretical models and hypotheses (e.g., multiple carrier trapping or the assumption of high carrier mobility in extended states). Progress in clarifying these processes greatly depends on the possibility of a quantitative investigation of the spectrum of the elementary excitations in a wide energy region, with time resolution of the order of the characteristic time of the investigated elementary processes. These requests widen when the elementary excitations in a light field (external driving force) are investigated, including the nonlinear and nonstationary medium behavior in states far from the thermodynamic equilibrium.

Amorphous semiconductors, including ChG, are attractive candidates for the fabrication of all-optical passive and active devices. In recent years a variety of both passive (fibers, planar waveguides, lenses, gratings) and active (nonlinear devices mainly based on Fabry–Perot interference, optical bistability and optical hysteresis) elements have been demonstrated (Andriesh, Bykovskii, Kolomeiko, Makovkin, Smirnov and Shmal'ko [1977], Andriesh, Bykovskii, Smirnov, Cernii and Shmal'ko [1978], Suhara, Handa, Nishihara and Koyama [1982], Hajto and Janossy [1983], Andriesh, Enaki, Cojocaru, Ostafeichuk, Cerbari and Chumash [1988], Haro-Poniatowski, Fernandez Guasti, Mendez and Balkanski [1989]),

Nasu, Kubodera, Kobayashi, Nakamura and Kamiya [1990], Heo, Sanghera and Mackenzie [1991], Bertolotti, Chumash, Fazio, Ferrari and Sibilia [1991], Andriesh, Chumash, Cojocaru and Enaki [1991], Asobe, Suzuki, Kanamori and Kubodera [1992], Bertolotti, Chumash, Fazio, Ferrari and Sibilia [1993], Chumash, Cojocaru, Bostan, Cerbari and Andriesh [1994].

The aim of this work is to illustrate the present state of knowledge and some unresolved problems of the nonlinear interaction of a strong laser radiation with ChG. The primary focus is nonlinear phenomena that are characteristic of the amorphous semiconductors and that, as a rule, have no analog in crystalline phase. Permanent photoinduced effects that result from laser excitation are not considered here (i.e., effects that remain in ChG after irradiation) because of previous work (see, e.g., Kastner [1985], Elliot [1986], Tanaka [1990] and the references therein).

§ 2 examines the peculiarities of the nonlinear transmission of CW laser radiation through thin ChG films. § 3 addresses nonlinear absorption of laser pulses into ChG films, and § 4 discusses optical hysteresis and nonlinear interaction of short laser pulses with the ChG. A physical model, taking into account the light interaction with nonequilibrium phonons, is considered in § 5 in order to explain the experimental results. The results of the numerical calculations of the phenomenological equations are compared with the experimental data. § 6 examines results of the Z-Scan spectroscopy investigation of ChG thin films under interband and intraband CW and picosecond irradiation. Nonlinear refraction, nonlinear absorption, and permanent photostructural changes on ChG thin films, suitable for planar waveguiding structures, are reviewed and possible applications of ChG refractive index changes are discussed.

§ 2. Nonlinear Transmission of CW Laser Radiation through Thin ChG Films

Several researchers studied some of the light-induced reversible changes of the ChG optical constants to examine the nonlinear transmission of focused CW laser radiation through thin film samples.

Toth, Hajto and Zentai [1977] observed a nonlinear change of the light transmission, when $GeSe_2$ and AsSe films were irradiated with a focused Ar-laser beam. With the purpose of excluding the contribution of irreversible changes of the optical parameters, due to photostructural changes, the samples were "stabilized" in the laser beam in advance or were annealed. The properties of the reversible nonlinear change of the ChG sample transparency depend on the laser input intensity.

The light transmission of ChG samples decreases (laser radiation with intensity from 2.5 W/cm² up to 15 W/cm²) with the irradiation time (t) according to a relation close to t^{-1}, whereas the value of the relative transmission change is almost proportional to the intensity of the exciting light. For example, when the Ar-laser exciting intensity (with wavelengths $\lambda_{1ex} = 4880$ Å and $\lambda_{2ex} = 5145$ Å) increased from 0 to 15 W/cm², the light transmission of an AsSe film (with $E_g = 1.86$ eV and thickness $d = 1.85$ μm) decreased 5 times, whereas the transmission of a GeSe₂ film ($E_g = 2.1$ eV, $d = 6.4$ μm) decreased 2.4 times. After switching off the exciting light the ChG film transmission recovers its starting value exponentially. The transmission changes of the ChG films occur with characteristic times of several seconds and no fast components are revealed (Toth, Hajto and Zentai [1977]). It is worth noting that, as a result of the ChG photostructural changes, the studied films are bleached (GeSe₂) or darkened (AsSe); in contrast, in an intensive CW laser radiation their transmission always decreases.

Some studies (Hajto, Zentai and Kosa Somogyi [1977], Hajto and Janossy [1983]) reported that with a CW-focused He–Ne laser ($\lambda = 632.8$ nm, with a fixed radiation intensity), the photocurrent, transmission, and reflection coefficients of GeSe₂ films (deposited on glass substrates or self-supported in the air) show periodic oscillations in time. The material returns to its initial transparent state if the laser is switched off. The interaction of the laser radiation with the air self-supported GeSe₂ thin films takes place at a considerably smaller light intensity (~40–50 W/cm²), compared with the films on glass substrates (~2 kW/cm²). The frequency (3–50 Hz) and amplitude of the light transmission oscillations noticeably depend on the incident radiation intensity: an increase in the laser radiation intensity is followed by an increase in the amplitude of the transmission oscillations and by a decrease of their frequency (fig. 1). It should be noted that the transmission oscillations are observed in strictly limited ranges of the laser intensities (from 1.39 kW/cm² up to 2.65 kW/cm² for the GeSe₂ film on the glass substrate). Near the laser threshold intensity the detected oscillations are distinguished by a high stability, and after about 10^4 cycles of oscillations no change in the ChG structure or any sign of matter transport are detected. A logarithmic time dependence of the amplitude and frequency of the oscillations was reported by Hajto, Janossy and Choi [1985]. It was not possible to find an oscillation regime of the light transmission in crystalline GeSe₂, indicating that the clue to understand the physical mechanism lying at its base is related to the amorphous nature of the ChG.

Another kind of nonlinear interaction of laser radiation with the ChG, which has been revealed as an optical bistable light transmission, was found for the

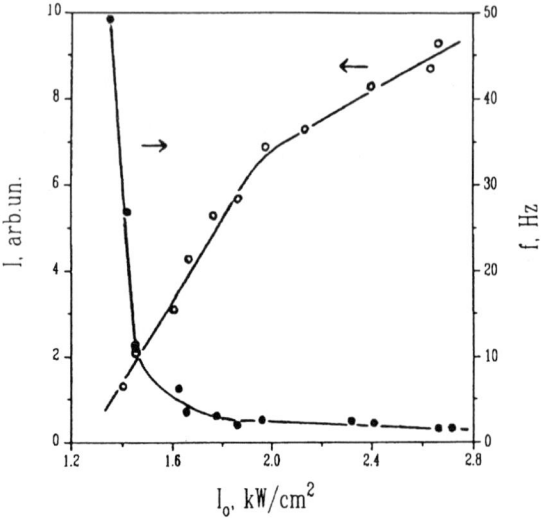

Fig. 1. Dependence of the light transmission (I) and oscillation frequency (f) on the input intensity (I_0) (Hajto, Zentai and Kosa Somogyi [1977]).

first time during the propagation of a focused helium–neon laser radiation (spot diameter <0.2 mm) through a self-supported $GeSe_2$ film (Hajto and Janossy [1983]). Changing the laser power (in the range from 12 mW to 16 mW, which corresponds to light intensities around 50 W/cm^2), some transmission discontinuities are noted as a function of the input light intensity (fig. 2). The jumps of the light transmission at the discontinuity points (i.e., the bleaching or darkening of the amorphous films) are observed at different light intensities depending on increasing or decreasing the input laser radiation intensity. As a result, without placing the ChG sample in any optical resonators, optical hysteresis of light transmission is obtained as a function of the input light intensity. This intrinsic optical bistable light transmission was also found in bulk $GeSe_2$ glass samples (Haro-Poniatowski, Fernandez Guasti, Mendez and Balkanski [1989]).

Despite considerable research (Toth, Hajto and Zentai [1977], Hajto, Zentai and Kosa Somogyi [1977], Gazso and Hajto [1978], Hajto [1980], Hajto and Apai [1980], Hajto and Fustoss-Wegner [1980], Fazekas [1981], Griffiths, Espinosa, Remeika and Philips [1982], Hajto and Janossy [1983], Hajto, Janossy and Firth [1983], Hajto, Janossy and Choi [1985], Xu, Cai and Xie [1988], Janossy [1988], Haro-Poniatowski, Fernandez Guasti, Mendez and Balkanski [1989], Hajto, Janossy, Choi and Owen [1989]), until now no reliable data are

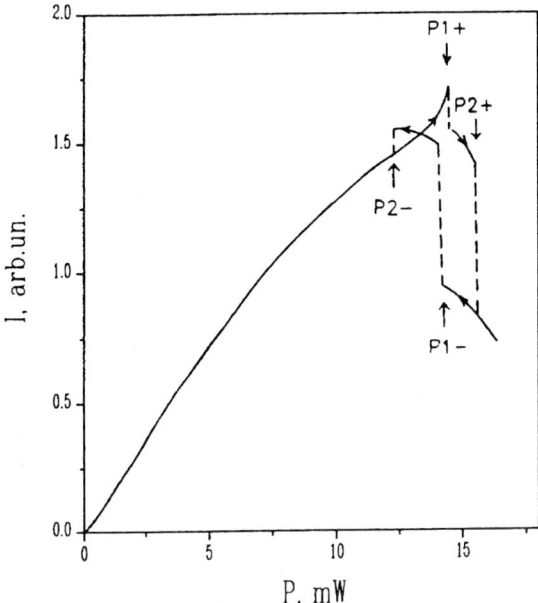

Fig. 2. Bistability behavior of the transmitted light intensity (I) in a-GeSe$_2$ film as a function of the incident laser power (P) (Hajto and Janossy [1983]).

available about the physical processes that adequately describe the nonlinear interactions of the CW laser radiation with noncrystalline semiconductors.

In the first experiments the kinetics of the transient reversible changes of AsSe and GeSe$_2$ light transmission were related to the laser heat-induced reversible structural changes of the medium (Toth, Hajto and Zentai [1977]). The authors established that thermal conductivity affects the kinetics of the light transmission variation. They assumed therefore that an increase of the sample temperature causes a red shift of the fundamental absorption edge of the amorphous semiconductors and, as a result, a decrease of the light transmission. This partly explains differences in the behavior of the self-supported GeSe$_2$ samples and those on the substrates. These authors established later (Hajto and Janossy [1983]), however, that the kinetics of the light transmission decrease possesses "a memory". The memory reveals itself from the finding that if, during the initial stage of the induced darkening, the laser light is switched off and the sample is kept in darkness, the initial light transmission value is not restored. When the laser radiation is switched on again, the ChG light transmission continues to decrease from the same value at which the radiation was switched off. This effect no longer exists if the light transmission change exceeds a critical value. The switching off of the radiation in such a case leads to the restoration of the initial transmission value, and the whole cycle of the sample photoinduced darkening in the laser field can be repeated. Thus the light-induced reversible darkening of the amorphous films in the CW laser field cannot be explained by a simple temperature increase in the illuminated spot (Hajto and Janossy [1983]). Moreover, the temperature oscillations accompanying the transmittance oscillations indicate a ~30 K rise in the dark phase. This could cause only a few percentage points change in the absorption coefficient.

The nature of the light-induced transmission oscillations was related by Hajto, Zentai and Kosa Somogyi [1977] to the possibility of repetitive crystallization of $GeSe_2$ from the melt in the laser beam. The unsatisfactory nature of such an explanation follows both from the simple estimation of the temperature rise, which is too small, and from the experimentally established facts that the amplitude and frequency of oscillations under the helium–neon laser treatment are very stable in time for more than 10^4 cycles of oscillations, material transport is absent, and no phase change occurs during the oscillations. Moreover, as Hajto [1980] and Gazso and Hajto [1978] noted it was impossible to achieve a regime of light transmission oscillations when changing the wavelength of the laser radiation from 6328 Å to 5145 Å or 4880 Å.

Another explanation of the ChG light transmission oscillations is suggested by Gazso and Hajto [1978]. Based on the assumption of the presence of different metastable and excited energy states in $GeSe_2$ films, which can appear under laser radiation, and explain the light absorption coefficient and refractive index changes in the illuminated spot. In addition, some thermal gradients (along the optical axis) and the play of recombination rates can lead to the appearance of a totally reflecting layer. Such a moving layer (from the front face) between the film boundaries can explain the time development of the oscillations. However, as Hajto and Apai [1980] mentioned later, this mechanism requires a simultaneous increase of the reflected light or of the light scattering during the period of transmittance decrease, a fact not revealed during the experiment.

The metastable absorption coefficient values (under the influence of an intensive laser radiation) changed more than twice (from 2.4×10^2 to 5.1×10^2 cm^{-1}) during the light transmission oscillations in $GeSe_2$ films. A decrease of the transmittance is accompanied by a simultaneous decrease of the reflectance and scattered light (independently from the scattering angle). The estimate of the refractive index change (from the measurement of the light transmission and reflection) showed that it reached its minimum value when the absorption coefficient was maximum, indicating the presence of an increase of the real light absorption during the transmission oscillations. On this basis Hajto [1980] and

Hajto and Apai [1980] proposed that the light transmission oscillations can be explained by the existence of light-induced metastable exciton states. Due to the strong electron–phonon coupling in the ChG, the formation of self-trapped excitons is possible. Thus the increased absorptivity during the oscillation period could be attributed to the appearance of self-trapped exciton states. The reverse process of absorption decrease could take place as a result of the thermo-release of the self-trapped excitons (due to the temperature increase of the laser irradiated region with a higher absorption coefficient). This assumption can easily be ascertained by a simultaneous registration of the photocurrent signal in the regime of light transmission oscillations. The oscillation of the light transmittance is accompanied by a simultaneous oscillation of the photocurrent signal (Hajto and Fustoss-Wegner [1980]). When the transmittance becomes lower (higher absorption), the photocurrent rapidly increases to a higher level and vice versa. Moreover, the self-trapped excitons should be metastable; that is, the material should not return to the stable-light state (within the switching time) if the laser is switched off. The increase of photocurrent when transmission decreases is clearly against the formation of excitons, which should subtract carriers from the current.

Attention was given to the importance of the interaction between carriers and lattice vibrations producing a local structural change in the glass network (a local change of the phase state) (Hajto and Fustoss-Wegner [1980]). At the same time it was assumed that in the glass a part of the atoms (or a group of atoms) might occur equally well in one of two equilibrium positions. The local bistable configurations will give two minima in the potential energy of the system as a function of some appropriate local atomic coordinates (Hajto and Fustoss-Wegner [1980]). We note that such an approach cannot explain the light intensity threshold for the ChG nonlinear transmission dependencies. The reverse mechanism, which explains the restoration of the glass in the initial bistable configuration, is also questionable.

To explain the oscillations of the light transmission through a $GeSe_2$ film, Fazekas [1981] suggested a new model of cooperative behavior (due to both Coulomb and laser-induced interactions) in a system of reasonably well-localized, charged, and neutral defects. It was assumed that the density of the defects was unusually high to ensure a sufficient Coulomb interaction between the intimate pairs. Such a microscopic model can qualitatively explain a sudden increase of the light absorption coefficient and of the photocurrent. However, no explanation is offered for the subsequent return of the medium to the initial, more transparent state.

A quantum-statistical, three-level cascade model for the investigation of the oscillation phenomenon in GeSe$_2$ films was proposed by Xu, Cai and Xie [1988] by considering a system of defects (D) immersed in an amorphous semiconductor. Three states were possible: neutral (D^0), negatively charged (D$^-$), and positively charged (D$^+$). It was supposed that when a laser beam passed through the film, the defect states would absorb photons and change their charge state. The density of the defects in GeSe$_2$ films could be estimated by some related experimental data. However, an exceedingly high density value (of the order of 10^{21} cm^{-3} in GeSe$_2$ films) was obtained (Xu, Cai and Xie [1988]).

Another explanation of the laser-induced transmission oscillations in GeSe$_2$ was proposed by Hajto, Janossy and Firth [1983]. Based on the assumption that the photostructural changes in these materials were much slower (some seconds) than the thermal effects. As a consequence, it was assumed that after the laser was switched on, a pseudothermal equilibrium (with a new corresponding light absorption coefficient in the irradiated region) was established much faster than the structural one. The temperature rise of the film followed the actual value of the absorption coefficient adiabatically. The main point to note is that this pseudoequilibrium is not necessarily stable against thermal fluctuations. Hajto, Janossy and Choi [1985] and Janossy [1988] assumed that the laser-induced transmission oscillations were conditioned by the instability of this pseudothermal equilibrium against the photostructural changes in GeSe$_2$, and as a result, the width of the optical band gap increased while the light absorption coefficient decreased. The influence of small changes in the refractive index (due to photostructural changes) together with the changes of the light absorption coefficient on the light transmission oscillation frequency and amplitude were analyzed by Hajto, Janossy, Choi and Owen [1989]. It is essential to note, however, that similar light transmission oscillations are found in AsSe, where, as a result of the photostructural transformations of the medium, the width of the optical band gap decreases and the feedback mechanism proposed by Hajto, Janossy and Firth [1983] is not justified.

As was pointed out by Griffiths, Espinosa, Remeika and Philips [1982] many aspects of the laser-induced light absorption oscillations in evaporated GeSe$_2$ films can be related to the reversible alterations of the molecular structure of the glass. Four distinct stages of ordering were observed by these authors. One new stage, called quasicrystallization, reverts to the original glassy structure when eliminating the laser flux. These new reversible Raman modes, which were experimentally observed in the spectrum of GeSe$_2$, when irradiated by a CW laser, can be regarded as the external driving forces that could produce not only the rearrangement or joining of the molecular clusters of the glassy matrix but, simultaneously, could change the light absorption mechanism (due to the strong carrier–phonon interaction in the glasses). Some aspects of this challenging problem will be discussed in § 5.

No less complicated, from the viewpoint of the physical interpretation, is the optical transmission bistability in $GeSe_2$ films (Hajto and Janossy [1983]). The critical laser powers for producing the optical discontinuities, the amplitudes of the optical discontinuities and of the optical hysteresis loops depend on experimental conditions, such as the sample temperature, laser spot diameter, and rate of the intensity increase. The phenomenological description of the optical bistability by the thermal model explains the discontinuities in the optical constants, the hysteresis, and the fact that the critical laser power densities depend on the laser beam diameter (Hajto and Janossy [1983]).

All the proposed physical models in the preceding works support neither a single qualitative or a quantitative description of the CW radiation nonlinear propagation in ChG thin films. Some papers propose combining simultaneously different physical mechanisms to give a qualitative description of the nonlinear phenomena in ChG, despite the lack of validity of using such mechanisms. Hajto and Janossy [1983], for example, assume that the nonlinear behavior of ChG thin films in a CW laser beam is conditioned by three mechanisms, that is, thermal effects, photostructural changes, and electronic nonlinearity. The electronic nonlinearity can explain the discontinuity in the optical constants (Fazekas [1981]), but cannot account for the memory effect and optical hysteresis. Models based on purely thermal processes explain the discontinuities, hysteresis phenomena, and the fact that the critical laser power densities depend on the laser beam diameter (Hajto, Zentai and Kosa Somogyi [1977], Hajto, Janossy and Firth [1983]). On the other hand, the memory effect and corresponding transmission oscillations cannot be explained within the limits of purely thermal processes.

We conclude that, at present, the physics of the nonlinear interaction processes of a CW laser radiation with ChG is far from being understood.

§ 3. Nonlinear Absorption of Laser Pulses in ChG

An understanding of the complex physical phenomena taking place in ChG, by setting them in an intensive laser field, largely depends on the possibility of their quantitative exploration with a sufficient short time resolution. This may occur during the use of short laser pulses and is often enhanced by the additional ability to perform quantitative spectroscopy over broad spectral ranges. Taking into account that the thermal conductivity of these glasses is of the order of $7 \times 10^{-3} \, J \, K^{-1} \, cm^{-1} \, s^{-1}$ (Stourac, Kolomiec and Silo [1968]), we consider only investigations where laser pulses with a duration shorter than 10^{-5} s were used. This section presents the results of several investigations in which the light nonlinear transmission in ChG samples under pulsed excitation were studied.

The nonlinear absorption of light pulses with a 25 ns duration from a ruby ($hv = 1.78$ eV) and a neodymium glass ($hv = 1.17$ eV) laser in amorphous and crystalline As_2S_3 was investigated by Krulikovskii, Lisitsa and Nasyrov [1977], Lisitsa, Nasyrov and Fekeshgazi [1977], Lisitsa, Nasyrov, Svechnikov and Fekeshgazi [1978], Nasyrov [1978], Nasyrov, Svechnikov and Fekeshgazi [1980], and Babinets, Vlasenko, Lisitsa, Mitsa, Pinzenik and Fekeshgazi [1988]. It was established that in samples with a thickness from 0.1 to 7 mm the nonlinear light absorption appears at laser pulse intensity higher than 1 MW/cm². These authors suggested that the nonlinear light absorption was conditioned by two-photon or two-step absorption. The coefficient of the two-photon absorption in a-As_2S_3 was estimated ($\beta \approx 0.14$ cm/MW for $hv = 1.78$ eV) from a fit of the experimental data. It is essential to note, however, that at this wavelength the coefficient of the linear absorption is $\alpha \approx 0.11$ cm^{-1}, and the suggestion of a two-quantum absorption mechanism is questionable. Moreover, the nonlinear absorption of the laser pulses with $hv < E_g$ depends on the local coordination of the atoms in ChG (Babinets, Vlasenko, Lisitsa, Mitsa, Pinzenik and Fekeshgazi [1988]). This fact seems to favor a linear or two-step optical transition rather than a two-photon one. Despite the interest of the involved physical mechanisms, these investigations also have practical importance, since the nonlinear light absorption may be the cause of the laser destruction of the ChG or of the film coating manufactured from them and used in high-power optics.

By reducing the laser pulse duration down to picoseconds or femtoseconds, it is possible to exclude or lower the influence of some processes, for example, the laser heating of the sample, and to investigate the mechanism of the nonlinear or nonstationary interaction of light with the ChG.

Optically induced transient absorption of laser pulses with subpicosecond duration (with quantum energy less than the optical band gap E_g) in a-As_2S_3 and c-As_2S_3 was reported for the first time by Fork, Shank, Glass, Migus, Bosch and Shah [1979]. The dynamics of this absorption with picosecond time resolution was investigated. Samples of amorphous and crystalline As_2S_3 were excited by strong subpicosecond pulses (0.5 ps) from a dye laser, operated at 0.61 μm wavelength and at 10 Hz repetition rate. The pulses were divided into two beams. The first beam induced the absorption into a-As_2S_3 and c-As_2S_3. The induced absorption was measured with the help of a subpicosecond continuum, which was used to probe the pump-excited sample during different time periods, both during and after excitation.

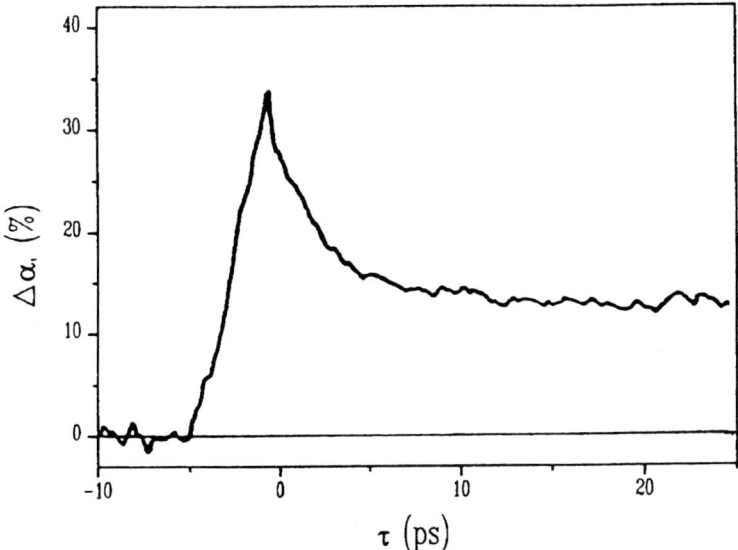

Fig. 3. Induced transient absorption ($\Delta\alpha$) versus probe light time delay (τ) in a-As$_2$Se$_3$ at 300 K (Fork, Shank, Glass, Migus, Bosch and Shah [1979]).

As a result of the a-As$_2$S$_3$ and c-As$_2$S$_3$ excitation with strong subpicosecond pulses (the intensity was higher than 1 GW/cm^2), an additional induced absorption was registered. It exhibited maximum amplitude during the excitation pulse and relaxed, in general, with several time constants (fig. 3). The induced absorption relaxed rather quickly in c-As$_2$S$_3$ (with a time constant about 14 ps, at 300 K). The fast components of the induced absorption relaxed in a-As$_2$S$_3$ in times shorter than 10^{-11} s, followed by a residual absorption with a relaxation time longer than 300 ps (but shorter than 0.1 s). A characteristic peculiarity of the induced transient absorption in a-As$_2$S$_3$ is the nonlinear dependence of the absorption coefficient from the light intensity, which shows a threshold-like dependence. Exciting a-As$_2$S$_3$ by strong pulses of subpicosecond duration, the light absorption coefficient increases more than four times. Moreover, in contrast with the spectrum of the metastable light absorption in a-As$_2$S$_3$ (induced by the one-photon CW argon laser excitation) the spectrum of the induced transient absorption possesses a weak dependence on the light wavelength in the spectral interval 0.65–1.5 μm.

The first investigations of the induced absorption relaxation in a-Si, a-Si:H, and As$_2$S$_{3-x}$Se$_x$ (0.25 < x < 0.75) with picosecond time resolution were reported by Ackley, Tauc and Paul [1979]. Experiments were conducted using a dye laser (similar to that used by Fork, Shank, Glass, Migus, Bosch and Shah [1979]), generating pulses of subpicosecond duration (1.5 kW peak power) in the 615 nm region, which was used for both the sample excitation and its probing. It should be noted that the a-Si, a-Si:H, and $As_2S_{3-x}Se_x$ samples were excited in nonequivalent conditions. In the cases of a-Si, a-Si:H the value of the exciting quantum energy considerably exceeded the optical band gap E_g, whereas for $As_2S_{3-x}Se_x$ a reverse condition took place. The induced absorption relaxation for the amorphous semiconductors in these two cases was found to be different (Ackley, Tauc and Paul [1979]). No induced transient light absorption was observed in As_2S_3, because the light absorption at the exciting wavelength was low and did not permit generation of a significant carrier concentration in the samples. The addition of Se to As_2S_3 leads to an increase of the light absorption linear coefficient in the 615 nm spectral region, and a long-living transient absorption appears in such conditions. It should be noted that the light absorption linear coefficient, was higher, i.e., the higher the laser radiation dose absorbed by the medium, the longer the relaxation time of the induced absorption. A fast relaxation component of the induced absorption, which is explained in Ackley, Tauc and Paul [1979] by the thermalization of hot carriers, is detected at small time delay values in all the ChG samples. The relaxation time decreases with the decrease of the sample temperature (at fixed wavelength of the exciting light). It may be related to the decrease of the light absorption linear coefficient due to the increase of the optical band gap, i.e., with the decrease in the exciting degree of the sample. In addition, the absorption coefficient changed from $10^{-2}\,cm^{-1}$ to $1\,cm^{-1}$, and this change did not depend on the ChG composition and temperature.

A mechanism based on the localization of the carriers on states in the gap and their energy redistribution is acknowledged by Fork, Shank, Glass, Migus, Bosch and Shah [1979] to explain the light-induced absorption in As_2S_3. Thus, the temporal and nonlinear behavior of a-As_2S_3 during its excitation by ultrashort light pulses was interpreted on the basis of a model in which carriers are excited due to two-photon absorption and then quickly relax to long-living states, which absorb at the probing light wavelength. Integrating the simultaneous nonlinear equations relating the pump and probe intensities, the authors determined the coefficients of the two- and three-photon absorption on the best least-squares fit with the experimental points.

An analogous model to explain the photoinduced absorption was proposed by Ackley, Tauc and Paul [1979]. Light with $hv > E_g$ excites electrons from the valence band in delocalized states of the conduction band with a subsequent relaxation to lower energy levels (S_1) or a capture on the localized states (S_2),

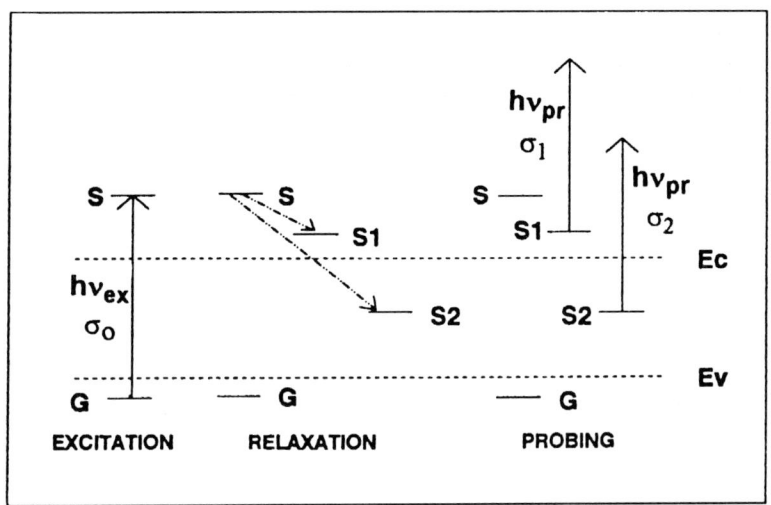

Fig. 4. Scheme of possible carrier transitions in ChG under the light field.

situated in the optical band gap of the amorphous semiconductor (fig. 4). From states S_1 and S_2 the carriers may be excited to higher states by the probing light (including the states of the same band, such transitions not being forbidden in amorphous semiconductors due to the nonfulfillment of the κ conservation rule). On the basis of their experimental results Ackley, Tauc and Paul [1979] concluded that the excited carriers thermalize in a time shorter than 1 ps. It was assumed that the thermalized (or localized) carriers, absorbing the light, are excited in the upper-lying energetic states of the same band, where the density of states is large. The subsequent absorption decrease was explained by a recombination of the carriers or their falling into deep traps where their absorption cross-section was smaller at the probe wavelength.

The difference discovered in the spectra of the transient and metastable absorption of the light with quantum energy $h\nu < E_g$ was explained by Fork, Shank, Glass, Migus, Bosch and Shah [1979] as a uniform carrier distribution on the localized states in the optical band gap of the amorphous semiconductors for the case of transient absorption and by a nonuniform distribution (where the deep lying states close to the middle of the optical band gap are mainly populated) in the case of metastable absorption.

Physical models that explain the transient light absorption in ChG, were proposed by Fork, Shank, Glass, Migus, Bosch and Shah [1979] and Ackley, Tauc and Paul [1979], and summarized by Vardeny [1983]. The scheme of such transitions is shown in fig. 4. The pumping light excites the amorphous semiconductor from the ground state (G) into the excited state (S), from which it relaxes into states S_1 or S_2. The probing beam measures the absorption in states S_1 or S_2 after a time delay t_d, which follows the excitation. If the absorption cross-sections σ_1 or σ_2 from the states S_1 or S_2 are higher than the absorption cross-section σ_0 from the state (G), it is possible to observe an induced absorption. If $\sigma_1, \sigma_2 < \sigma_0$, it is possible to observe a bleaching of the sample.

It is worth mentioning that the electron transitions proposed by Ackley, Tauc and Paul [1979] are possible. It is rather complicated to detect the absorption of the probing light (at the frequency of the exciting light), however, due to the previously noted transitions. This is contingent on the fact that the electron transition probability of the probing light field from the delocalized states or from the localized levels to higher levels of the conduction band is considerably smaller than the probability of the interband optical transitions. That is why the mechanism proposed by Ackley, Tauc and Paul [1979] cannot explain the light-induced absorption, at least when the amorphous semiconductor is irradiated by exciting (and probing) light with $h\nu \geqslant E_g$. In addition, the weak dependence of the induced absorption (effective cross-section) on the composition and temperature of the ChG sample is not explained by the proposed model. The differences in the dependencies of the induced absorption relaxation time in a-Si, a-Si:H, and $As_2S_{3-x}Se_x$ as opposed to the temperature are probably explained by different conditions of excitation rather than by different relaxation mechanisms, because the wavelength of the laser irradiation does not change. The change of sample temperature leads to a change of quantity $(h\nu - E_g)$, i.e., to a change of the linear light absorption coefficient and, as a result, to a different degree of the sample excitation.

The transient ChG light absorption is explained on the basis of the assumption of a cross-section change. However, with respect to the probability of the optical transition being determined not only by the light absorption cross-section but also by the population numbers of the initial and final states, the proposed physical models of Fork, Shank, Glass, Migus, Bosch and Shah [1979], Ackley, Tauc and Paul [1979], and Vardeny [1983] cannot explain unequivocally the measured characteristics. These investigations focus on the close resemblance between spectra of transient and metastable absorption, but there is a difference. In the picosecond case the transient absorption is absent at low intensities (lower than $1\,\text{GW/cm}^{-2}$).

More interesting investigations of the peculiarities of the picosecond and femtosecond spectroscopy of the amorphous semiconductors were performed primarily on a-Si and a-Si:H samples (see e.g., Hulin, Mourchid, Fauchet, Nighan and Vanderhaden [1991], Wraback, Tauc, Pang, Paul, Lee and Schiff [1991] and the references therein).

Recently, nonlinear light absorption of 100 femtosecond pulses in the case of interband and intraband excitation of As_2S_3 and As_2Se_3 thin films with the pump–probe technique was reported by Andriesh, Chumash, Cojocaru, Bertolotti, Fazio and Michelotti [1992], Fazio, Hulin, Chumash, Michelotti, Andriesh and Bertolotti [1993a], Andriesh, Chumash, Cojocaru, Bertolotti, Fazio, Michelotti and Hulin [1992], Fazio, Hulin, Chumash, Michelotti, Andriesh and Bertolotti [1993b]. Information about the laser system and experimental setup can be found in Fauchet, Gzara, Hulin, Tanguy, Mourchid and Antonetti [1988]. These materials were chosen because they showed similar band structures, just shifted in energy, to investigate interband and intraband excitation regimes with the same light energies. Different operation regimes (both pump and probe strongly interband absorbed; pump absorbed and probe not absorbed; both pump and probe not absorbed at intraband excitation) were investigated to reveal alteration of the nonlinear mechanisms due to both different excitation levels and different excitation frequencies. Different behavior of time evolution of induced nonlinear absorption in As_2S_3 and As_2Se_3 was revealed (fig. 5), that can be explained by different mechanisms of nonlinear interaction of strong light with semiconductor glasses.

The influence of photodarkening on the ultrafast dynamics of induced nonlinear absorption in As_2Se_3 was investigated by Andriesh, Chumash, Cojocaru, Bertolotti, Fazio and Michelotti [1992], Andriesh, Chumash, Cojocaru, Bertolotti, Fazio, Michelotti and Hulin [1992], Fazio, Hulin, Chumash, Michelotti, Andriesh and Bertolotti [1993a], and Fazio, Hulin, Chumash, Michelotti, Andriesh and Bertolotti [1993b]. Experimental measurements showed that the photodarkening did not change the relaxation speed but only the amplitude of the slow part of induced absorption.

Physical mechanisms that can contribute to nonlinear light absorption in amorphous semiconductors (including mechanisms proposed by Fork, Shank, Glass, Migus, Bosch and Shah [1979], Ackley, Tauc and Paul [1979], Vardeny [1983]) are discussed by Andriesh, Chumash, Cojocaru, Bertolotti, Fazio and Michelotti [1992], Andriesh, Chumash, Cojocaru, Bertolotti, Fazio, Michelotti and Hulin [1992], Fazio, Hulin, Chumash, Michelotti, Andriesh and Bertolotti [1993a], and Fazio, Hulin, Chumash, Michelotti, Andriesh and Bertolotti [1993b]). It was shown that in the case of interband transitions the mechanism of light absorption could explain the experimental peculiarities, taking into account the participation of nonequilibrium phonons. In the intraband excitation case, due to two-step and two-photon absorption of the exciting light, the

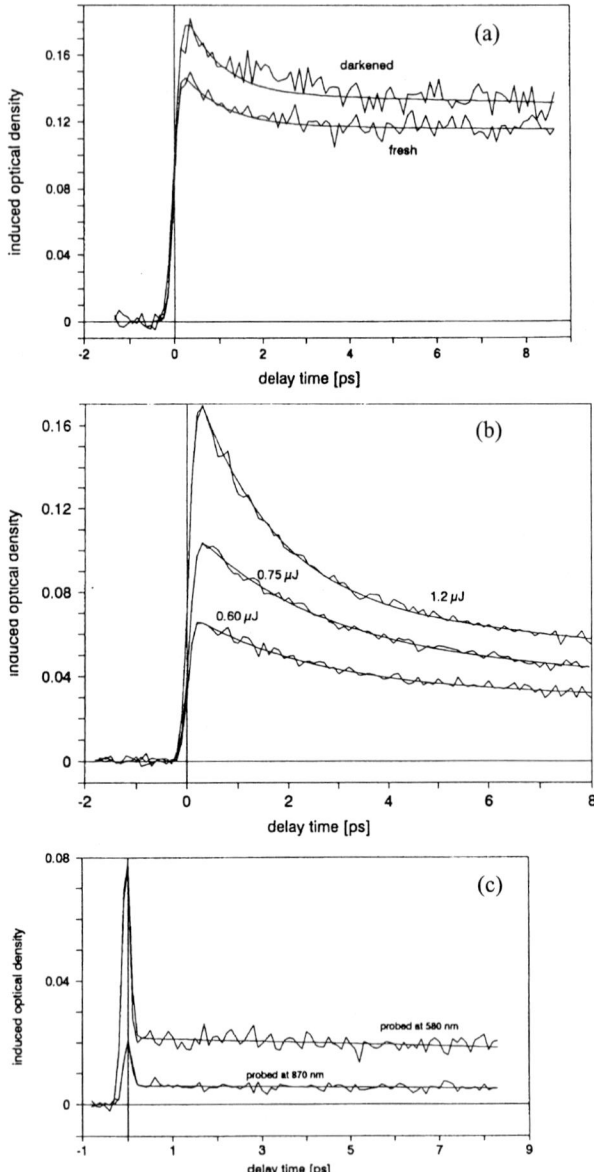

Fig. 5. (a) Time evolution of induced absorption in the case of As_2Se_3 for pumping at 2.0 eV and probing at 2.14 eV, for fresh and darkened samples. (b) Same as (a), for probe at 1.43 eV. Three measurements are reported for different pump energies: 1.2, 0.75 and 0.6 μJ. (c) Time evolution of induced absorption in the case of As_2S_3 for pumping at 2.0 eV and probing at 1.43 and 2.13 eV, for 100 fs pulses.

induced probe absorption is given both by two-step absorption by carriers in the delocalized states or localized on the traps and by nondegenerate two-photon absorption, obtained by the coupling of one pump and one probe photon (Andriesh, Chumash, Cojocaru, Bertolotti, Fazio and Michelotti [1992], Andriesh, Chumash, Cojocaru, Bertolotti, Fazio, Michelotti and Hulin [1992], Fazio, Hulin, Chumash, Michelotti, Andriesh and Bertolotti [1993a], Fazio, Hulin, Chumash, Michelotti, Andriesh and Bertolotti [1993b]).

The work of Mao and Yang [1982] occupies a separate place in investigations of nonlinear laser radiation interaction with ChG. It reports on optical absorption saturation in $AsSSe_xTe_{1-x}$. The transmission of $AsSSe_xTe_{1-x}$ bulk samples with a thickness of 1.5 mm increased in the field of strong laser pulses ($\lambda = 1.06\,\mu m$, duration 800 μs) from 1% to 30%. The author explained such a saturation of the optical absorption in chalcogenide glasses as a self-induced dynamic Burstein–Moss shift similar to the corresponding effect in crystalline semiconductors.

§ 4. Optical Hysteresis and Nonlinear Absorption of Laser Pulses in ChG

The interest in the nonlinear propagation characteristics of laser radiation in noncrystalline semiconductors is determined not only by the new fundamental physical mechanisms present in these materials, but also by a wide spectrum of possible applications in optoelectronic devices and photonic switching. Special attention is paid to the phenomena that can be observed at relatively low levels of light intensity, in relation to their possible use in active elements of integrated optics and optoelectronics. Optical bistability and optical hysteresis, are promising effects on the basis of which fast-response, all-optical switching and logical elements can be built (see, e.g., Gibbs [1985] and references therein, Bertolotti, Chumash, Fazio, Ferrari and Sibilia [1991], Bertolotti, Chumash, Fazio, Ferrari and Sibilia [1993]).

As mentioned in § 2, the bistable character of the light transmission in ChG was first registered in the propagation of focused helium–neon laser radiation through a-GeSe films. This section also describes the results of a systematic investigation of the optical hysteresis and nonlinear absorption at interband excitation of chalcogenide glass semiconductors As_2S_3, As_2Se_3, AsSe, $GeSe_2$, and $As_{22}S_{33}Ge_{45}$ with short laser pulses (300 K, 77 K), and discusses the physical mechanisms that can contribute to the nonlinear light absorption in ChG under pulsed excitation.

This section will summarize experimental work on thin film samples of chalcogenide glass semiconductors As_2S_3, As_2Se_3, AsSe, $GeSe_2$, and $As_{22}S_{33}Ge_{45}$

(0.2–5 μm thick), obtained by the vacuum thermal evaporation method on transparent substrates (polished K8 glass, mica, lavsan, etc.). When the input light (with $h\nu \geqslant E_g$) pulse intensity was relatively low, the transmission of the ChG films (As_2S_3, As_2Se_3, AsSe, $GeSe_2$, $As_{22}S_{33}Ge_{45}$, 0.2–5 μm thick) decreased with thickness according to the usual Beer law. However, increasing the incident light intensity over some threshold value (I_t) resulted in a nonlinear light transmission by the ChG films (Andriesh, Enaki, Cojocaru, Ostafeichuk, Cerbari and Chumash [1988]). The characteristic value of the threshold light intensity depends on the ChG film composition, excitation wavelength, temperature, and laser pulse duration. For example, for laser pulses with 7 ns duration at 570 nm the I_t values were equal; ~2 MW/cm^2 for As_2Se_3, ~3 MW/cm^2 for AsSe, and ~5 MW/cm^2 for $GeSe_2$; for laser pulses with microsecond duration, they were ~4 kW/cm^2 for As_2Se_3 and ~8 kW/cm^2 for AsSe (300 K), respectively.

As a result of the nonlinear light absorption, a change of the time profile of the laser pulses was recorded at the output of the amorphous semiconductors (Andriesh, Enaki, Cojocaru, Ostafeichuk, Cerbari and Chumash [1988], Andriesh, Enaki, Cojocaru, Ostafeichuk, Cerbari and Chumash [1989], Andriesh, Cojocaru, Ostafeichuk, Cerbari and Chumash [1989], Andriesh, Enaki, Cojocaru and Chumash [1990], Andriesh, Chumash and Cojocaru [1991], Andriesh, Bogdan, Enaki, Cojocaru and Chumash [1992]). The change of the time profile of a microsecond laser pulse at the output (I) of the As_2S_3 film is shown in fig. 6a (300 K). An analogous time profile change was registered for nanosecond pulses (Andriesh, Enaki, Cojocaru and Chumash [1990], Andriesh, Chumash and Cojocaru [1991], Andriesh, Bogdan, Enaki, Cojocaru and Chumash [1992]).

The kinetics of the induced darkening of ChG films in the field of laser pulses can be determined from the laser pulse oscillograms at the input and output of the sample. Figure 6b presents the curve of the transmission change (%) during the ChG film irradiation by a laser pulse of microsecond duration. It has been established that the type of ChG light transmission change depends on pulse amplitude, duration, and form (i.e., increasing steepness of intensity) of the input laser pulses.

The change of time profile of the laser pulses leads to hysteresis-like dependencies of the output light intensity (passed through the sample) on the corresponding value of the input (fig. 7). The extension of the optical hysteresis loop increases (changing its shape) with the input light intensity up to the damage of the sample surface. The loop of the optical hysteresis is inverted; i.e., it has a reverse direction, compared with the case of a Fabry–Perot resonator filled with

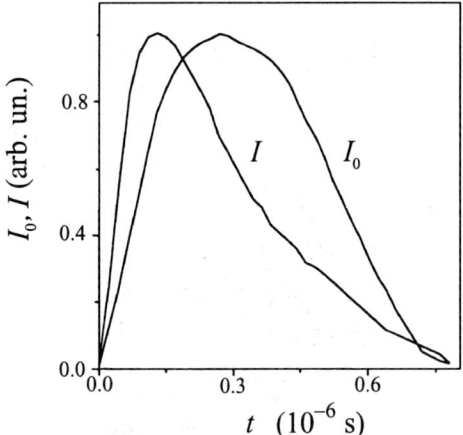

Fig. 6a. Oscillograms of incident (I_0) and transmitted pulses.

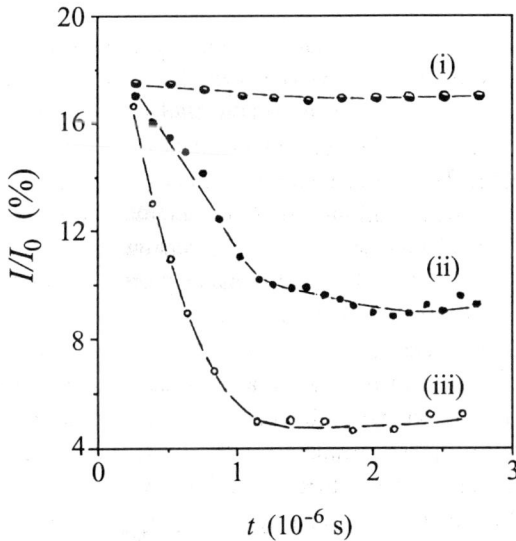

Fig. 6b. Kinetics of photodarkening of As_2Se_3 film (thickness 0.3 μm). Incident pulse intensity: (i) 5, (ii) 20 and (iii) 70 kW/cm^2. Exciting wavelength 605 nm, 300 K.

a saturating absorber, showing that with increases in the input light intensity the light absorption also increases (Gibbs [1985]).

The kinetics of the photoinduced darkening of ChG film in the field of laser pulses of microsecond and nanosecond durations was also measured by a low

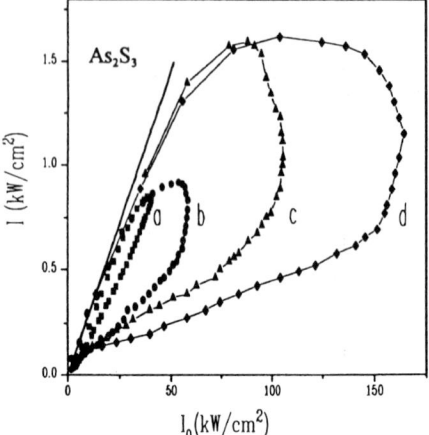

Fig. 7. Hysteresis dependence on light intensity transmitted through an As_2S_3 film. Pulse duration 0.4 µs, $h\nu = 2.57$ eV, 300 K. Incident peak pulse intensity: (a) 40, (b) 55, (c) 105 and (d) 160 kW/cm^2.

intensity probing light (CW radiation from a helium–neon laser or a pulse radiation of a microsecond laser). The wavelength of the probing light was the same as that of the exciting one. The results of the kinetics of the ChG photoinduced darkening in the fields of short laser pulses, investigated by these two methods, coincide. The characteristic times of establishing the new ChG light absorption quasiequilibrium state (corresponding to the horizontal part in fig. 6b) in the field of microsecond and nanosecond duration laser pulses were measured. They are less than 1 µs and 10 ns, respectively, and do not exceed the duration of the laser pulses.

The photoinduced increase of the ChG light absorption coefficient in the field of the laser pulses has a reversible character; that is, the medium fully restores to its initial transmission state after the ending of the laser pulse, and if the same place of the sample is irradiated with another laser pulse, the effect can be completely repeated. The restoration of the ChG initial absorption state was measured by a low intensity probing light at the exciting light wavelength. The time values after which the light transmission completely restored its initial value lie in the interval from several microseconds up to several dozens of microseconds, and depends on the ChG film composition, its thickness, the substrate material, and the excitation intensity. Figure 8 gives an example of restoration oscillograms of an AsSe-film initial transmission after its excitation by a 25 ns duration laser pulse. The curves can be approximated by two exponential dependencies with time constants $t_1 \approx 1$ µs and $t_2 \approx 1.5$ µs. A more detailed analysis of the restoration oscillograms, based on the non-archimedean

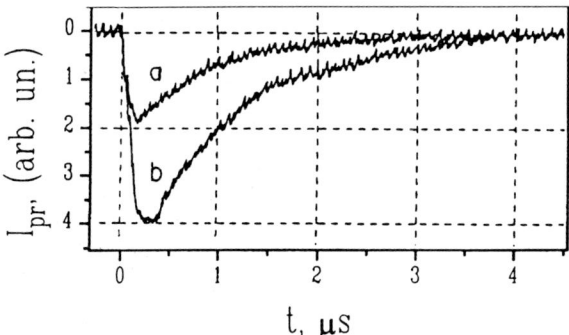

Fig. 8. Recovery of the initial light transmission state. The exciting pulse intensity was (a) 0.4 and (b) 0.8 MW/cm^2.

model, is given in Popescu, Andriesh, Chumash, Enaki, Cojocaru and Grozescu [1991].

The peculiarities of the nonlinear laser pulse absorption were also investigated by cooling the ChG films down to the liquid nitrogen temperature. An analogous changing of the laser pulse time profile, passing through a ChG film, was registered. The cooling of the samples leads to a considerable increase of the surface surrounded by the optical hysteresis loop (see the insertion in fig. 9). It was ascertained that by cooling the samples from 300 K to 77 K the light intensity threshold values I_t do not change (with a precision <20%).

The temperature behavior of the linear and nonlinear optical absorption in amorphous semiconductors points to the important role that phonons play in the process of the interband light absorption. In fact, the characteristic times of the ChG photoinduced transition into the new quasiequilibrium state (corresponding to the horizontal part of the curve in fig. 9) do not change during the sample cooling, from 300 K to 77 K. Thus, we can conclude that the quasistationary value of the light absorption coefficient in the laser pulse field is determined primarily by the light intensity in the medium.

The dependence of the ChG nonlinear light absorption characteristics on the laser radiation wavelength (in the region of the ChG optical absorption edge) was investigated. The wavelength was changed to values which were either higher than E_g (up to 0.4 eV) or lower than E_g (down to 0.2 eV). It was established that in this spectral range the nonlinear absorption characteristics were maintained. However, the pulse intensity threshold values I_t change with the wavelength inversely with the linear absorption coefficient of the medium.

As is known, the structure of freshly evaporated ChG films changes under band-gap illumination, and this structural transformation is associated with

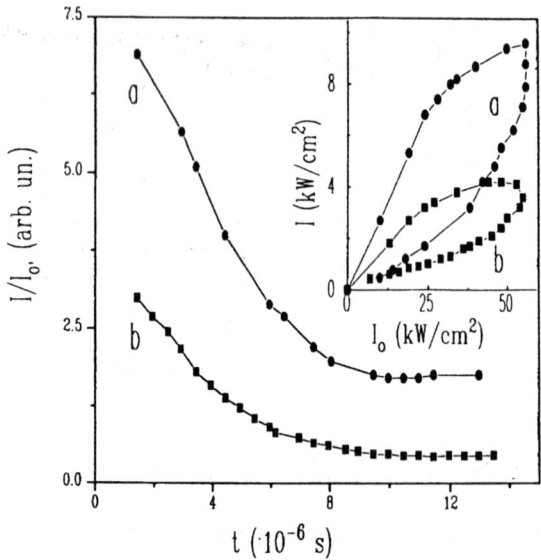

Fig. 9. Kinetics of the As_2Se_3 photodarkening for a laser pulse of 7 ns duration at (a) 77 and (b) 300 K; $h\nu = 2.18$ eV. Inset: the optical hysteresis loop at 77 K and 300 K.

certain changes in the optical properties (in particular with the shift of E_g), which are determined by the illumination dose (at low values of dose). Such photostructural transformations usually lead to two components in the E_g shift: one reversible and the other irreversible (Tanaka [1980a]). As a result of the heat treatment the reversible component recovers and the process may be repeated many times, whereas the irreversible component remains and cannot restore the initial structural state. This last component is probably connected with the disappearance of the macroscopic density fluctuation in the structure of the ChG films.

The ChG photoinduced darkening under laser pulses, described previously, is accompanied by photostructural transformations of the medium. These transformations (with the E_g shift) are noticed only in freshly prepared (previously not illuminated) samples and are irreversible after the exciting pulse ends. The photoinduced ChG darkening in the laser field (see Andriesh, Enaki, Cojocaru, Ostafeichuk, Cerbari and Chumash [1988], Andriesh, Enaki, Cojocaru, Ostafeichuk, Cerbari and Chumash [1989], Andriesh, Cojocaru, Ostafeichuk, Cerbari and Chumash [1989], Andriesh, Enaki, Cojocaru and Chumash [1990], Andriesh, Chumash and Cojocaru [1991], Andriesh, Bogdan, Enaki, Cojocaru and Chumash [1992], Andriesh, Chumash, Cojocaru, Bertolotti,

Fazio, Michelotti and Hulin [1992], Andriesh and Chumash [1993]) is of another physical nature than the ChG photostructural transformations. First, because it is reversible (the initial ChG light transmission sets in completely during 1–100 μs), and, second, the photoinduced darkening was found in materials which, as a result of the photostructural transformations, have a different sign of the E_g shift; that is, they give bleaching (GeSe$_2$, As$_{22}$S$_{33}$Ge$_{45}$) or darkening (As$_2$S$_3$, AsSe, As$_2$Se$_3$). It was shown experimentally that the peculiarities of the laser pulse nonlinear absorption are the same in annealed or in previously darkened (by other light sources) ChG films.

The saturation of the interband light absorption (the Burstein–Moss dynamic shift of the fundamental absorption edge (Gribkovskii [1975], Miller, Seaton, Prise and Smith [1981]) in semiconductors, connected with the generation of a sufficient number of electron–hole pairs, and consequently, with the Fermi level shifts into the conduction band (the absorption edge shifts towards higher photon energies), cannot explain the recorded behavior of the excited amorphous semiconductor, which is related to a considerable increase of the light absorption coefficient.

In the described experiments the behavior of the reflected light from the surface of the ChG films in a regime of laser pulse nonlinear transmission was also measured. It was ascertained that the oscillograms of the reflected pulses from the ChG repeated the incident ones. This points to a real increase of the ChG light absorption and it can not be explained by "metallization" effects, related to a carrier-concentration increase in the exciting region, and by melting of the material on the surface of this region.

A distinctive characteristic of the amorphous semiconductors is the presence of a high (10^{16}–10^{19} cm^{-3}) density of the localized states in the optical band gap. As a rule, in the literature the nonlinear light absorption peculiarities in the ChG transparent region are related to carrier occupation and redistribution on these states. In several investigations this mechanism is also proposed to explain the laser pulse nonlinear absorption at the ChG interband excitation (Vardeny [1983]). The hot carriers, created as a result of the ChG light excitation with $h\nu \geqslant E_g$, relax (during a period shorter than 10^{-12} s, 300 K), generally giving the excess energy to the phonons and being captured on the localized states. Such localized carriers may be excited by light, contributing to the light absorption. The laser pulse nonlinear absorption at ChG interband excitation, described in this chapter, however, cannot be explained on the basis of the occupancy and redistribution of the carriers on the localized states with their next excitation in the conduction band. It is conditioned by the fact that the probability of the transitions "localized-state band" is some orders lower than the interband optical transitions probability (Bonch-Bruevichi, Zveaghin, Kaiber, Mironov, Anderlain and Esser [1981]). Consequently, it is rather difficult to experimentally distinguish such a low contribution to the absorption (at the same light wavelength) from the interband absorption. This is even more difficult because the optical absorption spectrum of the investigated compounds and the energy dependence of the density of states in the optical band gap have a monotonous dependence on energy.

It is impossible to explain the laser pulse nonlinear transmission by the homogeneous heating of the ChG films. The temperature coefficient of the optical absorption edge shift in these compounds (dE_g/dT) is about 5×10^{-4} eV K^{-1} (Andreev, Kolomiets, Mazets, Manukian and Pavlov [1976]). Therefore, in order to change the absorption light coefficient in the absorption edge region by an amount equal to that revealed in the experiment (up to two times), it is necessary to heat the ChG sample to a temperature considerably higher than that of the material softening. Moreover, an analogous laser pulse nonlinear absorption was also registered at light quantum energies, considerably exceeding the ChG optical band gap (E_g), where the contribution of the temperature shift of the absorption edge to the light transmission change is smaller.

The contribution of the sample temperature increase (as a result of the absorption of a fraction of laser radiation energy) to the light pulse nonlinear absorption may be revealed from the absorption's dependence on the size of the excitation region, the ChG film thickness, and the substrate material (with different heat conductivities). The investigation of the laser pulse nonlinear transmission dependence on the excitation region size was carried out as follows: the ChG films were excited by laser pulses with a constant intensity, but obscured on different diameters (from 80 to 700 µm). As a result of the measurements, no dependence of the optical hysteresis transmission curves on the diameters of the exciting beam was revealed.

The influence of the substrate material (with a different heat conductivity κ) and of the sample thickness on the characteristics of the laser pulse transmission has been studied. Substrates of (a) glass ($\kappa = 0.7$ W m^{-1} K^{-1}), (b) mica ($\kappa = 0.24$ W m^{-1} K^{-1}), and (c) lavsan ($\kappa = 0.17$ W m^{-1} K^{-1}) were used. Figure 10 shows the curves of the optical hysteresis transmission of AsSe films evaporated on the indicated substrates. Within the experimental error (<20%), no influence of the substrate heat conductivity on the curves of the optical hysteresis transmission and no influence of the ChG film thickness on the nonlinear absorption (while changing the thickness from 0.1 µm to 3 µm) were revealed.

At the same time, as was pointed out previously, a significant influence of the

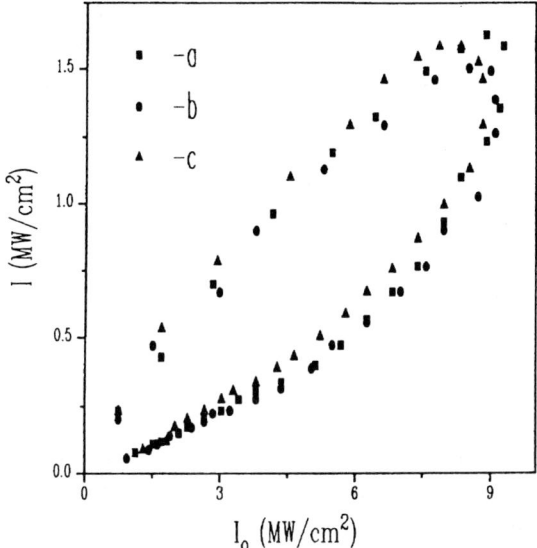

Fig. 10. Optical transmission dependence of AsSe films sputtered on (a) glass, (b) mica, and (c) lavsan substrates. Laser pulse duration 7 ns, pulse intensity 9 MW/cm^2, $h\nu = 2.05$ eV, 300 K.

ChG film composition, its thickness, of the substrate material (heat conductivity), and of the excitation intensity, was revealed on the transmission restoration kinetics after the end of the exciting pulse.

§ 5. Model of Nonlinear Light Absorption into a Highly Excited Noncrystalline Semiconductor: Phenomenological Approach

Structure inhomogeneities in amorphous semiconductors may lead to some spatial localization of the phonon vibrational modes: i.e., they may be considered as quasiresonators for the vibration energy (Malinovskii, Nesterihin, Novikov and Sokolov [1986], Orbach [1993]). The capture of nonequilibrium phonons by such quasiresonators increases their effective lifetime (Baumgartner, Engelhardt and Renk [1983], Orbach [1993]). Some properties of the vibrational mode localization in quasiresonators can be similar to the electron localization in quantum well structures. In the investigated ChG samples the interband light absorption coefficient depended on the temperature, conditioned by a strong electron–phonon interaction for ChG. Thermal phonons with a wavelength of the order of the distance between random inhomogeneities may also deform the random band potential in ChG. This, in turn, leads to a change of the light absorption coefficient as a result of the appearance (or disappearance) of additional optical transition channels between the earlier forbidden (or less probable) band states. Consequently, transitions with a simultaneous participation of light quanta and phonons play a considerable role in the amorphous semiconductor interband light absorption.

With these considerations in mind, the coefficient of the interband light absorption may be thought to consist of two parts:

$$\alpha = \alpha_0 + \beta N_a, \tag{5.1}$$

where α_0 does not depend on the temperature, and the second part depends on the concentration of N_a acoustic phonons. The second part of the light absorption coefficient in eq. (5.1) may change considerably by generating additional phonons as a result of the hot carrier (created by the light) energy dissipation, with their consequent capture on localized states in the ChG optical band gap. A part of the formed nonequilibrium phonons may be "captured" by the quasiresonators with a d_m size (where d_m is the average distance between the spatial inhomogeneities). Such quasiresonators may trap the standing acoustic modes with $\lambda_f \leq d_m$ (increasing their lifetime), and contribute to their further accumulation. Considerable deformation of the random (frozen) potential may occur at high concentration of nonequilibrium phonons (i.e., a considerable deformation of the quasiresonator "walls"). On the one hand, it leads to the phonons nonlinear decay from the quasiresonator regions, and, on the other, to the appearance (disappearance) of additional channels of interband light absorption.

The kinetic equations, describing the exciting and relaxation processes of electrons and phonons, can be written as follows (Andriesh, Enaki, Cojocaru and Chumash [1990], Andriesh, Bogdan, Enaki, Cojocaru and Chumash [1992], Andriesh and Chumash [1993]):

$$\frac{\partial E}{c\partial t} - \frac{\partial E}{\partial x} = \frac{\omega}{c}P; \tag{5.2a}$$

$$\frac{\partial P}{\partial t} = -\frac{P}{T_2} + p_{cv}^2 \frac{E}{\hbar}(N_v - N_c)(1 + \gamma n_a); \tag{5.2b}$$

$$\frac{\partial N_v}{\partial t} = -\frac{PE}{\hbar} - \frac{1}{\tau_0}(n+1)(N_v - N_v^0); \tag{5.2c}$$

$$\frac{\partial N_c}{\partial t} = \frac{PE}{\hbar} - \frac{1}{\tau_0}(n+1)N_c; \tag{5.2d}$$

$$\frac{\partial n}{\partial t} = \frac{\upsilon \delta}{\tau_0}(n+1)\left(N_c - N_v + N_v^0\right) - \frac{1}{\tau_{oa}}(n - n_0); \quad (5.2e)$$

$$\frac{\partial n_a}{\partial t} = \frac{\kappa}{\tau_{oa}}(n - n_0) - \frac{1}{\tau_{aa}} n_a \left(n_a - n_a^0\right); \quad (5.2f)$$

where E is the amplitude of the laser field, $P = p_{cv}/V$ is the matrix of the medium polarization for the interband transition per unit volume, p_{cv} is the transition dipole matrix between conduction and valence bands, N_c and $(N_v^0 - N_v)$ are the concentrations of electrons and holes in the conduction and valence band, respectively, created by the laser field, N_v^0 is the electron equilibrium concentration in the valence band; n_0 and n_a^0 are the occupation numbers of the equilibrium optical and acoustic phonons, respectively, $(n - n_0)$ and $(n_a - n_a^0)$ are the occupation numbers of the nonequilibrium optical and acoustic phonons, respectively; τ_{oa} is the decay time of an optical phonon into k acoustic phonons (one optical phonon is assumed to disintegrate into k acoustic phonons); τ_o is the hot carrier relaxation time as a result of the interaction with the optical phonons; T_2 is the transversal relaxation time of the medium polarization due to the electron–electron interaction; $\tau_{aa} = \tau n_a^0$; τ is the hot acoustic phonon relaxation time due to the phonon unharmonicity and to the interaction with the thermal body; δ/τ_o is the phonon generation rate by the hot carriers (δ is the optical phonon number generated by one hot carrier); υ is the volume of the semiconductor interacting with the light; and ω is the frequency of the exciting light.

Equation (5.2a) describes the laser energy loss during the interaction with the medium. The first term of eq. (5.2b) takes into account the medium polarization decay and the second member the transition of the carriers (under the laser field) from the valence to the conduction band and vice versa. The bracket $(1 + \gamma n_a)$ takes into account both direct transitions ($\gamma = 0$) and interband transitions with the participation of the local acoustic mode ($\gamma \neq 0$). Equations (5.2c) and (5.2d) describe the carrier concentration changes in the conduction and in the valence band under the action of the external laser field ($PE/\hbar$) and carrier relaxation with the emission of optical phonons (the second members). Equations (5.2e) and (5.2f) take into account the generation of optical and acoustic phonons. The first member of eq. (5.2e) considers the generation of optical phonons during the hot electron and hole decay. The generation rate is proportional to the concentration of the nonequilibrium electrons and holes in the semiconductor. The second member in eq. (5.2e) describes the optical phonon decay process into k acoustical phonons. In eq. (5.2f) the first member describes the generation of acoustic phonons as a result of the nonequilibrium optical phonon decay, whereas the second member describes their exit from the formed quasiresonators due to the phonon unharmonicity and the interaction with the thermal body.

Some possible stationary solutions of eqs. (5.2), describing the interband nonlinear light absorption in amorphous semiconductors under certain conditions, are now presented.

5.1. NO PARTICIPATION OF LOCAL ACOUSTIC PHONONS ($\gamma=0$)

If one neglects interband transitions with the participation of acoustic phonons (i.e., $\gamma=0$), the system of eqs. (5.2) describes the nonlinear bleaching of the medium, related to the saturation of the interband transition (Gribkovskii [1975], Miller, Seaton, Prise and Smith [1981]). Some cases are possible.

5.1.1. $n_a \approx n_a^0$; $n = n_0 \ll 1$

This is the case when the concentration of the nonequilibrium acoustic phonons is much smaller than the concentration of the equilibrium phonons, and the occupation numbers of the optical phonons are much smaller than 1; i.e., the energy of the optical phonon considerably exceeds $K_B T$, where K_B is the Boltzmann constant. The absorption coefficient in this case has the form:

$$\alpha = \frac{\alpha_0}{1 + 2\Omega^2 \tau_0 T_2}, \tag{5.3}$$

where

$$\Omega = \frac{E p_{cv}}{\hbar}; \qquad \alpha = -\frac{2\partial E}{E \partial x}; \qquad \alpha_0 = \frac{2\omega T_2 p_{cv}^2 N_v^0}{\hbar c}.$$

The absorption coefficient decreases with increasing laser field intensity. This occurs when the rate of the transitions, induced by the laser light, is greater than the reversible value of the carriers' relaxation time due to the interaction with the optical phonons $1/\tau_0$.

5.1.2. $n > n_0$

In this case a great number of optical and acoustic phonons is generated when the hot carriers relax at the bottom of the bands, and the light absorption coefficient results:

$$\alpha = \alpha_0 \frac{N_v - N_c}{N_v^0}, \tag{5.4}$$

where

$$N_v - N_c = \frac{1}{2}\left\{\left[N_v^0 - \frac{\tau_o}{v\delta\tau_{oa}}\right] - \frac{\hbar\omega N_v^0(n_0+1)}{vc\alpha_0\delta\tau_{oa}E^2}\right.$$
$$+ \left[\left(N_v^0 - \frac{\tau_o}{v\delta\tau_{oa}} - \frac{\hbar\omega N_v^0(n_0+1)}{vc\alpha_0\delta\tau_{oa}E^2}\right)^2\right.$$
$$\left.\left.+ \frac{4\hbar\omega(N_v^0)^2(n_0+1)}{vc\alpha\delta\tau_{oa}E^2}\right]^{1/2}\right\}. \quad (5.5)$$

As seen from eqs. (5.4) and (5.5), during excitation of the semiconductor by a strong laser pulse, two stages of the absorption process are possible. At the initial stage the interband light absorption coefficient decreases due to the nonequilibrium saturation of the interband transition ($N_c \approx N_v$ and $n \ll 1$, see eq. 5.3). By increasing the concentration of the nonequilibrium optical phonons (as a result of the hot carriers' energy dissipation), the rate of the induced relaxation of the hot electrons at the bottom of the conduction band increases (see eq. 5.2d). The probability of interband transitions increases at this stage, due to the increase of "free places" in the states of the conduction band (see eqs. 5.4, 5.5). In other words, the absorption coefficient α changes in time from the value corresponding to the interband transition saturated state (the initial stage), to its initial value α_0, as the number of the inverted electrons, getting into resonance with the external field, tends to zero ($N_v - N_c \approx N_v^0$, see eq. 5.5). This effect may exist both in crystalline and noncrystalline semiconductors under a strong electron excitation in the conduction band ($hv > E_g$).

5.2. PARTICIPATION OF LOCAL ACOUSTIC PHONONS ($\gamma \neq 0$)

As was noted above, the nonlinear change of the light absorption coefficient may be reached in the noncrystalline semiconductors at much lower irradiation intensities than in the crystalline ones, when the above-mentioned interband transition saturation effect plays a negligible role (under this condition $N_c \ll N_v \approx N_v^0$). This is possible when transitions with the participation of local acoustic phonons ($\gamma \neq 0$) play an important role. Here, the system of kinetic eqs. (5.2) simplifies. As the time of the transversal relaxation T_2 of the medium polarization is small ($T_2 \approx 10^{-13} - 10^{-14}$ s, i.e., the polarization attenuation is large and $P/T_2 \gg \partial P/\partial t$), the right side of eq. (5.2b) may be put equal to zero. In the experiments the duration of the laser pulses was greater than the characteristic times of the optical and hot acoustic phonons decay. In this case it is possible to neglect the intermediate stages of the phonons decay, which arise during the optical and hot acoustic phonon decay, and consider only the kinetic of the local acoustic phonons. In eqs. (5.2) this corresponds to putting zero on the right side of eq. (5.2e).

Under such assumptions, corresponding to the experimental conditions, eqs. (5.2) may be written as follows:

$$\frac{\partial I(t,x)}{c\partial t} - \frac{\partial I(t,x)}{\partial x} = (\alpha_0 + \beta N_a)I(t,x);$$

$$\frac{\partial N_c}{\partial t} = -\frac{N_c}{\tau_0} + (\alpha_0 + \beta N_a)\frac{I(t,x)}{\hbar\omega};$$

$$\frac{\partial N_a}{\partial t} = \frac{\Theta N_c}{\tau_0} - \frac{N_a(N_a - N_a^0)}{\tau N_a^0},$$

(5.6)

where $\Theta = k\delta$ is the number of created acoustic phonons as the result of one hot carrier decay; $I = cE^2/2$ is the laser pulse intensity in the medium; and $\beta = 2\omega T_2 p_{cv}^2 \gamma N_v^0/\hbar c$; N_a^0 and N_a are the concentrations of the equilibrium and nonequilibrium acoustic phonons.

The solution of eqs. (5.6) for the stationary case ($\partial N/\partial t = 0$; $\partial N_a/\partial t = 0$) is as follows:

$$\frac{\partial I(x)}{\partial x} = -\frac{1}{2}\left\{2\alpha_0 + \beta\left[N_a^0 + \frac{c\tau_{aa}N_a^0\beta\Theta I(x)}{\hbar\omega}\right]\right.$$

$$\left. + \beta\left[\left(N_a^0 + \frac{c\tau_{aa}N_a^0\beta\Theta I(x)}{\hbar\omega}\right)^2 + \frac{c\tau_{aa}N_a^0\alpha_0\Theta}{\hbar\omega}\right]^{1/2}\right\}I(x),$$

(5.7)

which corresponds to the case of a strong phonon subsystem excitation, when the phonon generation rate is equal to the rate of phonons leaving the localization region. As follows from eq. (5.7), the absorption coefficient $\alpha = -\partial I(x)/I(x)\partial x$ increases with increasing light intensity, which is related to the increase of the acoustic phonon concentration in the sample.

The analytical solution of eqs. (5.6), corresponding to the experimental conditions, is difficult to obtain. Numerical solutions and their comparison with the experimental curves are given below.

The kinetic eqs. (5.6) were solved for parameter values and initial conditions, corresponding to the experimental situation. The value β was determined from the measurement of the linear light absorption coefficient α_ℓ in ChG at temperatures $T_1 = 300$ K and $T_2 = 77$ K, which can be represented as follows:

$$\alpha_\ell(T_i) = \alpha_0 + \beta N_a^0(T_i),$$

(5.8)

where $i=1$ or 2, and α_0 does not depend on the temperature. From eq. (5.8) it follows that the parameter β is expressed through the concentration of the phonons N_a^0,

$$\beta = \frac{\alpha_\ell(T_1) - \alpha_\ell(T_2)}{N_a^0(T_1) - N_a^0(T_2)}. \tag{5.9}$$

Taking into account that the main contribution to the light absorption is given by phonons with wavelength $\lambda_f \approx d_m$ ($d_m \approx 1$ nm), then β may be determined from the temperature dependencies of $\alpha_\ell(T_i)$ and $N_a^0(T_i)$. The equilibrium concentration of the phonons was calculated based on the definition

$$N_a^0(T) \approx 4\pi k_m^2 \Delta k_m n_k^a(T) \approx \frac{2(2\pi)^4 n_k^a(T)}{d_m^3}, \tag{5.10}$$

where $n_k^a(T) = \left[\exp(\hbar\omega_f^k/K_BT) - 1\right]^{-1}$ are the population numbers of the local phonon modes, $\omega_f^k = v_s k_m$, v_s is the sound velocity in the medium; $k_m \approx 2\pi/d_m$ and $\Delta k_m \approx 2\pi/d_m$. Using eqs. (5.9) and (5.10), we have $v_s \approx 10^5$ cm s^{-1}, $N_a^0(T_1) \approx 9 \times 10^{22}$ cm^{-3}, $N_a^0(T_2) \approx 1.3 \times 10^{22}$ cm^{-3}, and a value $\beta \cong 10^{-20}$ cm^2 (Andriesh, Bogdan, Enaki, Cojocaru and Chumash [1992], Andriesh and Chumash [1993]).

Other parameters in eqs. (5.6) had the following values: quantum energy $h\nu = 3.5 \times 10^{-19}$ J (0.57 μm); light absorption coefficient $\alpha_\ell = 5.6 \times 10^4$ cm^{-1}; $\Theta \approx 10$–100; and $\tau_0 = 10^{-12}$ s. The relaxation time of the nonequilibrium phonons $\tau \approx 10^{-6} - 10^{-5}$ s was determined by fitting the numerical results with the experimental curves (fig. 8) (Popescu, Andriesh, Chumash, Enaki, Cojocaru and Grozescu [1991]).

The temporal laser pulse shape in the calculations was chosen to have the Gaussian form

$$I\left(t - \frac{x}{c}\right) = I_0 \exp\left[-\frac{(t - x/c)^2}{t_0}\right], \tag{5.11}$$

where I_0 is the pulse intensity at the maximum, and t_0 is the half-intensity laser pulse duration. The initial conditions are $N_c = 0$ and $N_a = N_a^0$.

The results of the numerical solution of eqs. (5.6) fit the experimental dependencies reasonably well. Figure 11 gives the experimental and calculated hysteresis dependencies of the pulse intensity, passed through an AsSe film (thickness 0.6 μm), versus the corresponding intensity value at the input for 7 ns laser pulse duration. Some differences between the calculated curves and the

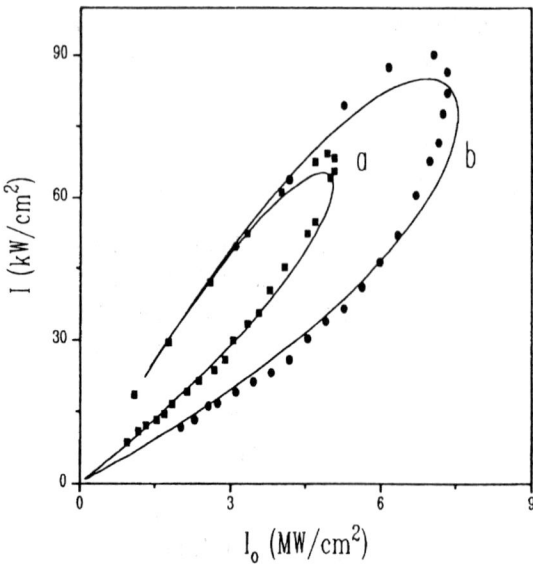

Fig. 11. Experimental (points) and calculated dependence of the AsSe light transmission for a laser pulse with 7 ns duration. Input light intensity: (a) 5.0 and (b) 7.5 MW/cm². Parameter values: $N_a^0 = 1.2 \times 10^{22}$ cm^{-3}, $\beta = 10^{-19}$ cm², $q = 50$, $\alpha_1 = 5.6 \times 10^4$ cm^{-1}, $\tau = 10\,\mu$s, $T = 300$ K.

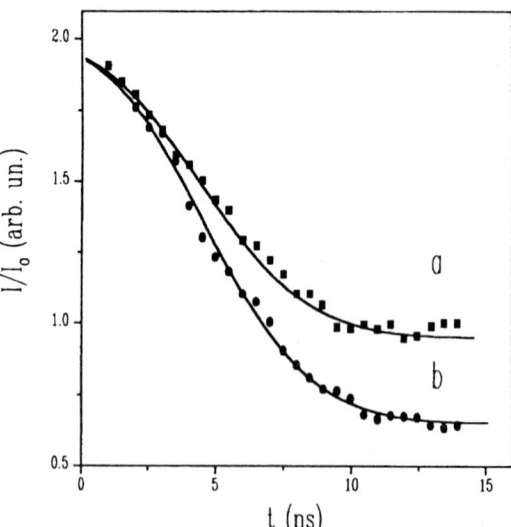

Fig. 12. Experimental (points) and calculated dependence of AsSe film darkening kinetics for a 7 ns laser pulse. Input light intensity and parameter values same as in fig. 11.

experimental points are determined by the deviation of the laser pulse shapes from the Gaussian shape and by the difficulty of incorporating the intermediate stage processes of the hot carrier energy decay into acoustic phonons.

The proposed model can be used as an explanation for the laser pulse nonlinear absorption of ChG, which can be supported by the fact that with the same parameter values the calculated curves coincide satisfactorily with the experimental ones, despite changes in laser pulse duration or amplitude in the wide range. Another example of the satisfactory correspondence of experimental and calculated dependencies of the AsSe film darkening kinetics during its irradiation with the nanosecond laser pulse is shown in fig. 12.

§ 6. Refractive Index Changes of Chalcogenide Glasses

According to the Kramers–Kronig relations, it is anticipated that the light absorption changes in ChG should lead to refractive index changes. In ChG, physical properties are permanently modified by photostructural transformations (Tanaka [1980a]). De Neufville, Moss and Ovshinsky [1973/74] investigated irreversible photostructural transformations in As_2Se_3 and As_2S_3 films, which gave rise to a large absorption edge shift to smaller energy and to an appreciable refractive index change. For example, in the transparent region of the samples the refractive index of As_2S_3 increases by $\Delta n \approx 0.1$ ($hv = 1.5$ eV), on annealing or illumination, and $\Delta n \approx 0.06$ ($hv = 1.0$ eV) for As_2Se_3. De Neufville classified the phenomena into reversible and irreversible processes, according to whether a heat treatment could or could not restore the initial structural states. Irreversible changes in the refractive index are considered as promising effects from the viewpoint of applications such as image recording, high-capacity information storage, and optical waveguiding. The reversible photostructural transformations in As_2Se_3 and As_2S_3 are accompanied by a negligible refractive index change ($\Delta n \leqslant 0.01$ or smaller) when compared with the irreversible ones.

A photoinduced increase in the refractive index of ChG was also observed by several groups. Ohmachi and Igo [1972] use the holographic storage technique in order to make quantitative measurements of refractive-index change in As–S–Ge glass films. Tanaka and Ohtsuka [1976] studied photoinduced changes in the refractive index, refractive-index dispersion and thickness in As_2S_3 films by means of laser heterodyne interferometry, and optical waveguiding, using a prism-coupling technique (Tanaka and Ohtsuka [1978], Tanaka [1979]). Two distinct changes, irreversible and reversible, of the refractive index are observed, which are recovered by an appropriate heat treatment. The irreversible change

($\Delta n \approx 0.133$, $\lambda = 633$ nm) associates only with the first appropriate illumination for the virgin films, whereas the reversible one ($\Delta n \approx 0.029$, $\lambda = 633$ nm) reproduces well under successive heat treatments and illuminations; on the other hand, the thickness decreases.

In addition, a dynamical (or transient) change in refractive index, $\Delta n \approx 0.003$, $\lambda = 633$ nm, appears under the band-gap illumination of about $10\,\text{mW/cm}^2$, which are recovered by additional illumination or after the completion of photoexcitation (Tanaka and Ohtsuka [1978], Tanaka [1978], Tanaka [1980b]). The mechanism of the dynamical changes in refractive index in As_xS_{100-x} films (with x between 18 and 43) was investigated, and it was suggested that these changes originated from the trapping of photoexcited carriers (Tanaka [1978], Tanaka [1980b]).

Systematic studies of the effect of ultraviolet exposure on the refractive index, extinction coefficient, and optical band gap as a function of wavelength and various ultraviolet exposure durations were made by Dawar, Shishodia, Chauhan, Joshi, Jagadish and Mathur [1990]. The initial decrease in refractive index and increase in transmission for short ultraviolet exposures can be qualitatively attributed to the fact that As_2S_3 films first expand and then contract when illuminated by ultraviolet light; i.e., for short exposures the films expand, and decrease in density and in refractive index.

Hajto [1980] estimated the changes of refractive index during nonlinear propagation of CW He–Ne laser radiation through $GeSe_2$ thin films from the measured transmittance and reflectance changes. The refractive index changes in the opposite direction as the absorption coefficient reaches the maximum, indicating that a real photoinduced absorption change occurs during the nonlinear light propagation in the form of transmittance oscillations.

Considerable interest has been shown in optical devices that can provide optical signal processing free from mechanical motions and electric noises. Switching, modulation, and deflection of guided light in the ChG films, using the photoinduced changes of the refractive index, were reported by several authors. Matsuda, Mizuno, Takayama, Saito and Kikuchi [1974] observed an optical "stopping effect", that is, a decrease in transmittance of a guided light beam induced by band-gap irradiation, in As_2S_3 films, and they ascribed the origin to photoexcitation of trapped carriers with the guided beam.

Principles and fabrications of photo-optical devices based on the photoinduced dynamical refractive index changes in As_2S_3 films are described by Tanaka and Odajima [1981] and Tanaka, Imai and Odajima [1985]). In these devices, propagation of a light beam in ChG optical waveguides with prism couplers is controlled with blue and red (band-gap and sub-band-gap) light illumination, which can modify the refractive index of the films. Although the switching time of the device is not fast, it allows signal processing of the optical energy to be carried out. In addition, the switching times of the devices depend primarily on the intensities of the blue and red illuminations.

Mazets, Pavlov, Smorgonskaya and Shifrin [1987] observed in the transparent region of glassy As_2Se_3 films large variations of the absorption coefficient and refractive index at high pulsed pumping levels above the absorption edge, associated with the filling of localized states in the band gap. The dependencies of refractive-index variation in As_2S_3, As_2Se_3, $(As_2S_3)_{0.5}(As_2Se_3)_{0.5}$, and $GeSe_2$ films on the pumping level above the absorption edge were investigated by Kalmikova, Mazets, Pavlov, Smorgonskaya and Shifrin [1988].

True and McCaughan [1991] measured a light-induced, refractive-index change in thin As_2S_3 films over a range of pump intensities. A relatively large refractive nonlinearity of the resonant type, but in a spectral region of low absorption, was measured. It was confirmed that for long time scales the refractive-index changes were thermal in origin, but for shorter time scales the thermal effect was no longer observed. True and McCaughan [1991] note that ChG can present a novel type of nonlinearity, when strongly absorbed light (at one wavelength) produces a relatively large change in the optical properties of the material at a different wavelength where the absorption is weak; that is, the advantages of resonant enhancement with the wavelength flexibility and low absorption of nonresonant nonlinearity can be combined. In addition, the speed of refractive index changes is potentially fast.

A large nonlinear refractive index n_2 value of the As_2S_3 glass fibers (two orders larger than that of silica glass fibers), taking into account the effect of group velocity dispersion and two-photon absorption, was investigated by Asobe, Kanamori and Kubodera [1993]. Applications of As_2S_3 glass fibers in ultrafast, all-optical switches have been reported. Switching speed and switching power were investigated experimentally and through calculations. Switching time of 12 ps and switching power of 5 W can be achieved using a 10 ps gate pulse and only a 1 m chalcogenide glass fiber (Asobe, Kanamori and Kubodera [1993]).

Recently the nonlinear optical properties of As_2S_3 ($E_g = 2.54$ eV) and As_2Se_3 ($E_g = 1.78$ eV) thin films, obtained by thermal evaporation on glass substrates in vacuum conditions to a thickness of 0.3–2.0 μm, were investigated by means of the Z-Scan technique (Bertolotti, Michelotti, Andriesh, Chumash and Liakhou [1992a,b], Michelotti, Bertolotti, Chumash and Andriesh [1992], Andriesh, Chumash, Cojocaru, Bertolotti, Fazio, Michelotti and Hulin [1992], Michelotti, Fazio, Senesi, Bertolotti, Chumash and Andriesh [1993]).

This technique can characterize nonlinear refraction and nonlinear absorption of a wide variety of materials (Sheik-Bahae, Said, Wei, Hagan and Van Stryland [1990]). Nonlinear refraction, usually referred to as n_2, gives rise to self-lensing and self-phase modulation. These effects are desirable to achieve optical switching and optical limiting. Knowledge of n_2 values is important to devise a positive component of n_2 (causing optical breakdown in materials or degradation in laser beam quality) from the negative one.

The study of the nonlinear optical behavior of thin films of the chalcogenide amorphous semiconductor As_2S_3, when excited with radiation below the band gap with a CW argon laser ($\lambda = 514.5$ nm, $h\nu = 2.41$ eV), and As_2Se_3, when excited with radiation above the band gap with a pulsed dye laser ($\lambda = 590$ nm, $h\nu = 2.1$ eV, $t = 4$ ps, $f_{rep} = 290$ kHz) is described further.

In the Z-Scan technique the transmission of a focused laser beam through a finite aperture in the far field is measured as a function of the displacement (z) of the film along the propagation direction with respect to the focal plane. The transmittance of the aperture in the linear regime is defined as

$$S = 1 - \exp\left[-2\frac{r_a^2}{\omega_a^2}\right], \qquad (6.1)$$

where r_a and ω_a are, respectively, the aperture radius and the beam radius at the aperture position. The transmittances with fully opened ($S = 1$) and partially closed ($S = 0.4$) apertures, when measured as a function of sample position for several input average powers, allow to determine changes of absorption ($S = 1$) and refractive index ($S = 0.4$) with light intensity.

In the case of As_2S_3 thin films in CW excitation, performing sequentially Z-Scan measurements on an initially virgin sample at the same point by increasing input optical power, at low light levels, enabled observation of the onset of a Z-Scan signal. This can be explained by positive refractive index and absorption coefficient changes. Lowering the light level to the starting condition, the original low intensity Z-Scan signal could not be recovered. This phenomenon was attributed to a change of the optical properties of the film due a photostructuring process in the chalcogenide film (Michelotti, Bertolotti, Chumash and Andriesh [1992], Andriesh, Chumash, Cojocaru, Bertolotti, Fazio, Michelotti and Hulin [1992], Michelotti, Fazio, Senesi, Bertolotti, Chumash and Andriesh [1993]).

As radiation energy close to the band gap is used, strong structural changes can give rise to a very large detectable change of the refractive index, which produces a permanent lens effect inside the sample in a region that is comparable with the focal plane laser spot size (Michelotti, Fazio, Senesi, Bertolotti, Chumash and Andriesh [1993]). This lens can give rise to an additional Z-Scan signal.

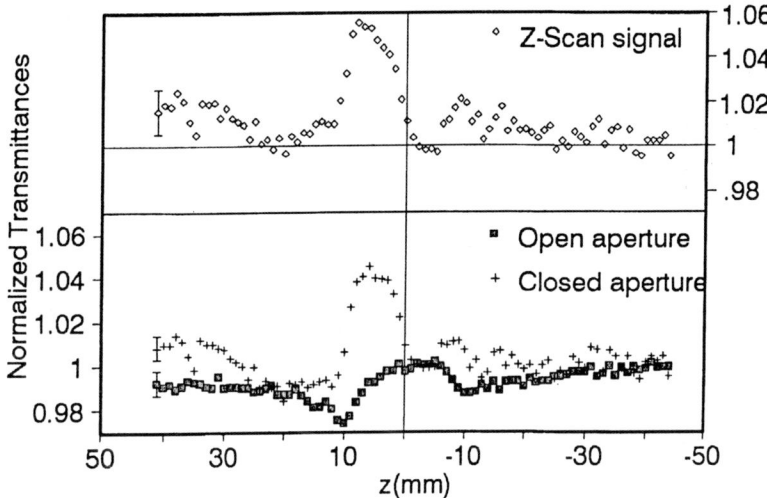

Fig. 13a. Z Scan As$_2$S$_3$ transmittance curves for CW Ar laser ($\lambda = 514.5$ nm) with $P_{in} = 0.1$ mW on a virgin sample.

In figs. 13a–c three Z-Scan measurements are shown for the same site on the sample, starting with a very low input power (fig. 13a, $P = 0.1$ mW) and coming back to the same power (fig. 13c) after performing a higher power scan (fig. 13b, $P = 40$ mW). In these figures the measurement obtained with $S = 1$ are indicated with squares, the measurement obtained with $S = 0.4$ with crosses, and the normalized measurement obtained dividing the second by the first one with diamonds.

In the first case (fig. 13a, $P_{in} = 0.1$ mW; focal plane intensity $I_0 = 0.013$ W/cm^2) a competition between a dynamical response of the material, due to nonlinearity, and a permanent photostructural change that occurred during the measurement time, was detected. The intensity on the sample I is such that during the measurement, slow optical transmission changes occurred due to photostructuring, which modulated our Z-Scan signal. Far from the focal plane ($I \ll I_0$) the changes were so slow that the Z-Scan was not sensitive to them. The closer the sample gets to the focal plane, the faster are the changes, because the intensity increases and the photostructural-change rate, being in first approximation proportional to the light intensity, increases with it. In this case bleaching or darkening, depending on the irradiation time, of the relative transmittance is observed (Michelotti, Fazio, Senesi, Bertolotti, Chumash and Andriesh [1993]).

In fig. 13b the transmission signal at an input power of 40 mW (focal plane intensity: $I_0 = 5.3$ W/cm^2) is shown. In this case the intensity is higher than that

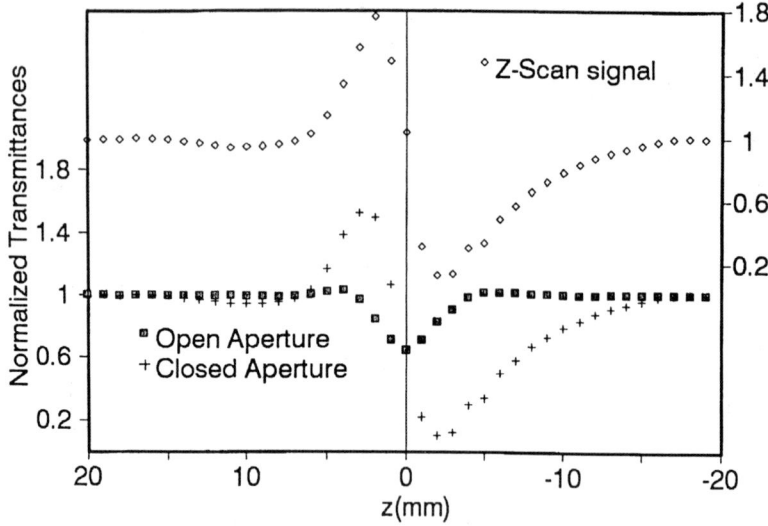

Fig. 13b. Z-scan As$_2$S$_3$ transmittance curves for CW Ar laser ($\lambda = 514.5$ nm) with $P_{in} = 40$ mW.

shown in fig. 13a, and photostructuring takes place in much shorter times than the time between one measurement and the next. Each time the measurements are performed in a completely photostructured sample (steady state situation) and a Z-Scan signal, which is the sum of the dynamical response of the material and the additional permanent lens effect, is observed.

In order to extract, in the previous case, the information on dynamical refractive index changes, the measurement was repeated at the same position on the sample, lowering the input power to the same level as that in fig. 13a. As the intensity was much lower, the background signal could be observed (see fig. 13c), which is due only to the permanent changes that we induced in the preceding scans. As in the previous case, the Z-Scan curve resulted in a strong positive change of refractive index and a relatively lower change in absorption coefficient, as reported in the literature (Lee and Paesler [1987]).

The background Z-Scan signal can be attributed to the creation of a small portion of the sample where the optical properties are changed due to photostructural changes. In fig. 13c the transmittance behavior is indicated with a solid line due to a 35 μm wide lens, whose focal length is $f_p = 40$ mm. The transmittance is calculated with a simple geometrical model taking into account the transverse finite dimensions (Michelotti, Fazio, Senesi, Bertolotti, Chumash and Andriesh [1993]).

Finally, if the scan of fig. 13c is subtracted from that of fig. 13b, only

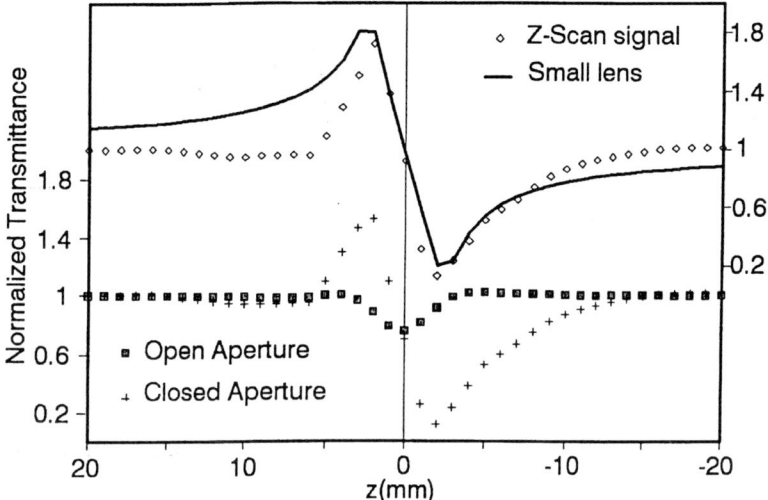

Fig. 13c. Z-scan As_2S_3 transmittance curves for CW Ar laser ($\lambda = 514.5$ nm) with $P_{in} = 0.1$ mW in photostructured sample. The solid line indicates the geometrical model fit.

the contribution due to dynamical optical nonlinearity can be obtained, which is present in the 40 mW measurement (fig. 14). In this case the sign of the nonlinearity change is negative (Bertolotti, Michelotti, Andriesh, Chumash and Liakhou [1992a]).

From the fitted parameters, obtained through nonlinear regression analysis, a maximum change of the refractive index in the focus equal to $\Delta n = -8 \times 10^{-3}$ is obtained, which corresponds to a value for n_2 equal to $n_2 = -1.6 \times 10^{-3}$ cm^2/W (Michelotti, Fazio, Senesi, Bertolotti, Chumash and Andriesh [1993]).

In the case of the As_2Se_3 thin films in pulsed excitation, performing Z-Scan measurements on a sample that was previously photostructured shining with highly absorbed light, a Z-Scan signal was observed that can be explained by negative refractive index and positive absorption coefficient changes.

In fig. 15 the normalized transmittances for $S = 1$ (triangles), $S = 0.4$ (diamonds), and their ratio (squares) are shown for a 1.6 μm thick As_2Se_3 sample under irradiation at $\lambda = 590$ nm, with a $t = 4$ ps pulsed dye laser (repetition rate $f_{rep} = 290$ kHz) (Bertolotti, Michelotti, Andriesh, Chumash and Liakhou [1992a,b]). In this case the percent absorption coefficient change is much larger than that of the refractive index, due to the preceding band-gap operation.

From the fitting parameters (line that fits the normalized transmittance data in fig. 15) a nonlinear refractive index value $n_2 = 6.8 \times 10^{-9}$ cm^2/W can be observed. This value is six orders of magnitude lower than the value for As_2S_3 with

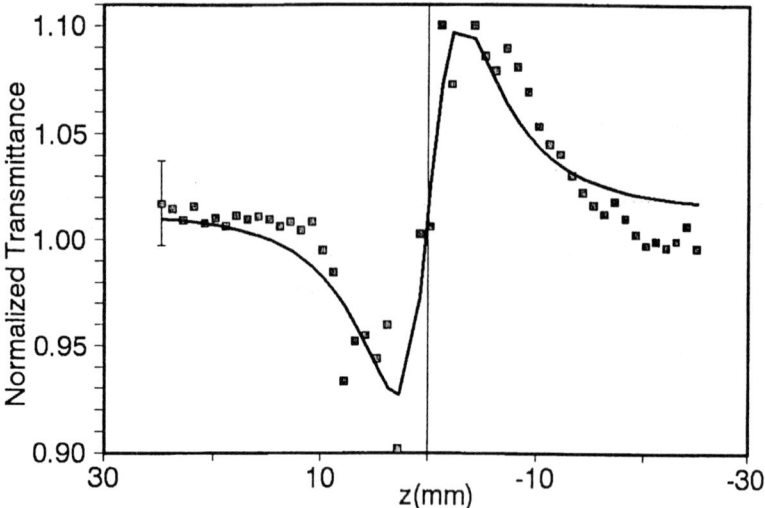

Fig. 14. Difference As_2S_3 Z-Scan signal obtained by subtracting curve 13c from curve 13b with theoretical fitting. Fit parameters: $Z_0 = 3.4$ mm, $\Delta F_0 = 0.42$ rad, $T_0 = 1.01$, $z_{orig} = 0$.

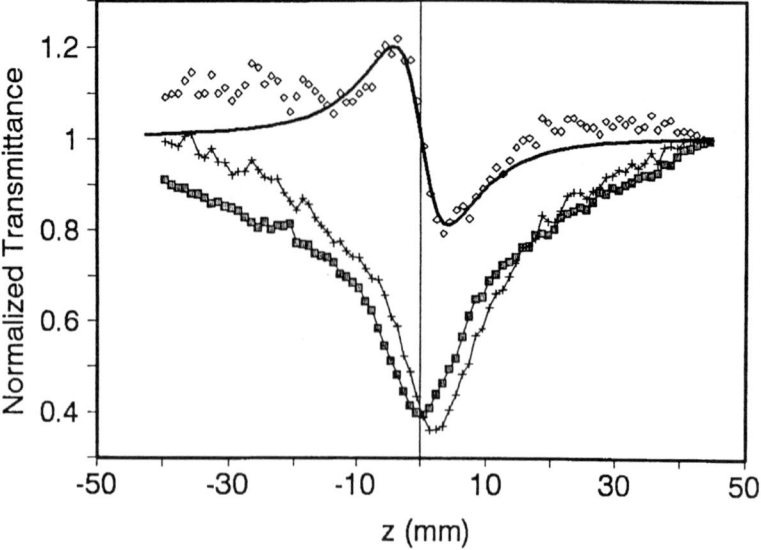

Fig. 15. Z-Scan As_2Se_3 transmittance curves for pulsed dye laser ($\lambda = 590$ nm) with average input power $P_{in} = 6$ mW in a photostructured sample. The solid line indicates the geometrical model fit.

excitation below the band-gap energy, showing the influence of trapped carriers on the nonlinear optical response of amorphous semiconductors.

The Z-Scan investigation of refractive and absorptive parts of the nonlinearity of two commercial ChG infrared windows (As_2S_3 and $Ge_{33}As_{12}Se_{55}$) and of an $Ag_2As_{39}S_{59}$ silver-doped sample around their two-photon edge has been reported recently by Rangel-Rojo, Kosa, Hajto, Ewen, Owen, Kar and Wherrett [1994]. The experimentally measured values of nonlinear refraction are $n_2 = 5.7 \times 10^{-5}$ cm^2/GW for As_2S_3 and $n_2 = 7.9 \times 10^{-5}$ cm^2/GW for $Ge_{33}As_{12}Se_{55}$. For the silver-doped sample the authors observed a large negative value of $n_2 = -3.7 \times 10^{-3}$ cm^2/GW, nearly two orders of magnitude greater than those for the undoped materials.

§ 7. Conclusions

The research that considers the nonlinear interaction of strong laser radiation with ChG, including both CW radiation and short or ultrashort pulse radiation, was described in this paper. Despite many investigations, no definite model for the physics of the nonlinear interaction of strong light with ChG can be found.

Nonlinear interactions of short and ultrashort laser pulses with ChG were briefly examined. The peculiarities of nonlinear light transmission include a nonlinear increase of the light absorption coefficient; a nonlinear refractive-index change, whose magnitude depends strongly on the excitation wavelength and excitation intensity; the threshold character of the nonlinear transmission; the change of the time profile of the laser pulses, leading to a hysteresis transmission dependence; the characteristic time constants of the sample photoinduced darkening, which does not exceed the laser pulse duration; the independence of the nonlinear transmission character on the light beam diameter, on the film thickness, and on the substrate material; the weak dependence from the laser pulse wavelength; the successive step character of restoration of light transmission to the initial state; the full restoration of light transmission initial state in a microsecond time interval; the temperature behavior of the parameters of the nonlinear interaction (the invariance of the light intensity threshold values and of the transient time into the nonlinear light absorption state); and the correlation with the photostructural transformations. It was shown that these characteristics cannot be explained satisfactorily within the limits of the existing physical models proposed in the literature. A new mechanism of nonlinear light absorption in amorphous semiconductors, taking into account the interaction with nonequilibrium phonon modes, is briefly reviewed. It is not really difficult to estimate that the number of created phonons due to the energy dissipation of hot carriers is of the same order as the number of the equilibrium phonons using laser pulse energy density in the range $1-10\,\text{mJ}\,\text{cm}^{-2}$. A reasonable agreement of the calculated with the experimental dependencies is obtained.

Large and reversible changes of the refractive index in ChG, shown by the Z-Scan technique, exhibit considerable interest for the fabrication of photo-optical devices based on the photoinduced dynamical refractive index changes (for light switching, modulation, deflection). In addition, the advantages of large refractive index changes with strong dependence from the wavelength and low absorption of nonresonant nonlinearity can be combined in ChG. The switching times of the devices primarily depend on the intensities of the interband and intraband illuminations and can be very fast, as was shown in femtosecond experiments.

Acknowledgement

We acknowledge the partial support of the NATO Programme for Priority Area on High Technology (Grant HTECH. EV 950772) for this research.

References

Ackley, D.E., J. Tauc and W. Paul, 1979, Phys. Rev. Lett. **43**, 715–718.
Andreev, A.A., B.T. Kolomiets, T.F. Mazets, A.L. Manukian and S.K. Pavlov, 1976, Sov. Solid State Phys. **18**, 53–57.
Andriesh, A.M., O.I. Bogdan, N.A. Enaki, I.A. Cojocaru and V.N. Chumash, 1992, Bul. Ross. Acad. Sci. Phys. Ser. **56**, 96–109.
Andriesh, A.M., Y.A. Bykovskii, E.P. Kolomeiko, A.V. Makovkin, V.L. Smirnov and A.V. Shmal'ko, 1977, Sov. J. Quantum Electron. **7**, 347–352.
Andriesh, A.M., Y.A. Bykovskii, V.L. Smirnov, M.R. Cernii and A.V. Shmal'ko, 1978, Sov. J. Quantum Electron. **8**(13), 293–296.
Andriesh, A.M., and V.N. Chumash, 1993, Bul. Acad. Sci. Moldova Ser. Phys. Tech. **2**(11), 49–66.
Andriesh, A.M., V.N. Chumash and I.A. Cojocaru, 1991, Inst. Phys. Conf. Ser. **115** (IOP, Bristol) pp. 235–238.
Andriesh, A.M., V.N. Chumash, I.A. Cojocaru, M. Bertolotti, E. Fazio and F. Michelotti, 1992, in: Intern. Top. Meet. on Photonic Switching, Techn. Digest, Minsk, p. 2F2.
Andriesh, A.M., V.N. Chumash, I.A. Cojocaru, M. Bertolotti, E. Fazio, F. Michelotti and D. Hulin, 1992, in: Photonic Switching, SPIE Proc. **1807**, 126–135.
Andriesh, A.M., V.N. Chumash, I.A. Cojocaru and N.A. Enaki, 1991, in: Laser and Ultrafast Processes, Vol. 4 (Vilnius Univ. Press) pp. 124–126.
Andriesh, A.M., I.A. Cojocaru, N.D. Ostafeichuk, P.G. Cerbari and V.N. Chumash, 1989, in: Proc. Int. Conf. on Noncrystalline Semiconductors 89, Vol. 1 (Uzhgorod Univ. Press) pp. 204–206.

Andriesh, A.M., N.A. Enaki, I.A. Cojocaru and V.N. Chumash, 1990, Bul. Acad. Sci. Moldova Phys. Tech. **2**, 3–11.
Andriesh, A.M., N.A. Enaki, I.A. Cojocaru, N.D. Ostafeichuk, P.G. Cerbari and V.N. Chumash, 1988, JETP Lett. **14**, 1985–1989.
Andriesh, A.M., N.A. Enaki, I.A. Cojocaru, N.D. Ostafeichuk, P.G. Cerbari and V.N. Chumash, 1989, Vestn. Beloruss. Ser. 1 **1**, 34–36.
Asobe, M., T. Kanamori and K. Kubodera, 1993, IEEE J. Quant. Electron. **29**, 2325–2333.
Asobe, M., K. Suzuki, T. Kanamori and K. Kubodera, 1992, Appl. Phys. Lett. **60**, 1153–1154.
Babinets, Iu.Iu., Iu.V. Vlasenko, M.P. Lisitsa, V.N. Mitsa, V.P. Pinzenik and I.V. Fekeshgazi, 1988, Sov. Quantum. Electron. **15**, 2040–2042.
Baumgartner, R., M. Engelhardt and F.F. Renk, 1983, Phys. Lett. **94A**, 55–58.
Bertolotti, M., V.N. Chumash, E. Fazio, A. Ferrari and C. Sibilia, 1991, Inst. Phys. Conf. Ser. **115**, 169–172.
Bertolotti, M., V.N. Chumash, E. Fazio, A. Ferrari and C. Sibilia, 1993, J. Appl. Phys. **74**, 3024–3027.
Bertolotti, M., F. Michelotti, A.M. Andriesh, V.N. Chumash and G. Liakhou, 1992a, in: Integrated Photonics Research, New-Orleans, OSA Technical Digest Ser. **10**, 20–21.
Bertolotti, M., F. Michelotti, A.M. Andriesh, V.N. Chumash and G. Liakhou, 1992b, in: IQEC'92, XVIII Int. Quantum Electron. Conf. (Technical Digest, Vienna) p. 176.
Bonch-Bruevichi, V.L., I.P. Zveaghin, R. Kaiber, A.G. Mironov, R. Anderlain and B. Esser, 1981, The Electronic Theory of Disordered Semiconductors (Moscow, Nauka) 384pp.
Chumash, V.N., I.A. Cojocaru, G. Bostan, P.G. Cerbari and A.M. Andriesh, 1994, in: Properties and Characteristics of Optical Glass III, SPIE Proc. **2287**.
Dawar, A.L., P.K. Shishodia, G. Chauhan, J.C. Joshi, C. Jagadish and P.C. Mathur, 1990, Appl. Optics **29**, 1971–1973.
De Neufville, J.P., S.C. Moss and S.R. Ovshinsky, 1973/74, J. Non-Cryst. Solids **13**, 191–223.
Elliot, S.R., 1986, J. Non-Cryst. Solids **81**, 71–98.
Fauchet, P.M., K. Gzara, D. Hulin, C. Tanguy, A. Mourchid and A. Antonetti, 1988, in: Ultrafast laser probe phenomena in bulk and microstructure semiconductors II, ed. R.R. Alfano, SPIE Proc. **942**, 92–99.
Fazekas, P., 1981, Philos. Mag. B **44**, 435–452.
Fazio, E., D. Hulin, V.N. Chumash, F. Michelotti, A.M. Andriesh and M. Bertolotti, 1993a, in: Proc. 3rd Nat. Conf. Photonic Techn. for Information Fotonica 93 (IIC, Genova) pp. 305–308.
Fazio, E., D. Hulin, V.N. Chumash, F. Michelotti, A.M. Andriesh and M. Bertolotti, 1993b, J. Non-Cryst. Solids **168**, 213–222.
Fork, R.L., C.V. Shank, A.M. Glass, A. Migus, M.A. Bosch and I. Shah, 1979, Phys. Rev. Lett. **43**, 394–398.
Gazso, J., and J. Hajto, 1978, Phys. Status Solidi a **45**, 181–186.
Gibbs, H.M., 1985, Optical Bistability: Controlling Light with Light (Academic Press, New York).
Gribkovskii, V.P., 1975, Teoria pogloshchenia i ispuscania sveta v poluprovodnikah (Nauka & Tehnika Press, Minsk).
Griffiths, J.E., G.P. Espinosa, J.P. Remeika and J.C. Philips, 1982, Phys. Rev. B **25**, 1272–1286.
Hajto, J., 1980, J. Phys. (Paris) C4 **41**, 63–69.
Hajto, J., and P. Apai, 1980, J. Non-Cryst. Solids **35–36**, 1085–1090.
Hajto, J., and M. Fustoss-Wegner, 1980, in: Proc. Conf. on Amorphous Semiconductors 80, Kishinev, ed. B.T. Kolomietz (Stiiuta-Press) pp. 189–195.
Hajto, J., and I. Janossy, 1983, Philos. Mag. B **47**, 347–366.
Hajto, J., I. Janossy and W.K. Choi, 1985, J. Non-Cryst. Solids **77–78**, 1273–1276.
Hajto, J., I. Janossy, W.K. Choi and A.E. Owen, 1989, J. Non-Cryst. Solids **114**, 304–306.
Hajto, J., I. Janossy and A. Firth, 1983, Philos. Mag. B **48**, 311–321.

Hajto, J., G. Zentai and I. Kosa Somogyi, 1977, Solid State Commun. **23**, 401–403.
Haro-Poniatowski, M., M. Fernandez Guasti, E.P. Mendez and M. Balkanski, 1989, Opt. Commun. **70**, 70–72.
Heo, J., J.S. Sanghera and J.D. Mackenzie, 1991, Opt. Eng. **30**, 470–479.
Hulin, D., A. Mourchid, P.M. Fauchet, W.L. Nighan Jr and R. Vanderhaden, 1991, J. Non-Cryst. Solids **137–138**, 527–530.
Janossy, I., 1988, Phys. Status Solidi b **150**, 783–789.
Kalmikova, N.P., T.F. Mazets, S.K. Pavlov, E.A. Smorgonskaya and E.I. Shifrin, 1988, Sov. J.T.P. Lett. **14**, 739–742.
Kastner, M.A., 1985, in: Physical Properties of Amorphous Materials, eds D. Adler, B.B. Schwartz and M.C. Steele (Plenum Press, New York) pp. 381–396.
Krulikovskii, B.K., M.P. Lisitsa and U. Nasyrov, 1977, Kvantovaya Electron. (Kiev) **12**, 74–79.
Lee, J.M., and M.A. Paesler, 1987, J. Non-Cryst. Solids **97–98**, 1235–1238.
Lisitsa, M.P., U. Nasyrov and I.V. Fekeshgazi, 1977, Ukr. Fiz. J. **22**, 674–676.
Lisitsa, M.P., U. Nasyrov, G.S. Svechnikov and I.V. Fekeshgazi, 1978, in: Proc. Conf. on Amorphous Semiconductors 78, Pardubice, Vol. 2, pp. 460–463.
Malinovskii, V.K., Iu.E. Nesterihin, V.N. Novikov and A.P. Sokolov, 1986, Avtometriia **2**, 3–9.
Mao, X., and P. Yang, 1982, J. Non-Cryst. Solids **52**, 315–320.
Matsuda, A., H. Mizuno, T. Takayama, M. Saito and M. Kikuchi, 1974, Appl. Phys. Lett. **24**, 314–315.
Mazets, T.F., S.K. Pavlov, E.A. Smorgonskaya and E.I. Shifrin, 1987, J. Non-Cryst. Solids **90**, 537–540.
Michelotti, F., M. Bertolotti, V.N. Chumash and A.M. Andriesh, 1992, in: Photonics for Computers, Neural Networks, and Memories, San Diego, SPIE Proc. **1773**, 423–432.
Michelotti, F., E. Fazio, F. Senesi, M. Bertolotti, V.N. Chumash and A.M. Andriesh, 1993, Opt. Commun. **101**, 74–78.
Miller, D.A.B., C.T. Seaton, M.E. Prise and S.D. Smith, 1981, Phys. Rev. Lett. **47**, 197–200.
Nasu, H., K. Kubodera, M. Kobayashi, M. Nakamura and K. Kamiya, 1990, J. Am. Ceram. Soc. **73**(6), 1794–1796.
Nasyrov, U., 1978, Sov. Phys. Semiconduct. **12**, 1210–1212.
Nasyrov, U., G.S. Svechnikov and I.V. Fekeshgazi, 1980, Ukr. Fiz. J. **25**, 424–428.
Ohmachi, Y., and T. Igo, 1972, Appl. Phys. Lett. **20**, 506–508.
Orbach, R., 1993, J. Non-Cryst. Solids **164–166**, 917–922.
Popescu, M., A.M. Andriesh, V.N. Chumash, N.A. Enaki, I.A. Cojocaru and A. Grozescu, 1991, J. Non-Cryst. Solids **137–138**, 973–976.
Rangel-Rojo, R., T. Kosa, J. Hajto, P.J.S. Ewen, A.E. Owen, A.K. Kar and B.S. Wherrett, 1994, Opt. Commun. **109**, 145–150.
Sheik-Bahae, M., Ali A. Said, T. Wei, D.J. Hagan and E.W. Van Stryland, 1990, IEEE J. Quantum Electron. **26**, 760–769.
Stourac, L., B.T. Kolomiec and V.P. Silo, 1968, Czech. J. Phys. B **18**, 92–96.
Suhara, T., Y. Handa, H. Nishihara and J. Koyama, 1982, Appl. Phys. Lett. **40**(2), 120–122.
Tanaka, K., 1978, Solid State Commun. **28**, 541–545.
Tanaka, K., 1979, Appl. Phys. Lett. **34**, 672–674.
Tanaka, K., 1980a, J. Non-Cryst. Solids **35–36**, 1023–1034.
Tanaka, K., 1980b, Solid State Commun. **34**, 201–204.
Tanaka, K., 1990, Rev. Solid State Sci. **4**, 641–659.
Tanaka, K., Y. Imai and A. Odajima, 1985, J. Appl. Phys. **57**, 4897–4900.
Tanaka, K., and A. Odajima, 1981, Appl. Phys. Lett. **38**, 481–483.
Tanaka, K., and Y. Ohtsuka, 1976, Opt. Commun. **19**, 134–137.

Tanaka, K., and Y. Ohtsuka, 1978, J. Appl. Phys. **49**, 6132–6135.
Toth, L., J. Hajto and G. Zentai, 1977, Solid State Commun. **23**, 185–188.
True, E.M., and L. McCaughan, 1991, Opt. Lett. **16**, 458–460.
Vardeny, Z., 1983, J. Non-Cryst. Solids **59–60**, 317–324.
Wraback, M., J. Tauc, D. Pang, W. Paul, I.K. Lee and E. Schiff, 1991, J. Non-Cryst. Solids **137–138**, 531–534.
Xu, L.B., S.F. Cai and C.X. Xie, 1988, Phys. Status Solidi a **150**, 797–803.

E. WOLF, PROGRESS IN OPTICS XXXVI
© 1996 ELSEVIER SCIENCE B.V.
ALL RIGHTS RESERVED

II

QUANTUM PHENOMENA IN OPTICAL INTERFEROMETRY

BY

P. Hariharan

CSIRO Division of Applied Physics, PO Box 218, Lindfield, New South Wales 2070, Australia

AND

B.C. Sanders

*School of Mathematics, Physics, Computing and Electronics
and Centre for Lasers and Applications, Macquarie University,
North Ryde, New South Wales 2109, Australia*

CONTENTS

		PAGE
§ 1.	INTRODUCTION	51
§ 2.	OPTICAL SOURCES	54
§ 3.	SECOND-ORDER INTERFERENCE	65
§ 4.	THE GEOMETRIC PHASE	73
§ 5.	FOURTH-ORDER INTERFERENCE	78
§ 6.	TWO-PHOTON INTERFEROMETRY	91
§ 7.	COMPLEMENTARITY	108
§ 8.	QUANTUM LIMITS TO INTERFEROMETRY	117
§ 9.	CONCLUSIONS	122
	REFERENCES	123

§ 1. Introduction

From the time of Young's classical experiment, interference has been regarded as a conclusive demonstration of the wave-like nature of light. However, under appropriate conditions, light also behaves as indivisible particles known as photons. The particle-like behavior of light is particularly noticeable at low light levels, when photo detectors register distinct events corresponding to the annihilation of individual photons. At the single-photon level, quantum effects cannot be ignored.

1.1. QUANTUM EFFECTS

What are quantum effects? The answer lies in an understanding of complementarity. Under appropriate conditions, light can be treated as a wave, subject to the behavior dictated by Maxwell's equations; in other situations, light seems to be composed of localized particles. As delocalization is necessary for interference to occur, these two views of light are incompatible. Nevertheless, the two views do apply to light from the same source, and this contention is supported by experimental evidence. This dichotomy is the essence of the quantum nature of light: not that light is a particle or a wave, but rather that light exhibits the characteristics of both a particle and a wave, leading to Dirac's famous statement that "... each photon interferes only with itself. Interference between different photons never occurs" (Dirac [1958]).

1.2. COMPLEMENTARITY

The paradox presented by something that is both corpuscular and undular is avoided by the existence of a complementarity principle in measurements. This principle of complementarity limits the possibility of performing a measurement which *simultaneously* demonstrates both undular and corpuscular behavior. We can, of course, use the same source of light first to observe particles and then to observe waves, and *vice versa*, but we cannot take the same "piece of light" and check one feature without altering, or demolishing, the other.

A way to resolve this paradox has been interference experiments at low light levels, using beams of light with low photon numbers. These experiments can be categorized according to the light source used, the manipulation of the beams and the detection technique. By permuting the various possibilities, experiments can be performed to explore a range of quantum phenomena.

1.3. SECOND-ORDER COHERENCE

The optical field can be characterized completely by the expectation values of various powers and products of the field variables, beginning with the second-order coherence (Wolf [1955]).

Second-order coherence corresponds to measurements of complex amplitudes for the field at two space–time points using interferometers, such as the Michelson and Mach–Zehnder interferometers, which produce interference fringes by mixing two fields with differing phases. The usual way to do this is by varying the optical path difference, but an alternative method is by operating on the geometric phase (Berry [1984, 1987]).

We first review experiments of this type which explore the quantum nature of light, ranging from early experiments involving low-intensity sources and fields from independent sources at low intensity levels, to measurements using nonclassical light sources, including single-photon states and photon-pair sources. The wave-like behavior of light is made clear by accumulating enough single-photon events; the corpuscular aspect is revealed by determining the path of the photon. For the photon to produce interference fringes, it must interfere with itself, a condition which requires the photon to traverse both paths; yet a direct measurement of the photon in either path should, by the principle of complementarity, reveal that the photon has localized itself to one path or the other. "Which path" (*welcher Weg*) measurements verify that the photon is indeed localized to one path or the other when an inspection is carried out, despite the fact that the photon can interfere with itself in the absence of an inspection.

1.4. NONCLASSICAL STATES OF LIGHT

A precise distinction between semiclassical and nonclassical states of light can be made by expanding the quantum field state in the Glauber coherent-state basis (Glauber [1963a,b]). If the corresponding c-number quasiprobability distribution, the Glauber–Sudarshan *P*-representation (Glauber [1963a,b], Sudarshan [1963]), satisfies the requirements of a probability distribution, namely that it is nonnegative and normalized to unity, then the state is referred to as a

classical state of light. A purely quantum state produces a Glauber–Sudarshan
P-representation which does not satisfy the criteria for a probability distribution.

1.5. FOURTH-ORDER COHERENCE

Measurements of second-order coherence cannot unambiguously distinguish a classical state of light from a quantum state. However, fourth-order coherence measurements, which correspond to intensity correlation measurements, permit quantum states of light to be distinguished from classical states. In the simplest conceptual form of these experiments, a single photon is directed at a beam splitter. A measurement at the two output ports of the beam splitter should, under ideal circumstances, detect the photon in either output port, whereas a classical field would be split between the two. The fourth-order coherence function is sensitive to the fact that the photon cannot be split into two regions of space, and there exists a range of values for fourth-order coherence measurements which are attainable for quantum states of light, but not for classical states of light. Besides providing unambiguous evidence for the existence of nonclassical light, fourth-order coherence experiments often present phenomena which appear counterintuitive.

1.6. ENTANGLED STATES

Correlated photon pairs produced by parametric down-conversion make possible a variety of experiments by which to probe the quantum mystery. The two photons are produced in an entangled state leading to a situation where observations on either photon separately reveal no interference, but observations involving coincidences of the two photons yield higher-order interference fringes.

Tests of Bell's inequality (Bell [1965]) play a crucial role in ruling out local realism as an alternative framework for describing quantum effects. Two-photon interferometry using entangled states makes it possible to carry out such tests without invoking polarization. Another series of experiments on two-photon interferometry vindicates Feynman's proposition that states interfere with each other only when they cannot be distinguished physically in the experimental setup. Yet another series of experiments has demonstrated the idea of a "quantum eraser", in which it is possible to appear to destroy "which-path" information without actually doing so.

1.7. BEAM-SPLITTING AND TUNNELING

As mentioned earlier, any description of light involving waves predicts that there will be some coincidences between photon detectors placed in the two output fields produced by a beam splitter. On the other hand, for single-photon states, quantum theory predicts, and experiments confirm, the probability of coincidences to be zero.

In a variant of these experiments, the beam splitter is replaced by two prisms with a very small air gap between them. It is then possible, with single-photon states, to observe particle behavior (anticoincidence) and wave behavior (tunneling) in the same apparatus.

A related question is the time taken by a photon to tunnel through such a barrier. Experiments using two-photon interference have revealed such puzzling effects as apparently superluminal tunneling velocities.

1.8. QUANTUM LIMITS

Finally, quantum effects set a limit to the precision attainable in measurements using interferometry. The complementarity of particle and wave aspects is responsible for an uncertainty principle linking the photon number and the measured phase. However, it does appear possible to perform measurements with precision beyond the standard quantum limit (SQL), either by injecting nonclassical light into the interferometer or by replacing passive optical elements in the interferometer by active nonlinear elements.

§ 2. Optical Sources

The earliest sources of light for interference experiments were thermal light sources, such as the sun and incandescent light. They have been supplemented by the laser, a source of coherent light, and, more recently, by atomic sources and nonlinear optical materials which provide nonclassical light.

Coherence functions have been used for many years to characterize classical light fields (Wolf [1955]), but Glauber [1963a,b] was the first to construct the quantum analog of classical coherence functions to characterize the coherence properties of nonclassical radiation. An important realization was that measurements of second-order coherence[1] cannot provide unambigous evidence of

---

[1] Glauber [1963a,b] refers to this function as the first-order correlation.

nonclassical light; measurements of fourth-order coherence, at the very least, are required to distinguish nonclassical from classical radiation.

2.1. QUANTUM DESCRIPTION OF RADIATION

By analogy with the coherence functions used to describe the interference properties of the classical radiation field, Glauber [1963a,b] introduced the normalized quantum coherence functions to describe the quantum field. The electric field at the space–time point $(\vec{r}_i, t_i)$ is replaced by the operator $\widehat{E}(\vec{r}_i, t_i)$, which can be separated into negative- and positive-frequency components, $\widehat{E}^-(\vec{r}_i, t_i)$ and $\widehat{E}^+(\vec{r}_i, t_i)$, respectively. What would be referred to, in classical terms, as the normalized coherence functions of order $2n$ are then defined as the correlation functions of order n:

$$g^{(n)}(\{\vec{r}_i, t_i \mid i = 1, \ldots, 2n\}) = \frac{\left\langle : \prod_{i=1}^{n} \widehat{E}^-(\vec{r}_i, t_i) \prod_{i=n+1}^{2n} \widehat{E}^+(\vec{r}_i, t_i) : \right\rangle}{\prod_{i=1}^{2n} \sqrt{\left\langle \widehat{E}^-(\vec{r}_i, t_i) \widehat{E}^+(\vec{r}_i, t_i) \right\rangle}},$$

(2.1)

where : : represents the normal-ordering operation, and the angular brackets refer to both ensemble averaging and to quantum state averaging. Whereas $0 \leqslant g^{(1)}(\{\vec{r}_i, t_i \mid i = 1, 2\}) \leqslant 1$ for both quantum and classical radiation fields, the second-order correlation function satisfies the criteria

$$1 \leqslant g^{(2)}(\{\vec{r}_i, t_i \mid i = 1, \ldots, 4\}) < \infty \quad \text{for a classical field,} \tag{2.2}$$

$$0 \leqslant g^{(2)}(\{\vec{r}_i, t_i \mid i = 1, \ldots, 4\}) < \infty \quad \text{for a quantum field.} \tag{2.3}$$

Hence, values of second-order correlation in the range

$$0 \leqslant g^{(2)}(\{\vec{r}_i, t_i \mid i = 1, \ldots, 4\}) < 1 \tag{2.4}$$

indicate unambiguously a nonclassical light source.

In addition to using the quantum correlation functions, it is important to describe the field in terms of quantum states. The field emitted by the source can be decomposed into arbitrary modes indexed by the three-vector $\vec{k}$, which, for a plane wave, is the wave vector.

We represent the n-photon state in mode $\vec{k}$ by the expression

$$|n\rangle_{\vec{k}} \equiv (n!)^{-1/2} \left(\hat{a}_{\vec{k}}^\dagger\right)^n |0\rangle, \tag{2.5}$$

in which $|0\rangle$ designates the zero-photon state or ground state, of the field mode, corresponding to the vacuum state, and the subscript $\vec{k}$ can be ignored where not required.

The nonclassical nature of the n-photon state is made evident when we calculate the second-order correlation function,

$$g^{(2)}(0) = 1 - \frac{1}{n}, \tag{2.6}$$

which, from eq. (2.4), is within the nonclassical regime. For $n=1$, the interpretation of eq. (2.6) is particularly straightforward: the photon is indivisible, and fractions of the photon cannot be detected at different space–time points.

The bridge between the classical and quantum descriptions of radiation is provided by the coherent state of light,

$$|\alpha\rangle = \exp\left(-\tfrac{1}{2}|\alpha|^2\right) \sum_{n=0}^{\infty} \frac{\alpha^n}{\sqrt{n!}} |n\rangle, \tag{2.7}$$

which has complex amplitude α and mean photon number $|\alpha|^2$. This coherent state of light is indistinguishable from classical coherent light: all orders of the correlation function defined by eq. (2.1) are equal to unity for the coherent state defined by eq. (2.7).

Classical radiation states can be expressed in the quantum framework as distributions of coherent states of light. Thus, the density matrix for the light field can be written as:

$$\hat{\rho} = \int d^2\alpha\, P(\alpha) |\alpha\rangle\langle\alpha|, \tag{2.8}$$

where $P(\alpha)$ is referred to as the Glauber–Sudarshan P-representation after the work of Glauber [1963a,b] and Sudarshan [1963], and the classical radiation states are diagonal in the overcomplete coherent-state basis. The P-representation for the coherent state $|\alpha_0\rangle\langle\alpha_0|$ is evidently $P(\alpha) = \delta^{(2)}(\alpha - \alpha_0)$, where $\delta^{(2)}$ is the second-order Dirac δ function. Nonclassical light corresponds to the states for which the Glauber–Sudarshan P-representation is not positive definite.

A multimode coherent state is a tensor product of single-mode coherent states,

$$|\vec{\alpha}\rangle = \prod_{i=1}^{n} |\alpha_i\rangle_i, \tag{2.9}$$

where each mode i has dimensionless complex amplitude α_i. A straightforward extension of the single-mode theory given above can then be made to the multimode case. For example, the Glauber–Sudarshan representation is generalized to:

$$\hat{\rho} = \int \prod_{i=1}^{n} d^2\alpha_i\, P(\vec{\alpha}) |\vec{\alpha}\rangle\langle\vec{\alpha}|. \tag{2.10}$$

A particularly important case is the weak field. If a coherent field is attenuated to a level at which the probability of detecting more than one photon in the field, with an ideal unit-efficiency detector, is negligible, then the multimode coherent state (2.9) can be expanded as the normalized state,

$$|\vec{\alpha}\rangle \approx \frac{\prod_{i=1}^{n} |0\rangle_i + \sum_{i=1}^{n} \prod_{j=1}^{n} \alpha_i |\delta_{ij}\rangle_j}{\sqrt{1 + \sum_{i=1}^{n} |\alpha_i|^2}}, \qquad (2.11)$$

where no more than one photon exists in the field. In the single-mode case, the weak coherent field can be expressed as:

$$|\alpha\rangle \approx \left[1 + |\alpha|^2\right]^{-1/2} \left[|0\rangle + \alpha|1\rangle\right], \qquad (2.12)$$

indicating that a coherent superposition exists between the absence of the photon and the presence of the photon in the field mode.

Interference involves the mixing of two fields. With an interferometer constructed with beam splitters, mirrors and passive phase-shifters, the output is a two-mode coherent state which, in the weak-field limit, can be viewed as representing the interference of a photon with itself. This self-interference of a photon has been interpreted as a sum over histories (Feynman, Leighton and Sands [1963]). In this picture, a photon can take either of two separate paths from the source to the detector. Associated with each path is a certain complex probability amplitude a_i ($i = 1, 2$), whose absolute square represents the probability of the photon taking this path. The intensity at the detector is then obtained by summing the probability amplitudes for the two paths and taking the square of its modulus $|a_1 + a_2|^2$, which gives the probability of detecting a photon at this point. Each photon therefore interferes only with itself; it is the basic quantum uncertainty of which path the photon takes through the apparatus that is responsible for the interference effect.

The quantum regime is attained for classical coherent light fields when the time interval between photodetection events, with an ideal, perfectly efficient detector, is much greater than the transit time of radiation through the system. If the state of the radiation field produced by the laser is represented by the coherent state $|\alpha\rangle$, then the attenuated laser field is given by $|\sqrt{\eta}\alpha\rangle$, where $|\alpha|^2$ is the photon flux prior to attenuation, and $\eta|\alpha|^2$ is the photon flux following attenuation. Attenuation does not destroy the coherence of the beam, or affect the coherence time. However, a reduction of the photon flux increases the integration time required to observe coherence effects, and can eliminate the possibility of detecting interference if the integration time required exceeds the coherence time.

2.2. INDEPENDENT SOURCES

To observe interference effects with independent sources, the measurements must occupy a time scale shorter than the mutual coherence time, and the average number of photons received during this time must be large enough to obtain an adequate signal-to-noise ratio.

If two independent laser beams are attenuated to a level where the mean photon number is less than one, the resulting field is brought into the quantum regime and can be regarded as the product of two coherent states corresponding to two independent modes $|\alpha_1\rangle_1$ and $|\alpha_2\rangle_2$. The state of the field is then given by the relation

$$|\alpha_1, \alpha_2\rangle_{1,2} = |\alpha_1\rangle_1 |\alpha_2\rangle_2 \approx \frac{|0\rangle_1 |0\rangle_2 + \alpha_1 |1\rangle_1 |0\rangle_2 + \alpha_2 |0\rangle_1 |1\rangle_2}{\sqrt{1 + |\alpha_1|^2 + |\alpha_2|^2}}. \quad (2.13)$$

The weak two-mode coherent state is therefore a coherent superposition of one photon in mode 1 or one in mode 2, as well as a contribution due to no photon in either mode. If the state represented by eq. (2.13) is conditioned on the detection of a photon, the vacuum state is eliminated from the superposition, yielding the entangled state,

$$|1;0\rangle_{1,2}^{\theta,\xi} \equiv \cos\theta \, |1\rangle_1 |0\rangle_2 + e^{i\xi} \sin\theta \, |0\rangle_1 |1\rangle_2, \quad (2.14)$$

where $\theta = \tan^{-1} |\alpha_2/\alpha_1|$, and $\xi = \arg(\alpha_2/\alpha_1)$.

2.3. TWO-ATOM SOURCES

Laser-driven atoms emit radiation through the process of resonance fluorescence. A case of particular interest is where two identical atoms are made to fluoresce coherently, so that each photon can be regarded as arriving at the detector by a superposition of the paths from the two sources. Vigué, Grangier, Roger and Aspect [1981] and Vigué, Beswick and Broyer [1983] produced two identical atoms travelling in opposite directions by photodissociation of a homonuclear molecule. Time-resolved studies of the fluorescence from such a diatomic source were carried out by Grangier, Aspect and Vigué [1985]. They excited the $X\,^1\Sigma_g^+$ state of a beam of Ca_2 molecules to the $^1\Pi_u$ dissociative state with a mode-locked pulsed Kr^+ laser beam having a wavelength of 406.7 nm. The fluorescence radiation (wavelength 422.7 nm) was emitted perpendicular to the plane of the atomic beam and the electric field of the Kr^+ laser beam. Time-resolved studies of the fluorescence revealed a modulation that could be taken as evidence for quantum interference, due to the indistinguishability of the paths from the two sources.

2.4. SOURCES OF NONCLASSICAL LIGHT

With a single-mode laser source, the arrival of photons at a photo detector exhibits a Poisson distribution. With a thermal source, the optical field can be regarded as the sum of the fields contributed by many independent coherent sources. Hence, with classical light sources, such as thermal sources and lasers, photo detections are more likely to occur at the same time, or very close together, than farther apart in time. This phenomenon is known as photon bunching (Mandel and Wolf [1965]) and has been attributed to the fact that photons are bosons. None of these sources can therefore generate any form of nonclassical light, such as a single-photon state.

2.5. SINGLE-PHOTON STATES

The creation of an n-photon state is not easy. One method for preparing an approximation to a single-photon state is by generating a pair of photons. Essentially, the process is one of conditional preparation: given that either two photons exist or no photon exists, the detection of one photon acts as a signal that a second photon is present in the field. The frequency and direction of propagation of the second photon are related to those of the first by conservation laws, and can be determined by analysing the first "gate" photon. The second photon field can then be regarded as being in a one-photon Fock state.

2.5.1. Atomic cascade

Two nearly simultaneous photons can be produced by an atomic cascade (Kocher and Commins [1967], Freedman and Clauser [1972]) using atoms of calcium, which are excited to the $6\,^1P_1$ state by means of a UV source, such as a hydrogen arc lamp. About 10% of these atoms then go into the $6\,^1S_0$ level, from which they return to the ground state *via* the $4\,^1P_1$ level. In this two-step process they emit, in rapid succession, two photons with wavelengths of 551.3 nm and 422.7 nm, respectively. A more efficient procedure is to excite the atoms selectively to the upper level of the cascade by two-photon absorption, using a Kr^+ laser ($\lambda = 406.7$ nm) and a dye laser tuned to resonance for the two-photon process ($\lambda = 581$ nm). The time interval between the emission of the two photons corresponds to the lifetime ($\tau_s = 4.7$ ns) of the intermediate state of the cascade (Aspect, Grangier and Roger [1981]).

2.5.2. Parametric down-conversion

The atomic cascade suffers from two drawbacks: the emission of the two photons is not perfectly simultaneous, and the correlation between their directions is not perfect. Parametric down-conversion overcomes these problems. In this process, a single UV photon decays spontaneously in a crystal with a $\chi^{(2)}$ nonlinearity into two photons (a *signal* photon and an *idler* photon) with wavelengths close to twice the UV wavelength (Harris, Oshman and Byer [1967], Klyshko [1967], Burnham and Weinberg [1970]). Down-conversion is facilitated by using a birefringent crystal to achieve phase matching. The two down-converted photons are highly correlated and are emitted with a negligible time separation (Hong and Mandel [1985], Friberg, Hong and Mandel [1985]). Since energy is conserved in the process, we have:

$$\hbar\omega_0 = \hbar\omega_1 + \hbar\omega_2, \tag{2.15}$$

where $\hbar\omega_0$ is the energy of the UV photon, and $\hbar\omega_1$ and $\hbar\omega_2$ are the energies of the two down-converted photons. Similarly, since momentum is conserved, we have:

$$\mathbf{k}_0 = \mathbf{k}_1 + \mathbf{k}_2, \tag{2.16}$$

where $\mathbf{k}_0$ is the momentum of the UV photon, and $\mathbf{k}_1$ and $\mathbf{k}_2$ are the momenta of the down-converted photons. It follows from eq. (2.15) that, while the frequencies of the individual down-converted photons may vary over a broad range, the sum of their frequencies is well defined. Similarly, it follows from eq. (2.16) that the photons in each pair are emitted on opposite sides of two cones, whose axis is the UV beam, and produce, as shown in fig. 2.1, a set of rainbow colored rings.

In a typical realization, Hong and Mandel [1986] used the UV beam from an argon-ion laser ($\lambda = 351.1$ nm) and a potassium dihydrogen phosphate (KDP) crystal to generate pairs of photons with wavelengths around 746 and 659 nm. These photons leave the crystal at angles of approximately $\pm 1.5°$ to the UV beam.

2.6. THE BEAM SPLITTER

Adam, Janossy and Varga [1955a,b] were the first to study the correlations between photons at a beam splitter, rather than the amplitude correlations

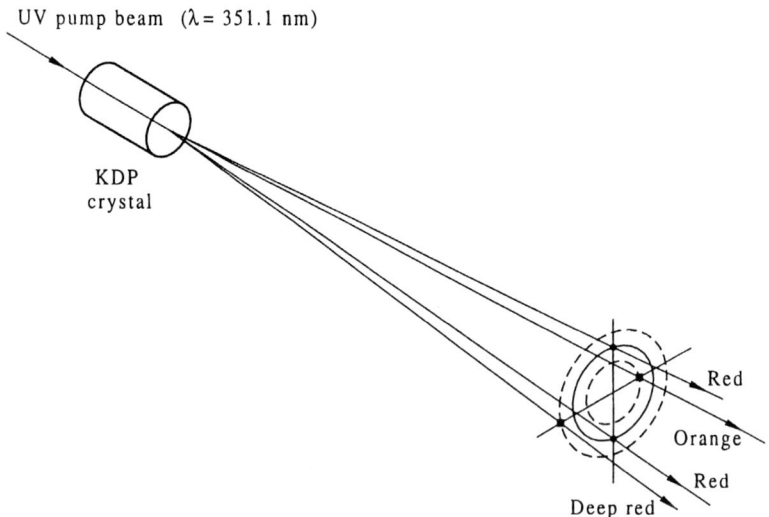

Fig. 2.1. Generation of photon pairs by parametric down-conversion of UV photons in a nonlinear crystal. Conjugate photons are emitted on opposite sides of the UV beam; typically, one photon has a slightly lower frequency than the other.

measured by normal interferometric techniques. Subsequently, Hanbury Brown and Twiss [1956] (see also Hanbury Brown [1974]) measured photon correlations in thermal light fields in their experiments leading to the intensity interferometer (see § 5), while Arecchi, Berné and Burlamacchi [1966] studied photon correlations with laser sources. For a thermal source, the correlation between photons detected at the two outputs of a beam splitter is positive; for a coherent field from a laser, there is no correlation between the two outputs.

On the other hand, nonclassical fields with definite photon number exhibit a very different behavior. With a single-photon state, quantum mechanics predicts a perfect anticorrelation between the counts at the two output ports of a beam splitter (Clauser [1974]). Grangier, Roger and Aspect [1986], as well as Diedrich and Walther [1987], observed such anticoincidences, indicating that each photon was either transmitted or reflected. The perfect anticorrelation of photons at the beam splitter can be regarded as evidence of the indivisibility of the photon.

The indivisibility of the photon provides the simplest example of entanglement. The output from the beam splitter can be regarded as the superposition of two histories, the first consisting of one photon at port 1 and no photon at port 2, and the second consisting of no photon at port 1 and one photon at port 2. This superposition state cannot be reduced because of the strong anticorrelation of the two modes that arises from the entanglement of the input field with the vacuum field represented by eq. (2.14).

It follows that any analysis of the effect of a beam splitter on an incident beam of light with definite photon number has to take into account the vacuum field at the unused input port (Fearn and Loudon [1987, 1989], Campos, Saleh and Teich [1989]). Such a treatment confirms the effects observed with classical fields as well as with photon-number states. When one of the input photon-number states is the vacuum and the other is a nonzero number state, the photon numbers at the output ports are described by a binomial distribution (Brendel, Schütrumpf, Lange, Martienssen and Scully [1988]). However, if the inputs at the two ports of a 50:50 beam splitter are identical single-photon states, the joint probability for detecting a photon at each of the two output ports vanishes. This implies that both incident photons must exit together at either of the two output ports, and is an example of quantum-mechanical interference of the probability amplitudes for a photon pair (Hong, Ou and Mandel [1987]).

2.7. SQUEEZED STATES OF LIGHT

The coherent state $|\alpha\rangle$ incorporates vacuum fluctuations which become observable with phase-sensitive detection schemes, such as heterodyne or homodyne detection. These vacuum fluctuations are responsible for the limit on the accuracy of measurements set by shot noise, known as the standard quantum limit (SQL). Yuen [1976] performed a detailed analysis of what he then called *two-photon coherent states* which revealed the possibility of reducing quantum noise using such states.

We can represent the electric field of a monochromatic light wave as the sum of two quadrature components in the form:

$$E = E_0 [X_1 \cos \omega t + X_2 \sin \omega t], \qquad (2.17)$$

where X_1 and X_2 are complementary operators satisfying the commutation relation $[X_1, X_2] = \frac{1}{2}i$, whose variances therefore obey the uncertainty relationship

$$\Delta X_1 \Delta X_2 \geqslant \tfrac{1}{4}. \qquad (2.18)$$

For normal coherent light the variances are equal. For a squeezed state the variances are unequal, although their product remains unchanged. Accordingly, it is possible to reduce phase fluctuations with squeezed light, as shown in fig. 2.2, at the expense of a corresponding increase in the amplitude fluctuations (Caves [1981], Walls [1983]).

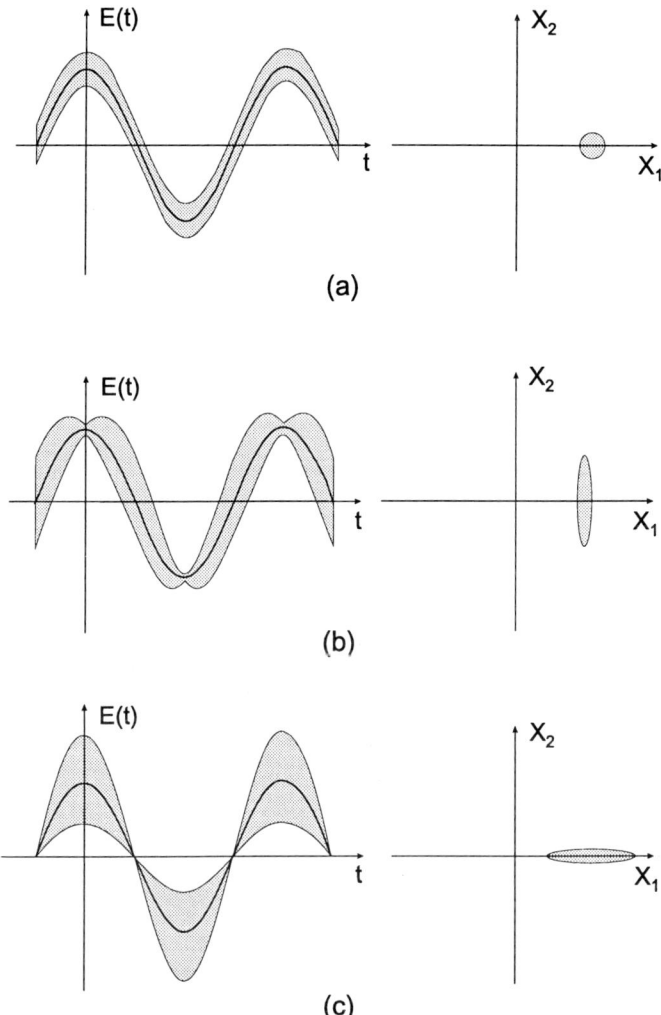

Fig. 2.2. Plot of electric field against time, showing the uncertainty for (a) a coherent state, (b) a squeezed state with reduced amplitude fluctuations, and (c) a squeezed state with reduced phase fluctuations (Caves [1981]).

The degree of squeezing can be measured with a balanced homodyne detector, which yields a phase-sensitive measurement of the noise. As shown in fig. 2.3, the squeezed light is combined with another strong beam from the same source, which constitutes a local oscillator, at a 50:50 beam splitter. The beams emerging from the beam splitter are directed to two photo detectors, and the difference of

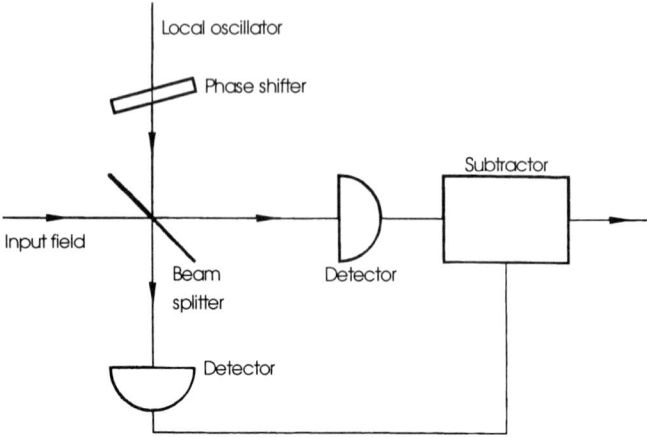

Fig. 2.3. Balanced homodyne detector used for measurements of noise due to a single field-quadrature component.

the two photocurrents is displayed. With this arrangement, intensity fluctuations in the local oscillator and the signal cancel out, and the output corresponds to the interference of the local oscillator with the signal. If the intensity of the local oscillator is much greater than that of the signal, the fluctuations in the output are essentially due to the signal. As the phase difference between the squeezed light signal and the local oscillator is varied, the detector becomes sensitive first to one quadrature amplitude and then to the other, and the output noise amplitude varies accordingly.

The concept of single-mode squeezed states can be extended to two or more correlated modes. Milburn [1984] and Caves and Schumaker [1985] considered two-mode states consisting of a two-mode coherent field with squeezed vacuum fluctuations. Whereas the single-mode squeezed state is generated by photon-pair creation and annihilation in one mode, the two-mode squeezed state is generated by photon-pair creation and annihilation in two modes. The two photons in the pair, each with a different frequency, are strongly correlated, and this correlation is responsible for the squeezed fluctuations.

The process of generating photon pairs is not restricted to parametric down-conversion and can be implemented with any order of nonlinearity. A number of physical phenomena can therefore be used, in principle, to generate squeezed states. The earliest and most common method has been degenerate four-wave mixing in a medium with a nonlinear susceptibility (Slusher, Hollberg, Yurke, Mertz and Valley [1985], Shelby, Levenson, Perlmutter, de Voe and Walls [1986], Maeda, Kumar and Shapiro [1987]). In such a process, energy is transferred from two strong pump beams to two weaker beams. As a result, correlations are established between the photons in the two weaker beams. When these two beams are combined, the resulting light exhibits the characteristics of squeezed states. A greater reduction in noise has been achieved by degenerate parametric down-conversion (Wu, Kimble, Hall and Wu [1986]). When the gain from parametric amplification becomes large in the down-conversion crystal, there is a transition from spontaneous to stimulated emission. Since the gain depends on the phase of the amplified light relative to the phase of the pump beam, vacuum fluctuations in one quadrature are squeezed (Kimble and Walls [1987]).

§ 3. Second-order Interference

The apparent contradiction between viewing light as particles and light as waves provided the impetus for studying the interference of a light beam with itself at power levels so low that the probability of two or more photons existing at the same time within the apparatus was negligible. If light is thought of as particles, and interference is a phenomenon involving the interaction of at least two particles, classical considerations suggest that interference effects should become weaker as the number of photons decreases and disappear completely when no more than one photon is in the apparatus at a time.

3.1. INTERFERENCE AT THE "SINGLE-PHOTON" LEVEL

These considerations led to a series of experiments involving photographic recordings of interference patterns at extremely low light levels, all of which showed that the quality of the pattern did not depend on the intensity (Taylor [1909], Gans and Miguez [1917], Zeeman [1925], Dempster and Batho [1927]). This result was also confirmed by counting photons with a photomultiplier (Janossy and Naray [1957]). A detailed review of these experiments involving interferometry at low light levels has been presented by Pipkin [1978].

While these experiments supported the predictions of quantum theory, all of them used conventional thermal light sources. The light from such a thermal source can be modeled as an ensemble of coherent states, each of which can be described by a classical electromagnetic field, even when it is highly attenuated. Accordingly, it could be argued that these experiments did not actually involve single-photon states.

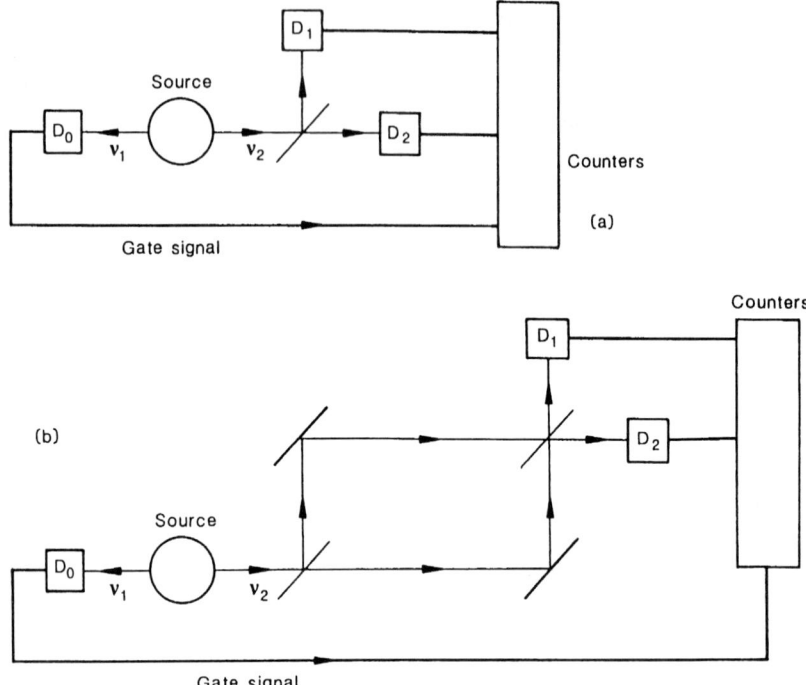

Fig. 3.1. Experimental arrangement used (a) to detect single-photon states and (b) to demonstrate interference with single-photon states (Grangier, Roger and Aspect [1986]).

3.2. INTERFERENCE WITH SINGLE-PHOTON STATES

Interference effects produced by light without a positive definite Glauber–Sudarshan P-representation cannot be explained in classical terms. A good example is interference with single-photon states, which was first studied using an atomic cascade (see § 2.5.1) by Grangier, Roger and Aspect [1986], and also by Aspect and Grangier [1987].

As shown in fig. 3.1a, the arrival of the first photon (frequency v_1) at the detector D_0 acted as a trigger for a gate generator, enabling the two detectors D_1 and D_2 on the two sides of the beam splitter for a time $2\tau_s$. During this period, the probability for the detection of a second photon (frequency v_2) emitted by the same atom is much greater than the probability of detecting a similar photon emitted by any other atom in the source.

While a classical wave would be divided between the two output ports of the beam splitter, a single photon cannot be divided in this fashion. We can therefore expect an anticorrelation between the counts on the two sides of the beam splitter at D_1 and D_2, measured by a parameter

$$\mathcal{A} \equiv \frac{N_{012}N_0}{N_{01}N_{02}}, \tag{3.1}$$

where N_{012} is the rate of triple coincidences between the detectors D_0, D_1 and D_2; N_{01} and N_{02} are the rate of double coincidences between D_0 and D_1 and D_0 and D_2, respectively, and N_0 is the rate of counts of D_0. For a classical wave, it follows from Schwarz's inequality that $\mathcal{A} \geqslant 1$, while the indivisibility of the photon should lead to arbitrarily small values of $\mathcal{A}$. As expected, the number of coincidences observed for the second photon, with a gate time of 9 ns, was only 0.18 of that expected from classical theory, but corresponded to that predicted by quantum theory. This source was then used in the optical arrangement shown in fig. 3.1b, with the detectors D_1 and D_2 receiving the two outputs from a Mach–Zehnder interferometer. The interferometer was initially adjusted and checked without the gating system in operation, and interference fringes with a visibility $\mathcal{V} > 0.98$ were obtained. In the actual experiment, with the gate on, the optical path difference was varied around zero in 256 steps, each of $\lambda/50$, with a counting time of 1 s at each step. The results of 15 such sweeps were then averaged to improve the signal-to-noise ratio. Analysis of the data showed that, even with the gate operating, values of the visibility $\mathcal{V} > 0.98$ were obtained.

The results of these experiments confirmed the predictions of quantum mechanics and Dirac's view that the photon interferes with itself. The self-interference of a photon can be understood, as discussed in § 2.1, through Feynman's concept of a sum over histories (Feynman, Leighton and Sands [1963]).

3.3. INTERFERENCE WITH INDEPENDENT SOURCES

Problems appear with Dirac's dictum when we consider interference effects produced by light beams from two completely independent sources (Magyar and Mandel [1963], Paul [1986]). Two independent waves can produce an interference pattern, provided that the phase difference between the waves is stable over the observation period. In the photon picture, however, the question is: "How can an interference pattern be produced if the photons in the two beams are created independently?" Since the photon that is detected does not always originate from the same source, the picture of a single photon being created and propagated along a superposition of two distinct paths no longer holds.

The first experiments demonstrating interference between two independent laser beams at very low light levels were performed by Pfleegor and Mandel

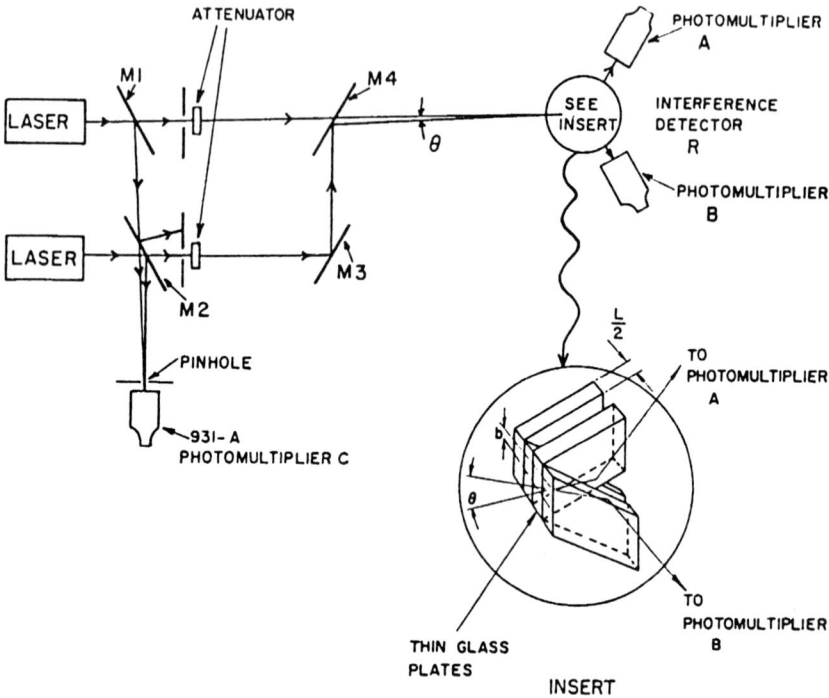

Fig. 3.2. Experimental system used to demonstrate interference with two independent laser sources (Pfleegor and Mandel [1968]).

[1967a,b, 1968]. As shown in fig. 3.2, the light beams from two independent He–Ne lasers were superimposed at a small angle to produce interference fringes on the edges of a stack of glass plates whose thickness was equal to half the fringe spacing. Two photomultipliers received the light from alternate plates, so that when interference fringes were present, a negative correlation was obtained between the number of counts registered by the photomultipliers. To minimize effects due to movements of the fringes, an additional photo detector was used to detect beats between the beams, and observations were restricted to 20 µs intervals, corresponding to periods during which the frequency difference between the two laser beams was less than 30 kHz. The transit time was approximately 3 ns, while the photon fluxes in the two beams were around 3×10^6 photons/s and the quantum efficiency of the photomultipliers was about 0.07, so that about 10 photons were detected in each 20 µs period. The average of 400 such measurements was taken in each experiment.

In this experiment, the positions of the fringe maxima are not predictable

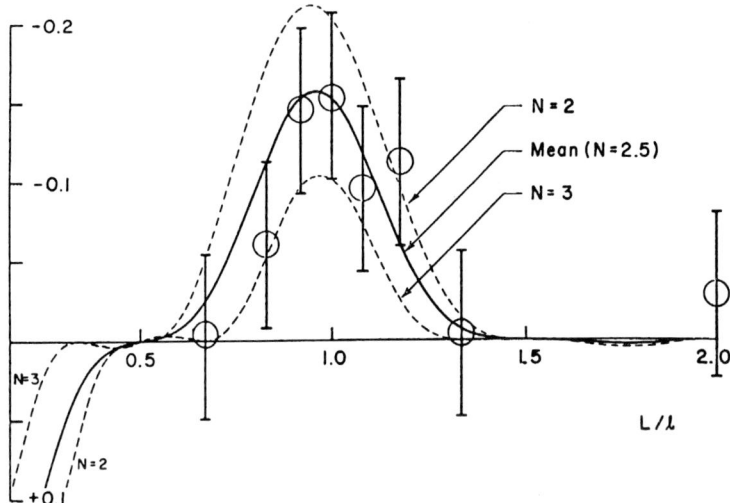

Fig. 3.3. Experimental results for the normalized correlation coefficient, together with the theoretical curves for $N=2$ and $N=3$ (Pfleegor and Mandel [1968]).

and vary from measurement to measurement. However, there should always be a negative correlation between the number of photons registered in the two channels, which should be a maximum when the fringe spacing l is equal to L, the thickness of a pair of plates. Figure 3.3 shows the variation in the degree of correlation of the two counts with the ratio L/l, together with the theoretical curves for $N=2$ and $N=3$, where N is the number of pairs of plates in the detector array. In subsequent experiments, the measurement procedure was automated, making it possible to record a much larger number of counts in each run, and to investigate the effects of varying the observation time and the number of interference fringes sampled.

Similar results were also obtained by Vain'shtein, Melekhin, Mishin and Podolyak [1981], in observations on the transient interference patterns formed by the beams from two lasers with two photomultipliers operating in the photon-counting regime.

These experiments confirmed that interference effects were associated with the detection of each photon, but their statistical accuracy was limited by the fact that observations could be made only over very short time intervals, during which a very small number of photons was detected. This problem has been overcome in more recent experiments involving observations of interference effects in the time domain.

3.4. INTERFERENCE IN THE TIME DOMAIN

Interference effects between two independent light sources were first observed in the time domain by Forrester, Gudmundsen and Johnson [1955], who mixed the Zeeman components of a visible spectral line at a fast photo detector. Subsequently, Javan, Ballik and Bond [1962] showed that beats could be obtained by superimposing the beams from two independent He–Ne lasers, or the beams corresponding to two axial modes of the same laser, at a photo detector. Observations on such beats have been used successfully to study interference effects with two sources at very low light levels (Hariharan, Brown and Sanders [1993]).

For such observations, there are significant advantages in using the beat produced by two axial modes of the same laser, since the frequency variations of the two modes due to thermal effects and mechanical variations of the cavity length are very nearly the same. A convenient low beat frequency can be obtained by applying a transverse magnetic field to a He–Ne laser that is oscillating in two longitudinal modes. The laser then oscillates on a single axial cavity mode composed of two orthogonally polarized components which exhibit a small frequency difference due to the magnetically induced birefringence of the gas in the laser tube (Morris, Ferguson and Warniak [1975]). These two Zeeman-split components can be regarded as equivalent to beams from two independent lasers, because the coupling between them is quite weak. In addition, with normal excitation, there is no coherence between the two upper states for the lasing transitions.

The experimental arrangement is shown schematically in fig. 3.4. The beat frequency was stabilized by mixing the two orthogonally polarized components in the back beam of the laser, with a polarizer, at a monitor photo diode and feeding the output to a frequency-to-voltage converter, which controlled the length of the cavity through a servo amplifier and a heating coil on the laser tube (Ferguson and Morris [1978]). A beat frequency of 80 kHz, with a frequency bandwidth estimated at 1 Hz, was obtained.

To make measurements, a set of neutral density filters was used to reduce the intensity of the output beams from the laser in accurately known steps over a range of $10^8:1$. The attenuated beams were incident, after passage through a polarizing prism that brought them into a condition to interfere, on a photo diode. The signal from this photo diode was taken through a band-pass filter to a homodyne detector that was fed with a reference signal from the monitor photo diode. Because the variations in the frequency of the beat signal were small and were tracked by the reference signal from the monitor photo diode,

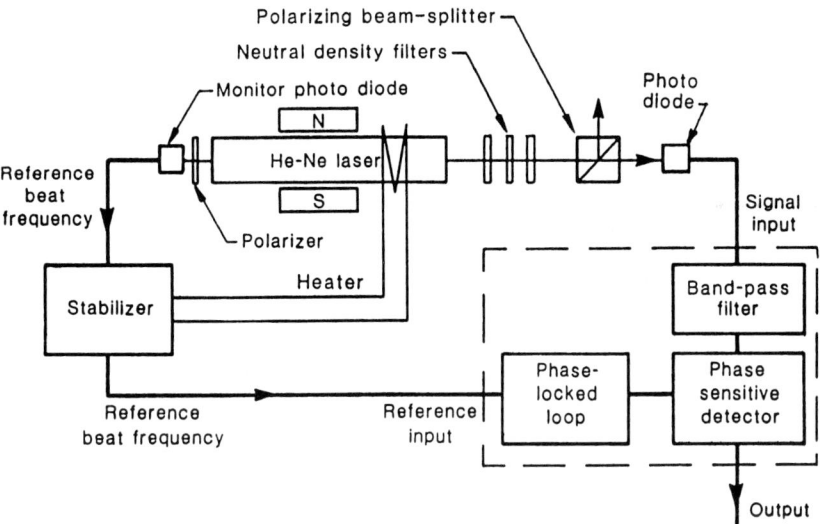

Fig. 3.4. Experimental system used to measure the amplitude of the beats produced by the two orthogonally polarized modes from a transverse Zeeman laser at very low light levels (Hariharan, Brown and Sanders [1993]).

measurements could be made with integrating times up to 100 seconds to obtain a good signal-to-noise ratio even at the lowest light levels.

Observations were made with the photo diode at a distance of 0.2 m from the laser, as the incident power was varied from $1.0\,\mu W$ down to $4.8\,pW$, corresponding to values of the incident flux ranging from 3.18×10^{12} photons/s to 1.53×10^{7} photons/s, respectively. At the lowest power level, the probability for the presence of more than one photon in the apparatus at any time, relative to that for the presence of a single photon, was less than 0.005.

Figure 3.5 shows the output from the homodyne detector plotted as a function of the power incident on the photo diode. The measurements showed no significant deviations from a straight line with a slope of unity, confirming that the interference phenomena remained unchanged down to power levels at which there was a very high probability that one photon was absorbed before the next one was generated.

These results were extended to cover interference involving more than two beams by Hariharan, Brown, Fujima and Sanders [1993] using a He–Ne laser operating in three axial modes. With such a laser, low-frequency beat signals are also obtained because the axial modes are not equally separated in frequency, due to the dispersion of the excited neon gas. The frequency of these beats

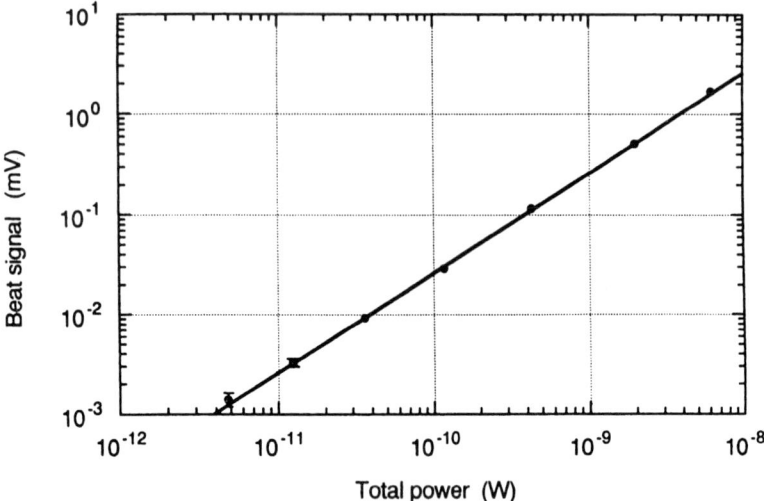

Fig. 3.5. Output signal from the homodyne detector as a function of the total power of the laser beams (Hariharan, Brown and Sanders [1993]).

corresponds to the second differences between the frequencies of the modes (Casabella and Gonsiorowski [1980]).

In the actual experiment, the photo detector was placed at a distance of 80 mm from the neutral density filters used to attenuate the beam. Measurements were made of the output from the homodyne detector as the power incident on the photo diode was varied from 75 nW down to 0.19 nW, corresponding to values of the flux ranging from 2.39×10^{11} photons/s to 6.04×10^8 photons/s. At the lowest power level, the ratio of the probability for the presence of at least one photon from all three modes to the total probability for the presence of at least one photon from any mode, was only 0.0009. However, the amplitude of the low-frequency beat was found to vary linearly with the power incident on the photo diode down to this power level.

These results also showed that the ratio of the beat amplitude to the incident power remained unchanged down to the lowest power level at which observations were made, even though, at this power level, the mean time interval between the arrival of successive photons at the photo detector was greater than the period of the beat (Hariharan, Brown, Fujima and Sanders [1995]).

3.5. SUPERPOSITION STATES

It follows that the interference phenomena observed in all these cases are associated with the detection of each photon and not with the interference of one photon with another. An analysis of the processes leading to interference effects made by Jordan and Ghielmetti [1964] and by Mandel [1964] suggested that for interference effects to become observable, the average number of photons in the same spin state falling on a coherence area in a coherence time, or the average occupation number per unit cell of phase space, would have to be appreciably greater than 1. An alternative explanation by de Broglie [1969] involved the assumption that the distribution of photons was determined by the superposition of weak electromagnetic waves from the two sources. However, it is now clear that the interference phenomena are produced by a sequence of photons, each one of which is in a superposition state that originates from the modes involved. The problem that remains is how the superposition state responsible for interference arises.

One explanation is that the superposition state is produced in the process of absorption at the photodetector, because it is impossible, in principle, to determine from which source the photon is emitted. The measurement therefore forces the photon into a superposition state in which it behaves as if it were associated with both light beams, and these two states of each photon interfere (Mandel [1976]). A field-theoretic analysis explaining how this happens has been presented by Walls [1977]. Alternatively, it is possible to regard the fields from each source as being in a superposition state of having one photon and no photon (Hariharan, Brown and Sanders [1993]). An explicit description of the interference effects produced by two independent laser beams using the techniques of quantum field theory has also been presented by Agarwal and Hariharan [1993].

§ 4. The Geometric Phase

An extension of the adiabatic theorem of quantum mechanics by Berry [1984] showed that the wave function of a quantum system may undergo a phase shift (a geometric phase) when the parameters of the system undergo a cyclic change. This phase change can be observed by interference if the cycled system is compared with another system that has not undergone any change.

4.1. THE GEOMETRIC PHASE IN OPTICS

Demonstrations of effects due to the geometric phase in optics followed. One example was the rotation of the plane of polarization of a linearly polarized light beam propagating in an optical fiber coiled into a helix (Tomita and Chiao [1986]). Another was the phase shift observed in a Mach–Zehnder interferometer in which the two beams traversed nonplanar paths arranged to have equal lengths, but opposite senses of handedness (Chiao, Antaramian, Ganga, Jiao, Wilkinson and Nathel [1988]).

Berry's paper also led to a reappraisal of earlier studies by Pancharatnam [1956] on the interference of polarized light, which could now be seen as manifestations of the geometric phase (Ramaseshan and Nityananda [1986], Berry [1987]).

Pancharatnam defined the phase difference between two beams in different states of polarization by considering the intensity produced when the two beams were made to interfere. He regarded the two beams as being "in phase" when the resultant intensity was a maximum. This approach made it possible to define how a beam changed its phase when its state of polarization was altered. It also led to the observation that a beam could be taken from one polarization state, without introducing any phase changes, through two other polarization states back to its original state, and exhibit a phase shift. The magnitude of this phase shift (the Pancharatnam phase) was equal to half the solid angle subtended by the circuit at the center of the Poincaré sphere. Several experiments have been described using interferometric techniques to measure this phase shift (Bhandari and Samuel [1988], Simon, Kimble and Sudarshan [1988], Chyba, Wang, Mandel and Simon [1988]).

4.2. OBSERVATIONS AT THE SINGLE-PHOTON LEVEL

Observations of the Pancharatnam phase have been made by Hariharan, Roy, Robinson and O'Byrne [1993] at light levels low enough to ensure that the probability of more than one photon being present simultaneously in the interferometer was negligible. They used a Sagnac interferometer in which the optical paths traversed by the two beams were always equal, and a phase difference could be introduced between them only by operating on the Pancharatnam phase. As shown in fig. 4.1, light from a He–Ne laser, linearly polarized at 45° to the plane of the figure by a polarizer P_1, was divided at a polarizing beam-splitter into two orthogonally polarized beams traversing the same closed triangular path in opposite directions. A second polarizer P_2, with its axis at 45° to the plane of the figure, brought the two beams leaving the interferometer into a condition to interfere at a photomultiplier. The phase difference between the beams was varied by a system consisting of a rotating

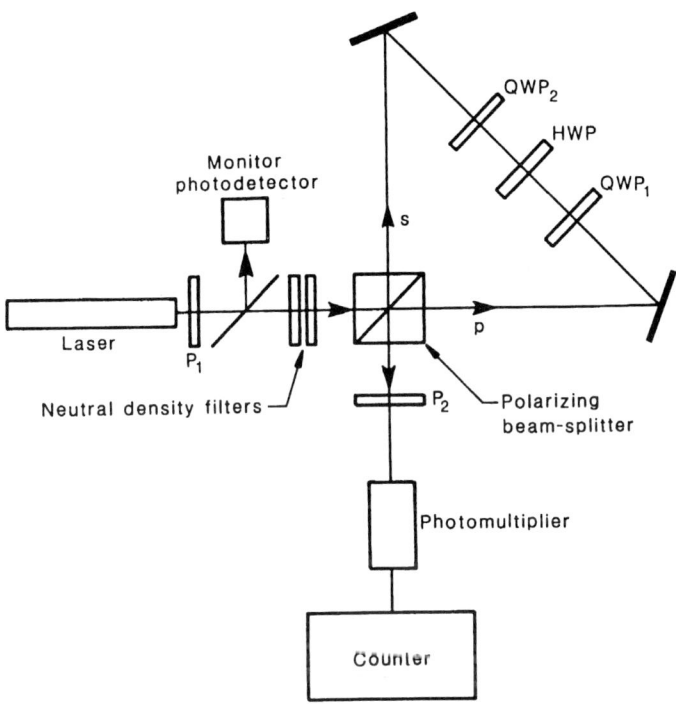

Fig. 4.1. Schematic of the experimental arrangement used for studies of the Pancharatnam phase at the single-photon level (Hariharan, Roy, Robinson and O'Byrne [1993]).

half-wave plate, HWP, located between two fixed quarter-wave plates, QWP$_1$ and QWP$_2$ (Hariharan and Roy [1992]).

The operation of this interferometer can be followed by means of the Poincaré sphere (Jerrard [1954]). Light (p-polarized) transmitted by the polarizing beam splitter, passes through the quarter-wave plate QWP$_1$, the half-wave plate HWP, and the quarter-wave plate QWP$_2$, in that order. As shown in fig. 4.2, the polarization state of this beam then traces out the path $A_1SA_2NA_1$ on the Poincaré sphere. If the half-wave plate HWP is set with its optic axis at an angle $+\theta$ to the optic axes of QWP$_1$ and QWP$_2$, the phase of the transmitted light is advanced by 2θ. Reflected (s-polarized) light traverses the interferometer in the opposite sense, and its polarization state traces out the path $B_1SB_2NB_1$ on the Poincaré sphere; its phase is therefore retarded by 2θ. These operations lead to a phase difference $\Delta\phi = 4\theta$ between the two fields when they reach the photomultiplier, without introducing any change in the optical paths.

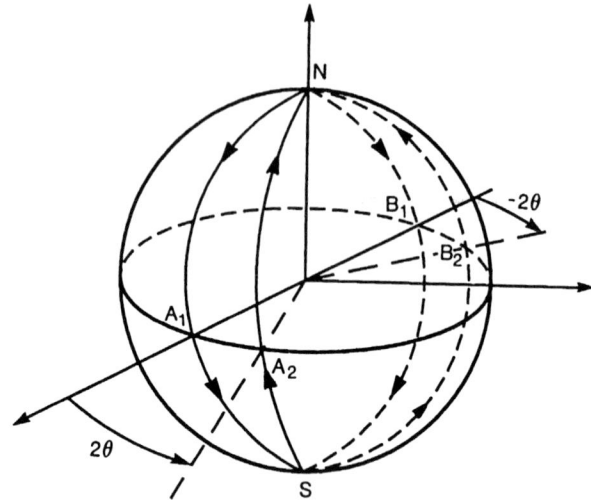

Fig. 4.2. Closed paths traversed by the polarization states of the two beams in the interferometer on the Poincaré sphere (Hariharan, Roy, Robinson and O'Byrne [1993]).

Measurements were made at an input power level <1 pW, corresponding to a photon flux $N_i < 3.2 \times 10^6$ photons/s, at which level

$$\frac{P(n>1)}{P(1)} = 0.005, \tag{4.1}$$

where $P(1)$ and $P(n>1)$ are, respectively, the probabilities for the presence of one photon and more than one photon in the apparatus.

The net counting rate, after subtracting the dark counting rate, exhibited the expected sinusoidal variation, corresponding to the relation $\Delta\phi = 4\theta$, over a wide range of values of θ. The visibility of the interference fringes was better than 0.97 and very close to that obtained in the classical regime.

4.3. OBSERVATIONS WITH SINGLE-PHOTON STATES

An experiment to demonstrate the existence of a geometric phase for single photons was performed by Kwiat and Chiao [1991]. They used a light source that produced pairs of photons with wavelengths centered at 702.2 nm by parametric down-conversion (see § 2.5.2). In the arrangement used by them (see fig. 4.3), the idler beam was transmitted through the filter F1 to the detector D1, while the signal beam entered a Michelson interferometer. The beam leaving the interferometer was incident on a second beam splitter B2, from which it was

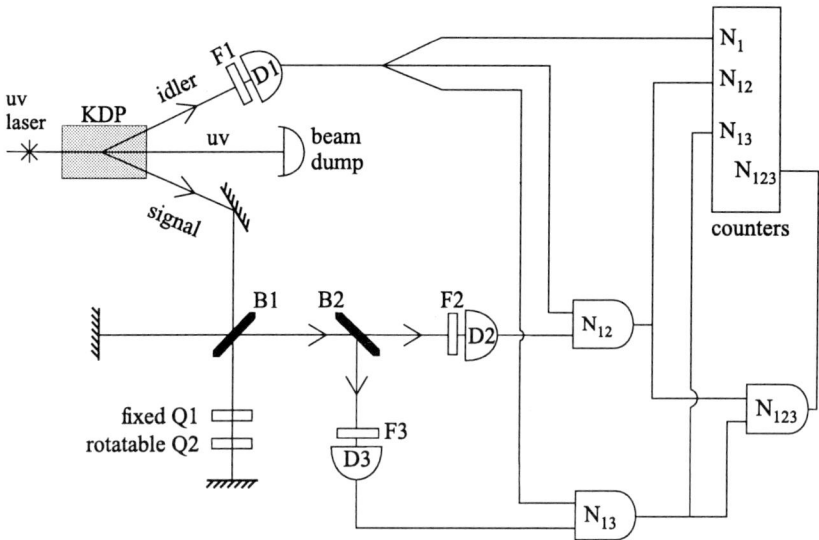

Fig. 4.3. Apparatus used to measure Berry's phase for single photons (Kwiat and Chiao [1991]).

transmitted to the detector D2 through the filter F2 or reflected to the detector D3 through the filter F3. The count rates for coincidences between D1 and D2 and between D1 and D3, as well as triple coincidences between D1, D2 and D3, were recorded.

One arm of the interferometer contained a fixed quarter-wave plate Q1, with its axis at 45°, as well as a quarter-wave plate Q2 that could be rotated. Since the beam traverses this system twice, a rotation of Q2 through an angle θ introduces an additional phase shift in this arm, $\Delta\phi = 2\theta$.

Data were recorded using filters with a bandwidth of 10 nm at F2 and F3, and an optical path difference of 220 µm, which is greater than the coherence length corresponding to this bandwidth (about 50 µm). As a result, the fringe visibility seen by the detectors D2 and D3, operating individually, was essentially zero. However, when a filter with a bandwidth of 0.86 nm was placed in front of D1, the count rate for coincidences between D1 and D3 varied with the angular setting θ of Q2, as shown in the lower part of fig. 4.4, with a visibility of 0.60±0.05. With a broad band filter at F1, the coincidence fringes disappeared, as shown in the upper part of fig. 4.4.

With this arrangement, it was possible to verify that the signal beam

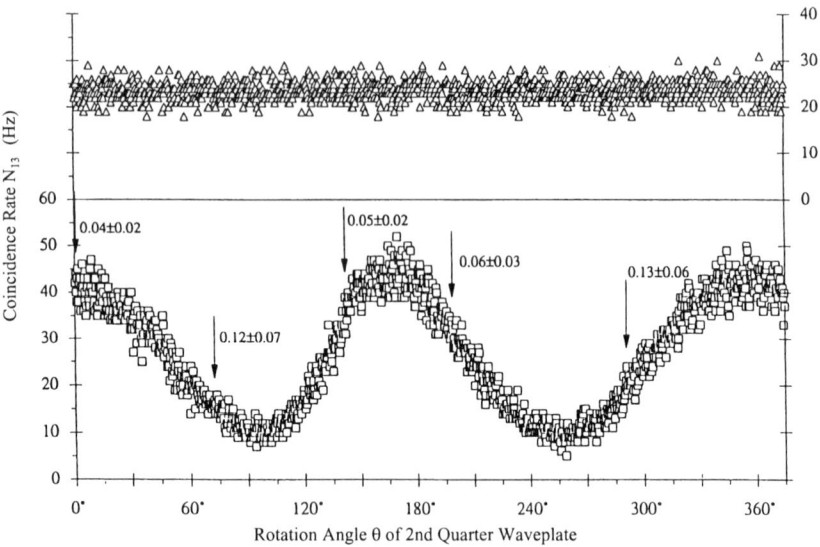

Fig. 4.4. Interference effects (lower trace, squares) with a slowly varying Berry's phase, observed as coincidences between D3 and D1, with an optical path difference of 220μm in the interferometer and a narrow band filter at F1. With a broad-band filter at F1, no interference is seen (upper trace, triangles) (Kwiat and Chiao [1991]).

was composed of photons in an $n=1$ Fock state by measurements of the anticorrelation parameter (see § 3.2):

$$\mathcal{A} \equiv \frac{N_{123}N_1}{N_{12}N_{13}}, \qquad (4.2)$$

where N_{123} is the rate of triple coincidences between D1, D2 and D3, N_1 is the rate of single counts by D1 alone, N_{12} is the rate of coincidences between D1 and D2, and N_{13} is the rate of coincidences between D1 and D3. The average value of $\mathcal{A}$ obtained with the two-photon source differed from that obtained with a thermal source by more than 13 standard deviations, confirming that the observations essentially involved photons in $n=1$ Fock states.

These observations suggest that the geometric phase observed in optics originates at the quantum level, but survives the correspondence principle limit into the classical level, although this question is still open to argument (Tiwari [1992]).

§ 5. Fourth-order Interference

Measurements of fourth-order coherence can be realized by using two spatially separated detectors, or by correlating photo detections which are separated in time.

Studies of fourth-order coherence began with the intensity interferometer (Hanbury Brown and Twiss [1956], Hanbury Brown [1974]). In this instrument, light from a star was focused on two photo detectors whose separation could be varied, and the correlation between the fluctuations in the output currents from the two detectors was measured.

The fluctuations in the output current from each detector then consist of two components. One is the shot noise associated with the current, while the other is due to fluctuations in the intensity of the incident light. The shot noise from the two detectors is not correlated, but the intensity fluctuations exhibit a correlation which depends on the degree of coherence of the fields at the two detectors. Since the fields are produced by a stationary thermal source, the normalized intensity correlation function depends only on the time difference, τ, and is given by the relation

$$\mathcal{R}(r_1, r_2, \tau) = \left|\gamma^{(1,1)}(r_1, r_2, \tau)\right|^2, \qquad (5.1)$$

where $\gamma^{(1,1)}(r_1, r_2, \tau)$ is the normalized second-order coherence function. When $\tau = 0$, the variation of the normalized value of the correlation with the separation of the detectors can be used to determine the angular diameter of a star. With a time delay τ produced by an optical path difference that is much greater than the coherence length of the radiation, the effects of such correlated intensity fluctuations can be observed as a spectral modulation (Alford and Gold [1958], Mandel [1962]).

5.1. NONCLASSICAL FOURTH-ORDER INTERFERENCE

Fourth-order interference provides a means for distinguishing classical and nonclassical light, since an optical field can exhibit nonclassical fourth-order interference effects even when the usual second-order interference effects cannot be observed (Mandel [1983], Ou [1988]). Such fourth-order interference effects can be observed with correlated photons produced by parametric down-conversion (Ghosh, Hong, Ou and Mandel [1986], Ghosh and Mandel [1987]).

We consider the detection of the field produced by the signal and idler modes from a two-photon source at a point x_1 (see fig. 5.1). Since the output from the down-converter is an approximation to the two-photon Fock state $|1,1\rangle_{AB}$ for

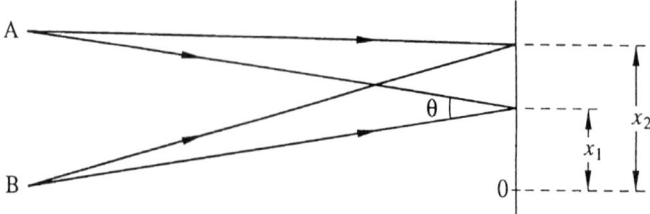

Fig. 5.1. Geometry of a fourth-order interference experiment (Ghosh and Mandel [1987]).

weak fields, the probability of detecting a photon between the arbitrary positions x_1 and $x_1 + \delta x_1$ can be shown to be:

$$\mathcal{P}_1(x_1)\delta x_1 = 2K_1\delta x_1, \tag{5.2}$$

where K_1 is a scale factor. This probability is independent of x_1, so that no interference fringes can be seen. A separate measurement made at x_2 with another photo detector yields a similar result. The reason for the absence of second-order interference is, of course, that no definite phase relationship exists between the two down-converted fields.

However, if we use two photo detectors at x_1 and x_2 to measure the joint probability $\mathcal{P}_{12}(x_1, x_2)\delta x_1 \delta x_2$, of detecting a photon within δx_1 and δx_2, we have (Ghosh, Hong, Ou and Mandel [1986])

$$\mathcal{P}_{12}(x_1, x_2)\delta x_1 \delta x_2 = 2K_1 K_2 \delta x_1 \delta x_2 \left[1 + \cos\left(\frac{2\pi(x_1 - x_2)}{L}\right)\right], \tag{5.3}$$

where $L = \lambda/\theta$ is the spacing of the second-order interference fringes corresponding to the geometry of fig. 5.1. We can regard the effects observed as interference between two different, two-photon probability amplitudes, because the system cannot distinguish between photons from A and B being detected at x_1 and x_2, respectively, or *vice versa*. As can be seen from eq. (5.3), the fourth-order interference fringes have a visibility of unity. On the other hand, with classical fields, it can be shown (Mandel [1983]) that the visibility of the fourth-order interference fringes cannot exceed 0.5, and the joint probability $\mathcal{P}_{12}(x_1, x_2)\delta x_1 \delta x_2$ never drops to zero.

Figure 5.2 is a schematic of the experimental arrangement used to observe the fourth-order interference fringes (Ghosh and Mandel [1987]). The beam from an argon-ion laser ($\lambda = 351.1$ nm) incident on a LiO$_3$ crystal generates photon pairs (see § 2.5.2) which are reflected through an interference filter so that they come together in a plane at a distance of 1.1 m at an angle of about 2°. The interference

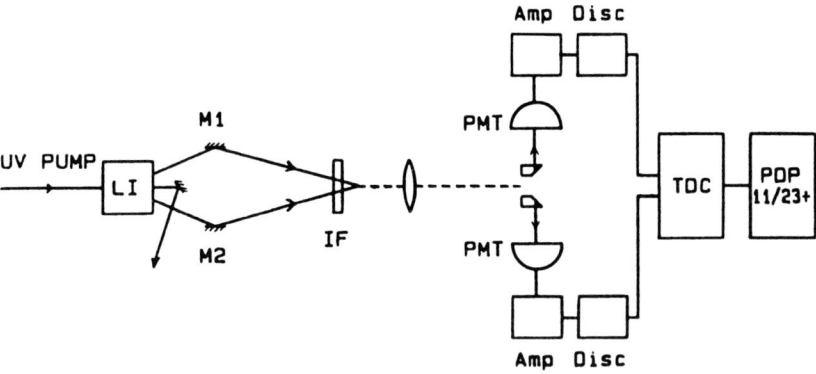

Fig. 5.2. Apparatus used to demonstrate fourth-order interference with a two-photon source (Ghosh and Mandel [1987]).

pattern formed in this plane is reimaged by a lens so as to give a spacing of the fourth-order interference fringes $L = 0.34$ mm. Two movable glass plates of thickness $\Delta x = 0.14$ mm collect the photons at x_1 and x_2 and direct them to two photomultipliers, whose outputs are fed to a counter. The number of coincidences was recorded over 10-hour periods for different values of the separation $(x_1 - x_2)$ of the plates. These values were corrected for accidental coincidences by making measurements with a delay extending from 35 to 75 ns, and subtracting the proportionate number expected within the 5 ns resolving time.

In practice, because of the finite width Δx of the detectors, the observed values of visibility are reduced by a factor:

$$\eta = \left[\frac{\sin(\pi \Delta x/L)}{\pi \Delta x/L}\right]^2. \tag{5.4}$$

Figure 5.3 shows the experimental values superimposed on a plot of the values of $\mathcal{P}_{12}(x_1, x_2)$ for a two-photon source corrected for this effect, and with the scale chosen to give the best fit with the measured coincidence rates (the solid curve). The corresponding curve from classical theory (the broken curve) is obviously a much poorer fit.

A more striking example of a nonclassical fourth-order interference effect can be observed when the inputs to the opposite sides of a beam splitter are one-photon Fock states (Fearn and Loudon [1987, 1989], Ou, Hong and Mandel [1987]). As shown in fig. 5.4, the superimposed beams leaving the beam splitter go to two photo detectors, D1 and D2, and measurements are made of the rate at which photons are detected in coincidence as the beam splitter is displaced in small steps $c\Delta\tau$ from the point where the two optical paths are equal. A sharp

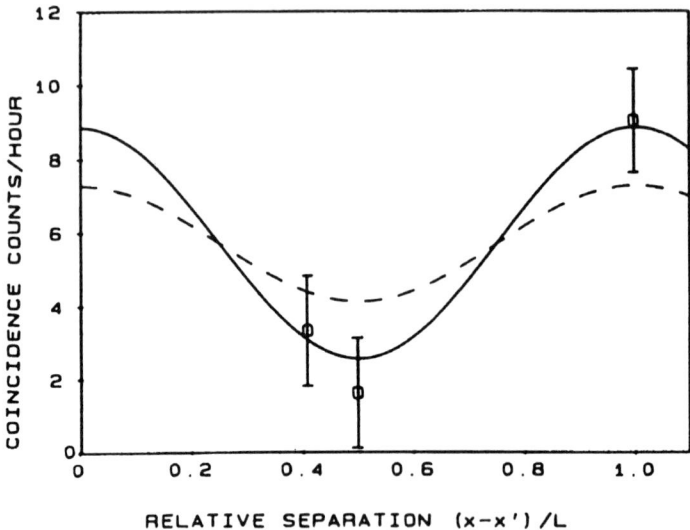

Fig. 5.3. Experimental results and predictions of quantum theory (solid curve) and classical theory (dashed curve) (Ghosh and Mandel [1987]).

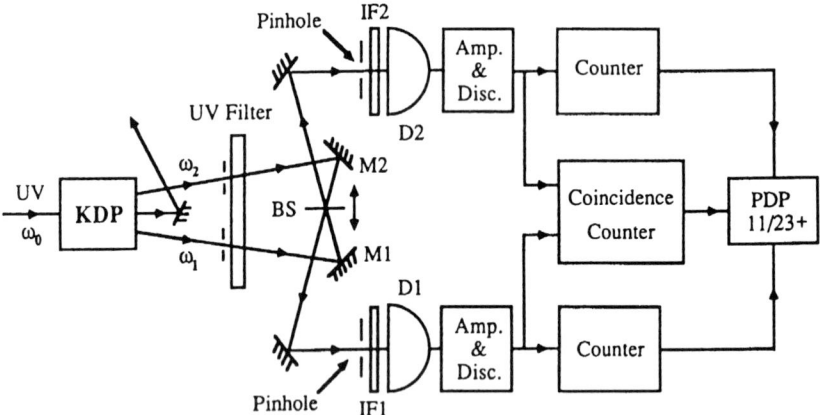

Fig. 5.4. Experimental arrangement used to demonstrate fourth-order interference effects with a varying optical path difference (Hong, Ou and Mandel [1987]).

reduction in the coincidence count rate occurs when the beam splitter occupies a symmetrical position (Hong, Ou and Mandel [1987], Rarity and Tapster [1989]).

We can label the field modes on the input sides of the beam splitter as $01, 02$ and on the output side as $1, 2$ and assume that the light is perfectly monochromatic. If the input state resulting from degenerate down-conversion is the two-photon Fock state $|1\rangle_{01}|1\rangle_{02}$, then the state on the output side of the beam splitter is:

$$|\psi_{\text{out}}\rangle = (R-T)|1\rangle_1|1\rangle_2 + i(2RT)^{1/2}\left[|2\rangle_1|0\rangle_2 + |0\rangle_1|2\rangle_2\right], \quad (5.5)$$

where R and T are the reflectance and transmittance of the beam splitter, with $R+T=1$. It follows that for a beam splitter with $R=\frac{1}{2}=T$, the first term on the right-hand side of eq. (5.5) is zero, corresponding to destructive interference of the two-photon probability amplitudes, and the entangled state $|2;0\rangle_{1,2}^{\pi/2,0}$ is obtained. This state is analogous to the single-photon entangled state defined by eq. (2.14), except that in this case, it is the photon pair that is entangled with the vacuum. Alternatively, we can regard the photon pair as being in a superposition state of adopting either path 1 or path 2. No coincidences should therefore be recorded.

However, the down-converted photons are never monochromatic, and the two-photon state can be represented more correctly by the linear superposition

$$|\psi\rangle = \int d\omega f(\omega, \omega_0 - \omega)|1\rangle_\omega |1\rangle_{\omega_0-\omega}, \quad (5.6)$$

where $f(\omega_1, \omega_2)$ is a weight function that is peaked at $\omega_1 = \frac{1}{2}\omega_0 = \omega_2$. The joint probability for the detection of photons at the detectors D1 and D2 at times t and $t+\tau$, respectively, is then:

$$\mathcal{P}_{12}(\tau) = K|G(0)|^2 \left\{ T^2|G_0(\tau)|^2 + R^2|G_0(2\Delta\tau - \tau)|^2 \right. \\ \left. - RT\left[G_0^*(\tau)G_0(2\Delta\tau - \tau) + \text{c.c.}\right]\right\}, \quad (5.7)$$

where $G(\tau)$ is the Fourier transform of the weight function,

$$G(\tau) = \int f(\tfrac{1}{2}\omega_0 + \omega, \tfrac{1}{2}\omega_0 - \omega)\exp(-i\omega\tau)d\omega, \quad (5.8)$$

$G_0(\tau) \equiv G(\tau)/G(0)$, and K is a constant characteristic of the detectors.

While the coincidence measurement corresponds to an integration of the probability $\mathcal{P}_{12}(\tau)$ over the resolving time of a few nanoseconds, this time is so much longer than the correlation time that we may integrate $\mathcal{P}_{12}(\tau)$ over all

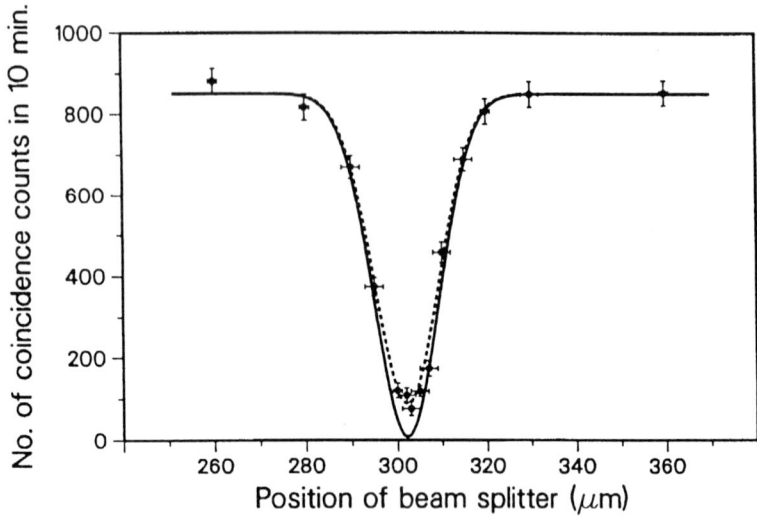

Fig. 5.5. Measured number of coincidences as a function of the displacement of the beam splitter, superimposed on the theoretical (solid) curve derived from eq. (5.10) with $R/T = 0.95$ and $\Delta\omega = 3 \times 10^{13}$ rad/s. The dashed curve was obtained by multiplying the factor $2RT/(R^2 + T^2)$ in eq. (5.10) by 0.9 (Hong, Ou and Mandel [1987]).

values of τ to obtain the expected number of observed coincidences. We then have:

$$N_c = C \left[R^2 + T^2 - 2RT \frac{\int_{-\infty}^{\infty} G_0(\tau) G_0(\tau - 2\Delta\tau) \, d\tau}{\int_{-\infty}^{\infty} G_0^2(\tau) \, d\tau} \right], \quad (5.9)$$

where C is a constant, which, when $f(\frac{1}{2}\omega_0 + \omega, \frac{1}{2}\omega_0 - \omega)$ is a Gaussian with bandwidth $\Delta\omega$, reduces to

$$N_c = C(T^2 + R^2) \left[1 - \frac{2RT}{R^2 + T^2} \exp(-\Delta\omega\delta\tau)^2 \right]. \quad (5.10)$$

It follows that when $\Delta\tau = 0$, $N_c = C(R - T)^2$, which vanishes when $R = \frac{1}{2} = T$, whereas when $\Delta\tau \gg G_0(\tau)$, one has $N_c = C(T^2 + R^2)$.

Figure 5.5 shows the number of coincidences observed, after subtracting accidentals, as a function of the displacement of the beam splitter. The rate of coincidences drops to a few percent of its normal value when the two optical paths are equal, because of destructive interference of the two-photon probability amplitudes. The width of the dip in the coincidence rate yields a measure of the length of the photon wave packet which agrees with the value derived from the width of the passband of the interference filters F1 and F2.

The occurrence of an almost complete null at the center of the dip confirms that this is a nonclassical effect, since according to classical theory, the visibility cannot exceed 0.5 (Mandel [1983]). The drop in the number of coincidences is associated with an increase in the number of photon pairs leaving the beam splitter in the same direction. This behavior arises from the Bose–Einstein commutation properties of the photon-creation and annihilation operators (Fearn and Loudon [1989]).

An extension of this experiment (Ou and Mandel [1988b]) involves the use of interference filters with pass bands centered on different, nonoverlapping frequencies, ω_1 and ω_2. If the complex frequency responses of these filters can be described by Gaussian functions with an rms width σ, the measured coincidence detection probability is:

$$\mathcal{P}_{12} \propto \sqrt{2}\pi\sigma^3 \exp\left[-\frac{(\omega_1 + \omega_2 - \omega_0)^2}{2\sigma^2}\right] \\ \times \left[T^2 + R^2 - 2TR\exp\left(\frac{-\sigma^2 \Delta\tau^2}{2}\right)\cos(\omega_1 - \omega_2)\Delta\tau\right], \quad (5.11)$$

which is a maximum when the two center frequencies are chosen to satisfy the condition $\omega_1 + \omega_2 = \omega_0$. If $T = R = \frac{1}{2}$, eq. (5.11) describes an interference pattern whose visibility is unity at the center, but falls off exponentially to either side. At the center, where $\Delta\tau = 0$, the probability of coincidences $\mathcal{P}_{12} = 0$.

In the actual experiment, the pass bands of the two interference filters were centered on conjugate wavelengths of 680 and 725 nm, corresponding to a frequency difference $(\omega_1 - \omega_2)/2\pi = 27 \times 10^{12}$ Hz. Figure 5.6 shows the observed two-photon coincidence rate as a function of the position of the beam splitter and the corresponding time delay $\Delta\tau$ between the signal and idler photons. The coincidence rate exhibits interference effects with a spatial period of 5.5 μm, corresponding to a temporal period of 37 fs, which is almost exactly the period of the beat frequency. This beat frequency is observed even though neither of the photo detectors individually registers a beat.

This experiment also reveals a violation of classical theory. In addition, it demonstrates that even though there are fundamental quantum limits in attempting to localize the position of a photon to better than a few wavelengths in space, or better than a few periods in time (Newton and Wigner [1949]), this limit does not apply to the average time interval between photons, which can always be determined with subperiod precision.

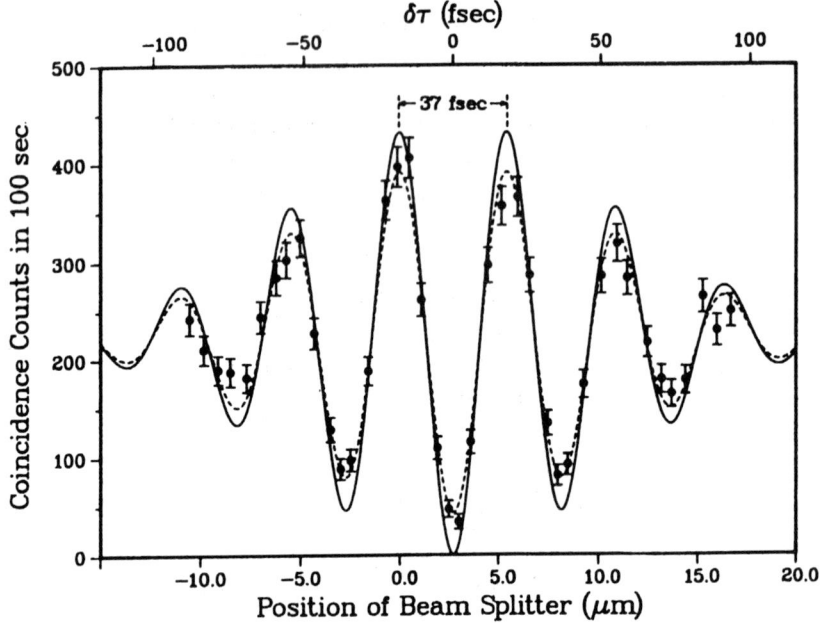

Fig. 5.6. Measured number of coincidences in 100s as a function of the position of the beam splitter, or the time delay $\Delta\tau$, between the signal and idler photons, along with the theoretical (solid) curve obtained from eq. (5.11) and the (dotted) curve obtained when the interference term was multiplied by 0.8 (Ou and Mandel [1988b]).

5.2. INTERFERENCE IN SEPARATED INTERFEROMETERS

Fourth-order interference effects also arise when pairs of photons enter one or more interferometers, and the coincidence rate is monitored at the output ports (Kwiat, Vareka, Hong, Nathel and Chiao [1990], Ou, Zou, Wang and Mandel [1990a], Rarity, Tapster, Jakeman, Larchuk, Campos, Teich and Saleh [1990]).

In the arrangement used by Ou, Zou, Wang and Mandel [1990a] (see fig. 5.7), the two photons traveled to two photo detectors *via* two unbalanced Michelson interferometers, which were adjusted so that the difference in the propagation time between the longer and shorter paths was the same in both channels and was much greater than the coherence time of the individual photons. Under these conditions, the count rate registered by the two detectors showed no dependence on the optical path difference in either of the interferometers. However, measurements of the two-photon coincidence rate, as a function of the position of one of the mirrors, revealed interference fringes with a spatial period corresponding to the wavelength of the pump beam. A visibility of

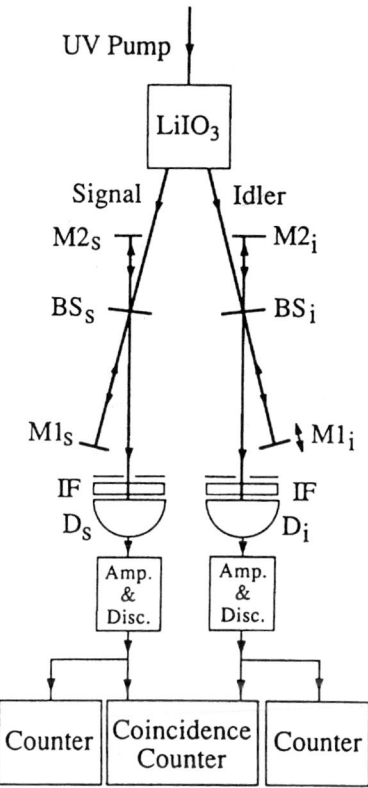

Fig. 5.7. Experimental arrangement used to observe fourth-order interference effects with photon pairs in two separated channels (Ou, Zou, Wang and Mandel [1990a]).

0.5 was obtained in this experiment, but this was due to the limited time-resolution of the detector system, and subsequent measurements with higher time-resolution gave coincidence fringes with a visibility of 0.87 (Brendel, Mohler and Martienssen [1991]).

Unusual interference patterns have also been observed with nondegenerate photon pairs (Larchuk, Campos, Rarity, Tapster, Jakeman, Saleh and Teich [1993]). In their experiments, pairs of photons whose center wavelengths differed by approximately 40 nm were used as the inputs to single and dual Mach–Zehnder interferometers (MZI). In the single MZI configuration, the paths of both down-converted beams overlapped completely within the interferometer. In the dual MZI configuration, their paths did not overlap; this is equivalent to sending each beam into a separate interferometer. Second-order interference was observed by counting the number of photons at each of the output ports, while

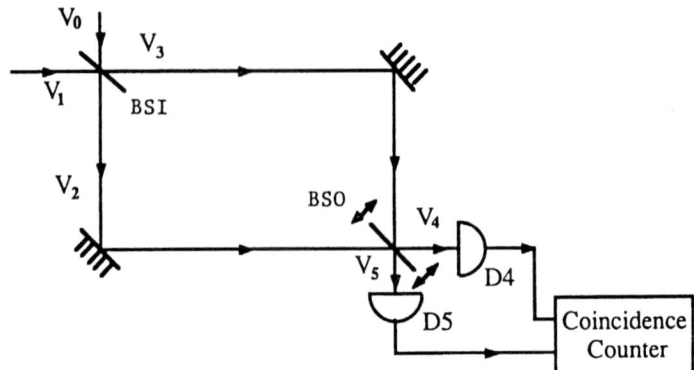

Fig. 5.8. Experimental arrangement used to study interference by two-photon superposition states formed by pairs of photons at a beam splitter (Ou, Zou, Wang and Mandel [1990b]).

observations of fourth-order interference were made by recording coincidences. Observations were made as the optical path difference was varied from zero to very large values.

The second-order interference pattern at both the output ports was the same for photon pairs as that predicted by classical theory for the center frequencies, ω_1 and ω_2, and as expected, disappeared when the optical path difference exceeded the second-order coherence length. However, the fourth-order interference patterns were found to be quite different. For small path differences, the coincidence rates exhibited interference fringes corresponding to the difference frequency $\omega_d = |\omega_1 - \omega_2|$, as well as the sum frequency $\omega_s = |\omega_1 + \omega_2|$, when the beams overlapped, and also when they did not overlap. When the beams did not overlap, interference fringes corresponding to the center frequencies, ω_1 and ω_2, were also observed. A striking observation was the existence of interference effects at the sum (pump) frequency at path-length differences that were greater than the second-order coherence length. This is a nonlocal quantum effect, confirming the high degree of entanglement of the down-converted photons.

In another fourth-order interference experiment (Ou, Zou, Wang and Mandel [1990b]), two photons produced simultaneously provided the two inputs to a Mach–Zehnder interferometer, as shown in fig. 5.8, and the photons emerging at the two outputs were counted. The rate of coincidences was found to exhibit interference fringes with high visibility when the optical path difference was varied, despite the fact that the two average output intensities did not vary with the optical path difference. The effects observed can be attributed to the fact that when two similar photons simultaneously enter a beam splitter at ports 0 and 1, two photons always emerge together either at port 2 or at port 3 (Hong, Ou and Mandel [1987]), so that the output from these ports is in a superposition state. The resulting fourth-order interference fringes have a visibility of unity and are a consequence of the interference of photon pairs, rather than single photons.

5.3. THE GEOMETRIC PHASE

As described in § 4, earlier observations of the geometric (Pancharatnam) phase were made either at low light levels with classical sources, or with single-photon states. In both cases, there is no difference in the effects predicted by a classical treatment or a quantum-mechanical treatment, since the measurements only involve second-order interference.

The effects produced by the geometric (Pancharatnam) phase in fourth-order interference have been studied by Brendel, Dultz and Martienssen [1995] using the experimental arrangement shown in fig. 5.9. In this setup, photon pairs generated by down-conversion of blue light ($\lambda = 458$ nm) from an argon-ion laser in a beta barium borate (BBO) crystal traversed a Michelson interferometer. One arm of this interferometer contained two quarter-wave plates, one of which was fixed at an azimuth of 45°, while the other could be rotated. A rotation of the second quarter-wave plate through an angle θ introduced geometric (Pancharatnam) phases $\Delta\phi = \pm 2\theta$, respectively, for the two orthogonal polarizations.

Experiments were carried out using two BBO crystals cut, respectively, for type-I and type-II phase matching, so that the photons of a pair could be prepared either in the same state of polarization (type-I) or in orthogonal states of polarization (type-II).

The photon pairs emerging from the interferometer were incident on a second beam splitter BS_2 which directed them to two photo detectors D_1 and D_2. With type-I phase matching, BS_2 was a normal beam splitter, while with type-II phase matching, BS_2 was a polarizing beam splitter.

Measurements with this system showed that the effects observed depended on the initial states of polarization of the two photons in a pair and the optical path difference. With near-zero optical path differences, second-order interference fringes were observed, and the effects of the dynamic phase and the geometric phase were equivalent. With large optical path differences and coincidence detection, no interference was observed due to the geometric phase with type-II phase matching. However, with type-I phase matching, interference fringes with a visibility of 0.78 were obtained with a period equal to half that expected with a classical light field.

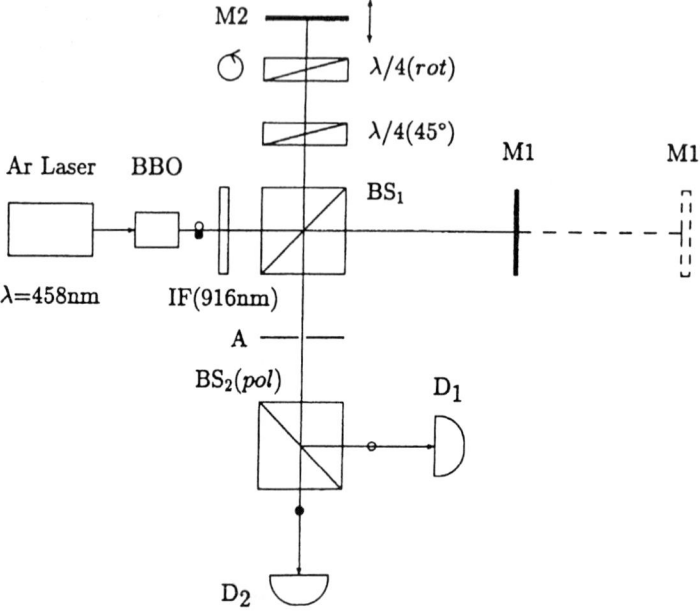

Fig. 5.9. Experimental setup used to demonstrate the effects of the geometric phase in fourth-order interference (Brendel, Dultz and Martienssen [1995]).

These results imply that pairs of photons with parallel polarizations acquire twice the geometric phase of single photons and behave like single particles with spin 2. On the other hand, pairs of photons with orthogonal polarizations acquire geometric phases with opposite signs and behave like a single particle with total spin 0. It follows that the equivalence between the dynamical phase and the geometric phase observed with second-order interference does not always exist with fourth-order interference.

5.4. TESTS OF QUANTUM THEORY

According to an alternative interpretation of quantum theory (de Broglie [1969]), the wave function describes a real physical wave, so that waves associated with different particles may interfere. An experiment to test a modified version of this theory (Croca, Garuccio, Lepore and Moreira [1990]) was performed by Wang, Zou and Mandel [1991] using the experimental arrangement shown in fig. 5.10.

If we assume 50:50 beam splitters, no idler photons will reach D_2, but signal photons will be detected one-fourth of the time. Idler photons will only reach D_1, and this will happen one-fourth of the time. Due to second-order interference,

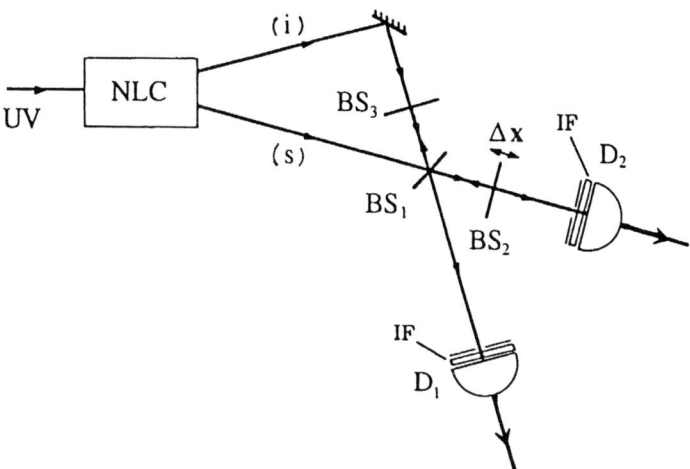

Fig. 5.10. Schematic of the experimental arrangement used to test the de Broglie guiding-wave theory (Wang, Zou and Mandel [1991]).

the rate at which signal photons reach D_1 will depend on the difference in the lengths of the optical paths from BS_1 to BS_2 and BS_3, resulting in interference fringes with a visibility of 0.5.

However, if we consider the rate of coincidences between D_1 and D_2, quantum theory predicts that this will be a constant, corresponding to the detection of an idler photon at D_1 and a signal photon at D_2. On the other hand, the theory of Croca, Garuccio, Lepore and Moreira [1990] predicts coincidence fringes with a visibility of 0.5 arising from the interference of the guiding wave for the signal photon reflected from BS_3 with the guiding wave for the idler photon.

The results of this experiment revealed no such interference effects, supporting the quantum theory (however, see Holland and Vigier [1991]).

§ 6. Two-photon Interferometry

In a classic paper, Einstein, Podolsky and Rosen [1935] presented a paradox which brought out the incompatibility of quantum theory and the assumption of local realism. They were led to conclude that the quantum-mechanical description of a system was incomplete and postulated the existence of "hidden variables", the specification of which would predetermine the results of any measurements. Subsequently, Bell [1965] proposed a test, based on a *Gedankenexperiment* of Bohm [1951] involving spin-half particles, that could distinguish between quantum theory and the entire class of theories based on local realism.

At the quantum level, polarization is associated with the spin of the photon. Although the photon is a spin-one particle, only two polarization states are allowed; hence, the angular momentum of the photon corresponds to that of a pseudo spin-half system. Polarization-correlation experiments therefore provide a convenient way to realize spin-half particle experiments. This led to a generalization of Bell's theorem, and a proposal for an experiment involving measurements of the polarization correlations of photon pairs produced by an atomic cascade (Clauser, Horne, Shimony and Holt [1969]).

6.1. ENTANGLED STATES AND BELL'S INEQUALITY

If we consider a pair of photons described by the entangled polarization singlet-like state,

$$|H;V\rangle_{1,2} = 2^{-1/2} \left(|H\rangle_1 |V\rangle_2 - |V\rangle_1 |H\rangle_2 \right), \tag{6.1}$$

where H and V denote single photons with horizontal and vertical polarizations, respectively, and the subscripts identify their propagation directions, a measurement involving photons travelling in one of these directions will show no preferred polarization. However, if the polarization of photon 1 is measured in some basis, the polarization of photon 2 can be predicted with certainty. We then find that while quantum mechanics and theories based on local realism agree in situations of perfect correlations or anticorrelations, quantum mechanics gives different predictions for polarizers at intermediate angles (Clauser, Horne, Shimony and Holt [1969]).

Initially, experimental tests of Bell's inequality, based on polarization correlations, were made using pairs of photons produced by an atomic cascade (Aspect, Dalibard and Roger [1982]). However, in these experiments the correlation of the polarizations was not complete because of the imperfect angular correlation of the photons.

Better results have been obtained in experiments using pairs of photons produced by parametric down-conversion (Shih and Alley [1988], Ou and Mandel [1988a]). In the arrangement used by Ou and Mandel [1988a] (see fig. 6.1), linearly polarized photons (wavelength about 702 nm) with their electric vector in the plane of the diagram were produced by degenerate parametric down-conversion. The idler photons passed through a half-wave plate that rotated their plane of polarization by 90°, while the signal photons traversed a compensating

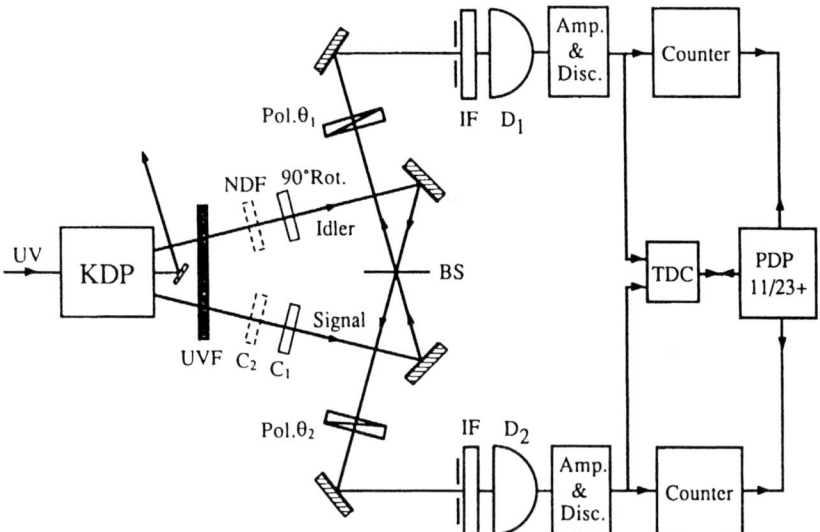

Fig. 6.1. Apparatus used to demonstrate violations of Bell's inequality using photon pairs produced by parametric down-conversion (Ou and Mandel [1988a]).

glass plate producing an equal time delay. The mixed signal and idler photons emerging from the two sides of the beam splitter, after passing through linear polarizers set at adjustable angles θ_1 and θ_2 and identical interference filters, were incident on two photodetectors D_1 and D_2. The coincidence counting rate provided a measure of the joint probability $\mathcal{P}(\theta_1, \theta_2)$ of detecting two photons for various settings θ_1, θ_1 of the two linear polarizers.

According to quantum theory, the probability of detecting a coincidence in this arrangement is:

$$\mathcal{P}(\theta_1, \theta_2) \propto \sin^2(\theta_2 - \theta_1), \tag{6.2}$$

which only depends on the difference in the angular settings of the two polarizers. The actual coincidence count rates obtained for various values of θ_1, with θ_2 fixed at 45°, are presented in fig. 6.2, along with curves corresponding to the predictions of quantum mechanics and classical theory. As can be seen, the observed relative modulation obtained from the best fit curve is about 0.76, which is below the value of 1.00 predicted by quantum mechanics, probably because of imperfect alignment of the signal and idler beams, but greater than the figure of 0.50 expected from classical theory. However, this result corresponds to a violation of Bell's inequality by about 6 standard deviations.

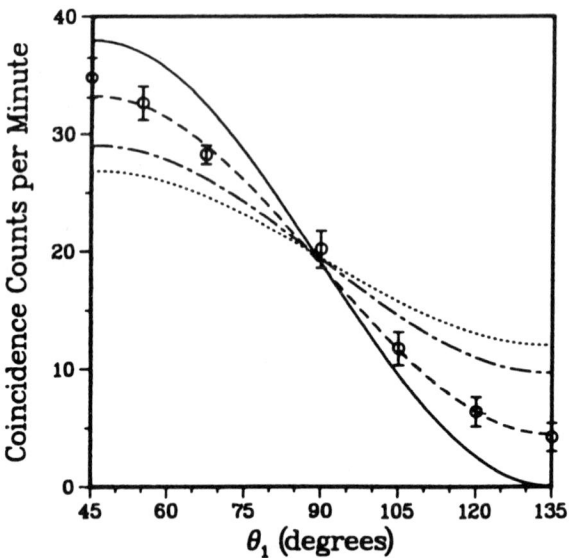

Fig. 6.2. Measured coincidence counting rate for different values of the polarizer angle θ_1, with θ_2 fixed at 45°. The solid curve and the dash–dotted curve correspond to the predictions of quantum mechanics and classical theory, respectively. The dashed and dotted curves are obtained by multiplying these curves by a factor of 0.76 to allow for reduced modulation due to imperfect alignment of the beams (Ou and Mandel [1988a]).

6.2. INTERFEROMETRIC TESTS OF BELL'S INEQUALITY

The generation of correlated photon pairs by parametric down-conversion has also made possible tests of Bell's inequality using two-photon interferometry, which are not based on polarization (Horne, Shimony and Zeilinger [1989]).

A general arrangement for two-photon interferometry is shown in fig. 6.3. In this arrangement, pairs of photons, one having a wavelength λ_1 and the other a wavelength λ_2, are selected by four pinholes in a diaphragm placed downstream from the nonlinear crystal to produce four beams, A, B, C, D, with wave vectors k_A, k_B, k_C and k_D, where

$$|k_A| = |k_D|, \tag{6.3}$$

$$|k_B| = |k_C|, \quad \text{but} \quad |k_B| \neq |k_A|, \tag{6.4}$$

and

$$k_A + k_C = k_B + k_D = k, \tag{6.5}$$

where k is the wave vector of the beam incident on the crystal.

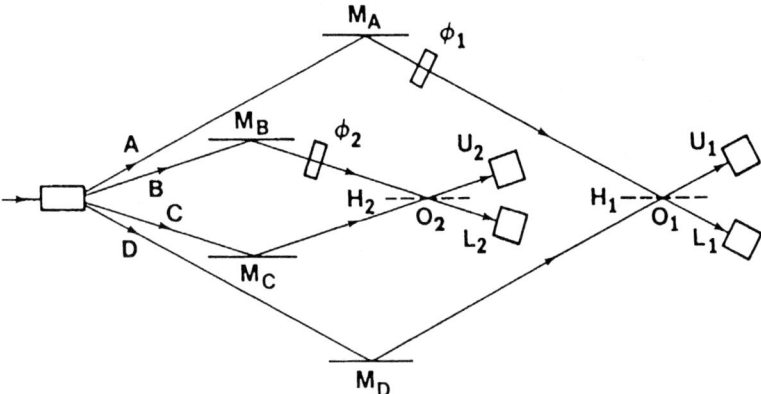

Fig. 6.3. Schematic of an interferometer used to demonstrate two-photon interference (Horne, Shimony and Zeilinger [1989]).

Each pair in the ensemble of photon pairs is in the quantum state

$$|\psi\rangle = 2^{1/2} \left[|A\rangle_1 |C\rangle_2 + |D\rangle_1 |B\rangle_2 \right], \tag{6.6}$$

which is a coherent superposition of the probability amplitudes corresponding to two distinct pairs of correlated paths. In one case, a photon with wavelength λ_1 goes into beam A, and a photon with wavelength λ_2 goes into beam C; in the other, a photon with wavelength λ_1 goes into beam D, and a photon with wavelength λ_2 goes into beam B. A variable phase difference ϕ_1 can be introduced between the beams A and D before they are recombined by the 50:50 beam splitter H_1 and proceed to the photo detectors U_1 and L_1. Similarly, a variable phase difference ϕ_2 can be introduced between the beams B and C before they are recombined by the 50:50 beam splitter H_2 and proceed to the photo detectors U_2 and L_2. It follows that the two interfering beams at H_1 have the same wavelength λ_1, while the two interfering beams at H_2 have the same wavelength λ_2; however, the wavelengths at H_1 and H_2 are different.

The quantum-mechanical probabilities for the joint detection of both photons by the detector pairs (U_1, U_2), (L_1, L_2), (U_1, L_2), and (U_2, L_1), are then proportional to the absolute squares of the corresponding probability amplitudes defined by eq. (6.6), and are given by the relations:

$$\begin{aligned} \mathcal{P}(U_1, U_2 \,|\, \phi_1, \phi_2) &= \mathcal{P}(L_1, L_2 \,|\, \phi_1, \phi_2) \\ &= \tfrac{1}{4}\eta^2 \left[1 + \cos(\phi_2 - \phi_1 + \phi_0) \right], \end{aligned} \tag{6.7}$$

and

$$\begin{aligned} \mathcal{P}(U_1, L_2 \,|\, \phi_1, \phi_2) &= \mathcal{P}(L_1, U_2 \,|\, \phi_1, \phi_2) \\ &= \tfrac{1}{4}\eta^2 \left[1 - \cos(\phi_2 - \phi_1 + \phi_0) \right], \end{aligned} \tag{6.8}$$

where η is the quantum efficiency of the photo detectors, and ϕ_0 is a phase factor determined, once and for all, by the placement of the mirrors and beam splitters.

On the other hand, the probabilities for detecting single photons by the four detectors are:

$$\mathcal{P}(U_1 | \phi_1, \phi_2) = \mathcal{P}(L_1 | \phi_1, \phi_2) = \mathcal{P}(U_2 | \phi_1, \phi_2) = \mathcal{P}(L_2 | \phi_1, \phi_2) \\ = \tfrac{1}{2}\eta. \tag{6.9}$$

It follows that while the count rate for single photons, which is defined by eq. (6.9), is constant and independent of ϕ_1 and ϕ_2, the count rates for coincidences, which are defined by eqs. (6.7–6.8), will vary sinusoidally with the phase shifts ϕ_1 and ϕ_2. These interference fringes observed with spatially separated two-photon states are a quantum-mechanical phenomenon arising from their entangled nature.

An experimental arrangement involving two-photon interferometry, similar to that shown in fig. 6.3, was used by Rarity and Tapster [1990] for a test of Bell's inequality based on the entanglement of the momenta of the photons in a pair. In this case, given the direction of one photon at one of the detectors U_2, L_2, quantum theory indicates that the direction taken by the other photon of the pair is dependent on the setting of the remote phase plate ϕ_2 to an extent that cannot be explained by any theory based on local realism.

To verify this proposition, measurements were made of the coincidence rates between the four detectors for selected values of the variable phase differences. The correlation coefficient

$$\mathcal{E}(\phi_1, \phi_2) = \frac{\mathcal{P}(U_1, U_2 | \phi_1, \phi_2) + \mathcal{P}(L_1, L_2 | \phi_1, \phi_2) - \mathcal{P}(U_1, L_2 | \phi_1, \phi_2) - \mathcal{P}(L_1, U_2 | \phi_1, \phi_2)}{\mathcal{P}(U_1, U_2 | \phi_1, \phi_2) + \mathcal{P}(L_1, L_2 | \phi_1, \phi_2) + \mathcal{P}(U_1, L_2 | \phi_1, \phi_2) + \mathcal{P}(L_1, U_2 | \phi_1, \phi_2)}, \tag{6.10}$$

can be taken as a measure of the distribution of coincidences between detectors on the same side and opposite sides of the beam splitters for these phase settings.

A generalization of Bell's inequality (Clauser and Shimony [1978]) then states that the combination of four such measurements, at various phase settings given by the relation

$$\mathcal{S} = \mathcal{E}(\phi_1, \phi_2) - \mathcal{E}(\phi_1, \phi_2') + \mathcal{E}(\phi_1', \phi_2) + \mathcal{E}(\phi_1', \phi_2'), \tag{6.11}$$

should always lie within the bounds

$$-2 \leqslant \mathcal{S} \leqslant 2, \tag{6.12}$$

if we assume local realism. However, quantum theory indicates that for appropriately chosen values of the phase angles ($\phi_1 = 0$, $\phi_1' = \tfrac{1}{2}\pi$, $\phi_2 = \tfrac{1}{4}\pi$, $\phi_2' = \tfrac{3}{4}\pi$),

$$\mathcal{S} = 2\sqrt{2}. \tag{6.13}$$

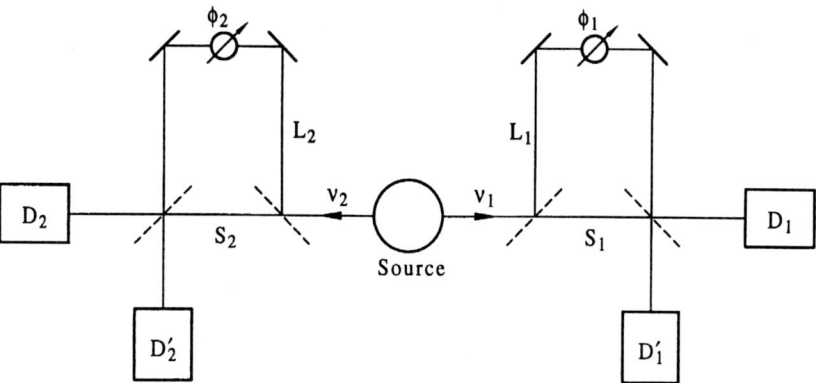

Fig. 6.4. Simplified schematic of the optical system for a Bell's inequality experiment based on energy and time correlations (Franson [1989]).

In actual measurements, a value of $S = 2.21 \pm 0.022$ was obtained. This lower value could be attributed to a reduced visibility $\mathcal{V} = 0.78$ of the interference fringes due to misalignment of the apparatus, but corresponded to a violation of Bell's inequality by 10 standard deviations.

Another experimental test of Bell's inequality was proposed by Franson [1989] and carried out by several groups (Franson [1991a], Brendel, Mohler and Martienssen [1992], Kwiat, Steinberg and Chiao [1993], Shih, Sergienko and Rubin [1993]). Figure 6.4 is a schematic of the basic arrangement. In the actual experiments, each of the photons from a down-converted pair was sent into an unbalanced interferometer, presenting a short (S) and a long (L) path to the final output.

Examination of the singles count rates when the imbalances were greater than the coherence length of the down-converted photons revealed no interference effects. However, when the difference of the path-length differences in the two interferometers was less than the coherence length of the down-converted photons, observations of the coincidence rates revealed interference effects arising from the impossibility of distinguishing between the two processes which led to coincidences. These interference effects could be observed even when the extra optical path traversed by one of the photons was quite long (Franson [1991b], Rarity and Tapster [1992]).

With detectors fast enough to exclude the possibility of one photon taking the short path, and the other taking the long path, high-visibility fringes could be obtained corresponding to observations of the quantum state

$$|\psi\rangle = \tfrac{1}{2}\left(|S_1, S_2\rangle - e^{i\phi}|L_1, L_2\rangle\right), \tag{6.14}$$

where ϕ is proportional to the sum of the relative phases in the two interferometers. As shown in fig. 6.5, sinusoidal fringes with a visibility greater than 0.8 were obtained (Kwiat, Steinberg and Chiao [1993]), whereas the maximum possible without violating Bell's inequality would be 0.71.

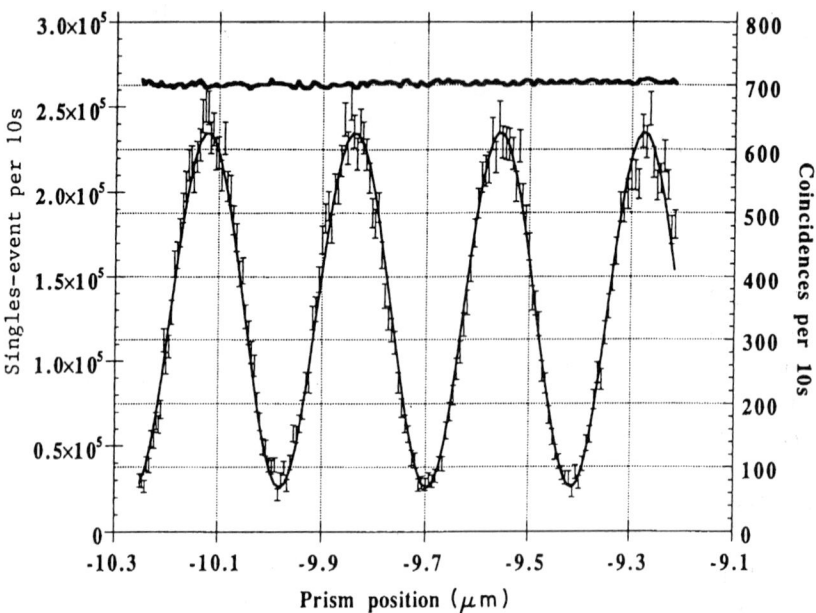

Fig. 6.5. Coincidence fringes obtained as the phase in interferometer 1 is varied. The constant single-event rate is also shown for comparison (Kwiat, Steinberg and Chiao [1993]).

A significant loophole in all these experiments has been the lack of detectors with unit quantum efficiency, necessitating the assumption that the fraction of the pairs detected is representative of the entire ensemble (Clauser, Horne, Shimony and Holt [1969], Santos [1992]). Some progress towards solving this problem has been made by the development of photo detectors with high quantum efficiencies (Kwiat, Steinberg, Chiao, Eberhard and Petroff [1993]). A possibility is the use of a nonmaximally entangled state (in which the magnitudes of the probability amplitudes of the contributing terms are not equal), which can lead to a significant reduction in the required detector efficiency (Eberhard [1993]). An experiment leading to a loophole-free test of Bell's inequality has been proposed by Kwiat, Eberhard, Steinberg and Chiao [1994].

6.3. OTHER TESTS OF LOCAL REALISM

Another solution of the problem of demonstrating that quantum mechanics violates local realism, which does not involve Bell's inequality, has been developed by Hardy [1992a,b, 1993] and by Jordan [1994]. Figure 6.6 shows the setup used by Torgerson, Branning, Monken and Mandel [1995] in an experiment based on this approach.

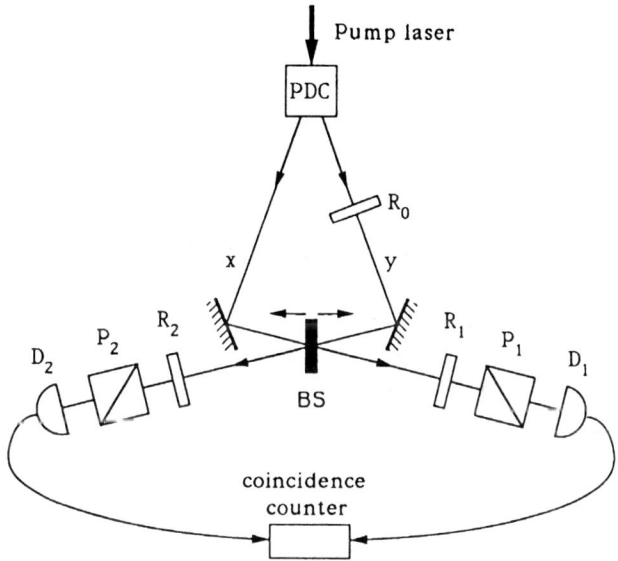

Fig. 6.6. Setup used to demonstrate violation of local realism (Torgerson, Branning, Monken and Mandel [1995]).

In this arrangement, pairs of photons with linear (x) polarizations were produced by parametric down conversion. A rotator R_0 inserted in the idler beam converted it to the orthogonal (y) polarization. The signal and idler beams were then mixed at a beam splitter, and the two outputs were taken to similar analyzers. Each of these consisted of a rotatable half-wave plate (R_1 or R_2) followed by a fixed linear polarizer (P_1 or P_2) and a photo detector (D_1 or D_2).

We consider measurements of the number of two-photon coincidences made with a series of polarizer settings when the signal and idler optical path lengths are equal. The angles θ_1 and θ_{10} define two possible settings of the polarizer in arm 1; similarly, θ_2 and θ_{20} define two possible settings of the polarizer in arm 2. The angles $\bar{\theta}_i = \theta_i + \frac{1}{2}\pi$ ($i = 1, 2, 10, 20$) define the orthogonal settings. If $\mathcal{P}_{12}(\theta_1, \theta_2)$ is the joint probability of detecting a photon in arm 1 with the polarizer set at θ_1 and a photon in arm 2 with the polarizer set at θ_2, quantum mechanics shows that, for a nonabsorbing beam splitter with $|T|^2 + |R|^2 = 1$ and $|T| \neq |R|$, there exist polarizer angles $\theta_1, \theta_2, \theta_{10}, \overline{\theta}_{10}, \theta_{20}$ and $\overline{\theta}_{20}$, such that

$$\mathcal{P}_{12}(\theta_1, \overline{\theta}_{20}) = 0, \tag{6.15}$$

$$\mathcal{P}_{12}(\overline{\theta}_{10}, \theta_2) = 0, \tag{6.16}$$

$$\mathcal{P}_{12}(\theta_{10}, \theta_{20}) = 0, \tag{6.17}$$

$$\mathcal{P}_{12}(\theta_1, \theta_2) > 0. \tag{6.18}$$

The value of $\mathcal{P}_{12}(\theta_1, \theta_2)$ is greatest when (Torgerson, Branning and Mandel [1995]):

$$\tan \theta_1 = \left(\frac{|T|}{|R|}\right)^3 = \cot \theta_2, \quad \tan \theta_{10} = -\frac{|T|}{|R|} = \cot \theta_{20}. \tag{6.19}$$

However, according to the point of view adopted by Einstein, Podolsky and Rosen [1935], eq. (6.18) contradicts eq. (6.17).

Experimental measurements confirmed that the value of $\mathcal{P}_{12}(\theta_1, \theta_2)$ was clearly non zero. Even though the values for $\mathcal{P}_{12}(\theta_1, \overline{\theta}_{20})$, $\mathcal{P}_{12}(\overline{\theta}_{10}, \theta_2)$ and $\mathcal{P}_{12}(\theta_{10}, \theta_{20})$ were not exactly equal to zero, the data contradicted local realism by about 45 standard deviations.

6.4. TWO-PHOTON INTERFERENCE

Another class of two-photon interference experiments makes use of the down-converted light beams from two nonlinear crystals which are optically pumped by mutually coherent beams from the same laser.

In one arrangement (see fig. 6.7), the signal beams s_1 and s_2 from the two down-converters are combined by one beam splitter (BS_A) and allowed to fall

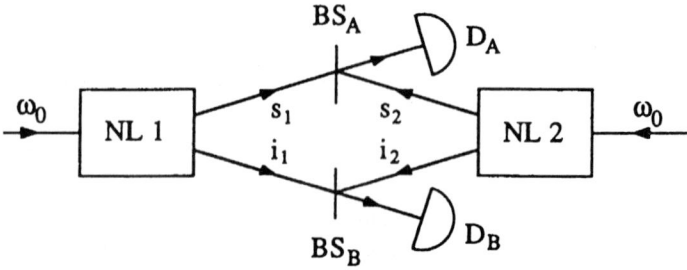

Fig. 6.7. Experimental arrangement used to observe interference effects produced by down-converted light beams from two nonlinear crystals (Ou, Wang, Zou and Mandel [1990]).

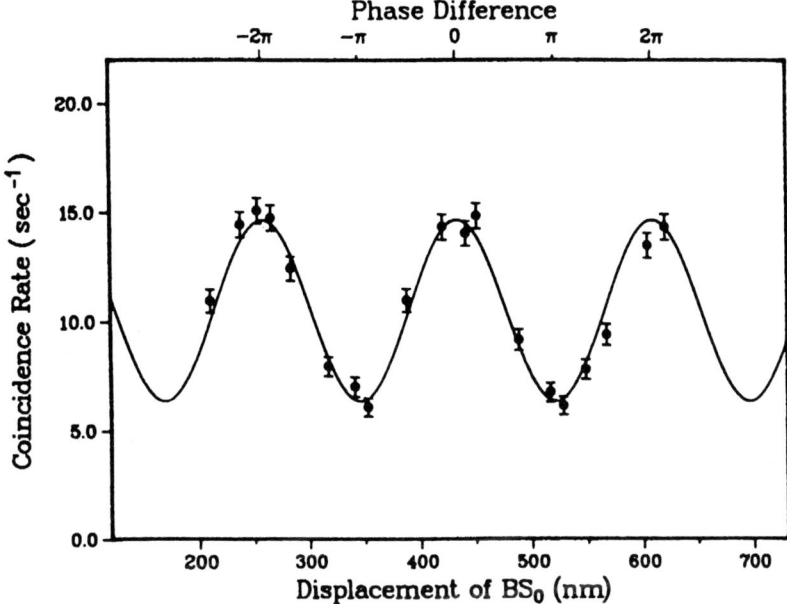

Fig. 6.8. Measured two-photon coincidence rate as a function of the displacement of the beam splitter (Ou, Wang, Zou and Mandel [1990]).

on one photo detector (D_A), while the two idler beams i_1 and i_2 are combined by another beam splitter (BS_B) and taken to another photo detector (D_B) (Ou, Wang, Zou and Mandel [1990]). Measurements of the counting rates of the individual photo detectors showed no change as the optical path difference was varied, confirming that the mutual coherence of the pump beams did not produce any mutual coherence, either between the two signal beams s_1 and s_2 from the two down-converters, or between the two idler beams i_1 and i_2. However, as shown in fig. 6.8, measurements of the coincidence rate for simultaneous detection of photons by both D_A and D_B, as a function of the optical path difference, revealed interference effects.

A modification of this arrangement, shown in fig. 6.9, uses a single nonlinear crystal traversed by the pump beam in opposite directions (Herzog, Rarity, Weinfurter and Zeilinger [1994]). Down-converted photons can be generated on either of the two passes, and it is possible to make the idler modes from the two processes overlap at one photo detector, while the signal modes overlap at the other. Since the two production processes are indistinguishable, interference effects are observed in the singles rates, as well as in the coincidence rates, when any one of the mirrors is translated. An interesting aspect of this experiment is

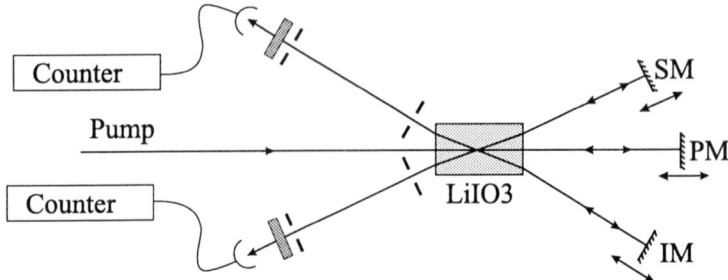

Fig. 6.9. Arrangement using a single nonlinear crystal to generate two sets of down-converted light beams (Herzog, Rarity, Weinfurter and Zeilinger [1994]).

that the distances to the mirrors can be much greater than the coherence lengths of the down-converted beams; one interpretation of the results is, therefore, a variable enhancement (or suppression) of the down-conversion process.

Nonclassical effects can also show up in certain second-order interference experiments in which only one photon is detected (Mandel [1982]). Figure 6.10 is a schematic of the optical system for such an interference experiment with beams from two parametric down-converters (Zou, Wang and Mandel [1991]). In this arrangement, both the nonlinear crystals, NL_1 and NL_2, were optically pumped by mutually coherent beams derived from the same laser by means of a beam splitter. However, while the two signal beams s_1 and s_2 were combined by means of another beam splitter and taken to a photo detector (D_s), the idler beam i_1 was allowed to pass through the nonlinear crystal NL_2 and fall, along with the second idler beam i_2, directly on the other photo detector D_i.

When the optical path difference was varied by translating the beam splitter BS_0, the photon counting rate at D_s was found to oscillate, indicating that s_1 and s_2 were mutually coherent (see curve A in fig. 6.11). These oscillations could be observed as long as i_1 and i_2 were collinear, but if either i_1 or i_2 was misaligned, or if i_1 was blocked so that it could not reach NL_2, the interference disappeared (see curve B in fig. 6.11).

If, instead of blocking i_1, an attenuator or beam splitter with a complex amplitude transmittance t was placed between NL_1 and NL_2, the visibility of the interference pattern registered by D_s was found to be proportional to $|t|$. However, the average rate of photon counts was the same in both cases, implying that the degree of mutual coherence of the two beams could be controlled without affecting their intensities.

In addition, the introduction of a delay τ, by varying the length of the path of the idler i_1 between the two nonlinear crystals NL_1 and NL_2, was found to

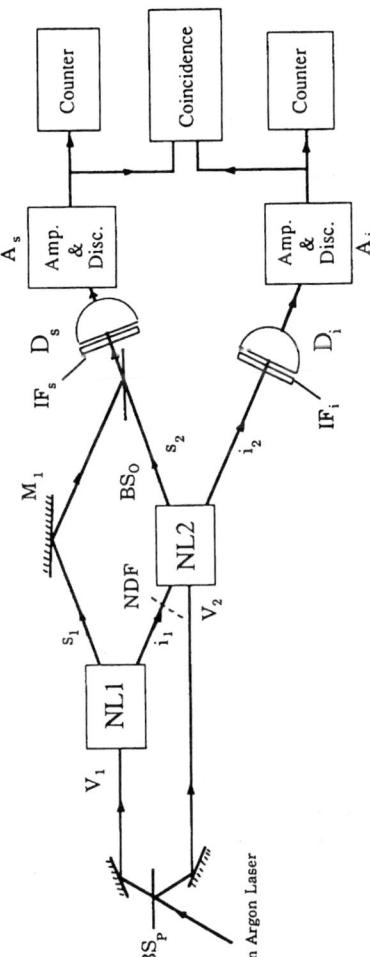

Fig. 6.10. Experimental setup used to observe interference effects produced by the signal beams from two parametric down-converters (Zou, Wang and Mandel [1991]).

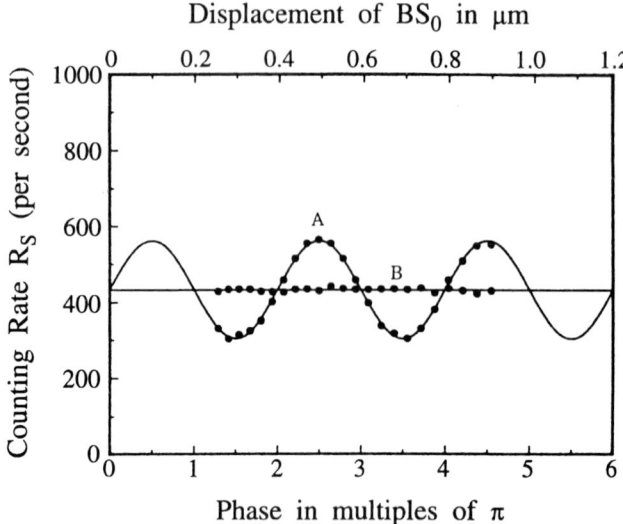

Fig. 6.11. Counting rate of the detector D_s as a function of the displacement of the beam splitter BS_0: (A) with the idler beams i_1 and i_2 aligned and (B) with the idler beam i_1 blocked (Zou, Wang and Mandel [1991]).

affect the visibility of the interference effects produced by the signal beams, s_1 and s_2, exactly as if the delay had been introduced in one of the signal paths (Zou, Grayson, Barbosa and Mandel [1993]). A phase shift of the idler beam i_1 introduced through the geometric (Pancharatnam) phase (see § 4.1) also had the same effect on the interference pattern produced by the signal beams (Grayson, Torgerson and Barbosa [1994]).

Figure 6.12 shows the variation of the visibility of the interference effects as a function of the time delay; as can be seen, when $\tau > \tau_c$, where $\tau_c \approx 1$ ps is the coherence time, the visibility of the interference effects drops to zero. However, it is well known that even when $\tau \gg \tau_c$, interference effects can still be seen in the spectral domain (Mandel [1962]). Such effects were observed in this experiment by inserting a scanning Fabry–Perot interferometer before the detector D_s (Zou, Grayson and Mandel [1992]). As shown in fig. 6.13, the expected modulation of the spectrum could be observed even with a differential delay $\tau \approx 5$ ps $\approx 5\tau_c$. This modulation disappeared when the idler beam i_1 was blocked.

All these effects can be understood in terms of the indistinguishability of the paths taken by the beams through the interferometer. In the arrangement shown in fig. 6.7, when a coincidence is registered, there is no way to determine whether the pair of photons involved originated in NL_1 or NL_2. Similarly, in the

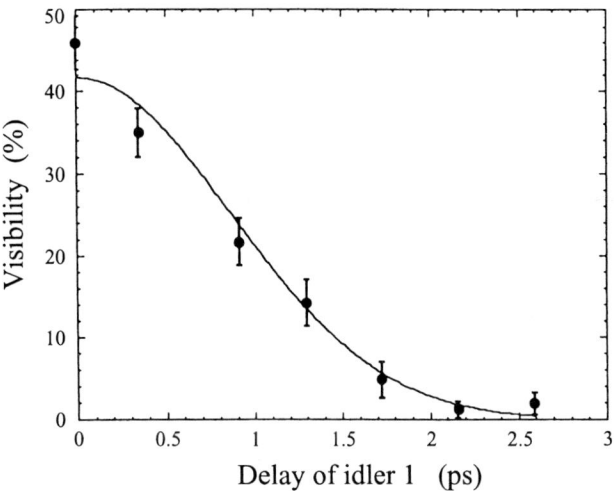

Fig. 6.12. Variation of the visibility of the interference effects produced by the signal beams s_1 and s_2 with the time delay inserted in the idler beam i_1 (Zou, Grayson, Barbosa and Mandel [1993]).

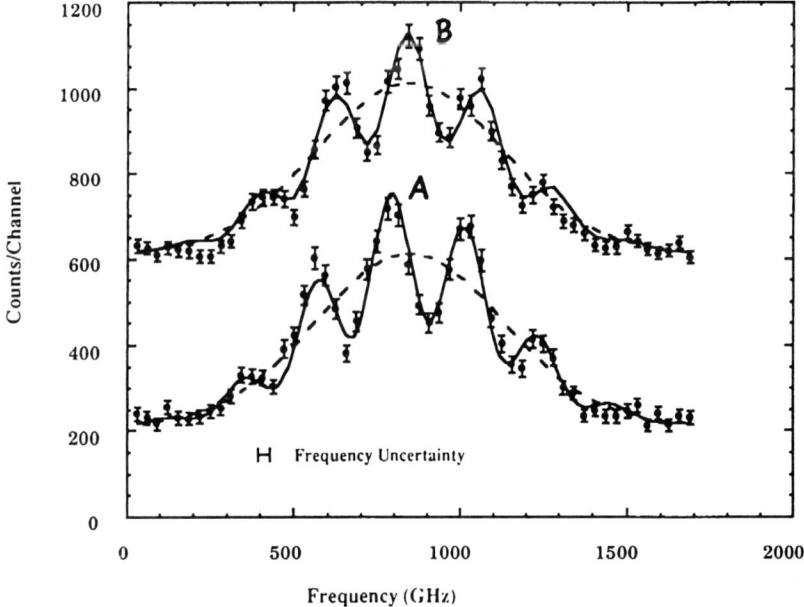

Fig. 6.13. Variation of the count rate as a function of the optical frequency with a time delay $\tau \approx 3\tau_c$ inserted in (A) s_1 and (B) i_1. The dashed curve shows the original unmodulated spectrum (Zou, Grayson and Mandel [1992]).

arrangement shown in fig. 6.10, there is no way to determine the origins of the photons reaching the photo detector D_s, as long as both i_1 and i_2 are incident on the detector D_i.

6.5. TWO-PHOTON TESTS OF BELL'S INEQUALITY

An experimental arrangement that could overcome the problems encountered with earlier interferometric tests of Bell's inequality has been proposed by Pavičić [1995]. This arrangement is based on fourth-order spin-correlated interferometry using two independent pairs of spin-correlated photons.

Ou, Hong and Mandel [1987] showed that a pair of orthogonally polarized photons incident on a symmetrically positioned beam splitter produce a singlet-like state. On the other hand, similar photons with parallel polarizations never appear on opposite sides of the beam splitter (Hong, Ou and Mandel [1987]). Subsequently, these observations were extended to show that the fourth-order interference interaction between a beam splitter and two incoming unpolarized photons imposes polarization correlations on the emerging photons. For an appropriate position of the beam splitter, incoming unpolarized photons emerge with orthogonal polarizations. More specifically, they appear entangled in a singlet state, similar to that described by eq. (6.1), when they exit on different sides of the beam splitter (Pavičić [1994], Pavičić and Summhammer [1994]).

In the arrangement shown in fig. 6.14 (Pavičić [1995]), a subpicosecond laser pulse pumps two nonlinear crystals, NL1 and NL2, to produce simultaneous pairs of signal and idler photons with the same frequency, which are converted to orthogonal polarizations by the 90° rotators. These photon pairs are incident on the two beam splitters, BS1 and BS2, which therefore act as sources of independent singlet pairs. Two of the photons, one from each pair, interfere at the beam splitter BS. As a result, the other two photons from these pairs appear to be in a singlet state, although they are completely independent and have never interacted. Even when no polarization measurements are carried out on the first two photons, one finds polarization correlations between the latter two photons. One of the subsets of these two photons contains only photons in the singlet state, and we can therefore consider them preselected by their pair-companions which interfered at BS.

It can then be shown that, with the polarizers P1 and P2 removed, and the polarizers P1' and P2' oriented at angles $\theta_{1'}$ and $\theta_{2'}$, respectively, the probability of coincident detection of four photons by the detectors D1, D2, D1' and D2' is given by the relation

$$\mathcal{P}(\theta_{1'}, \theta_{2'}) = \tfrac{1}{8}\left[1 - \mathcal{V}\cos^2(\theta_{1'} - \theta_{2'})\right], \tag{6.20}$$

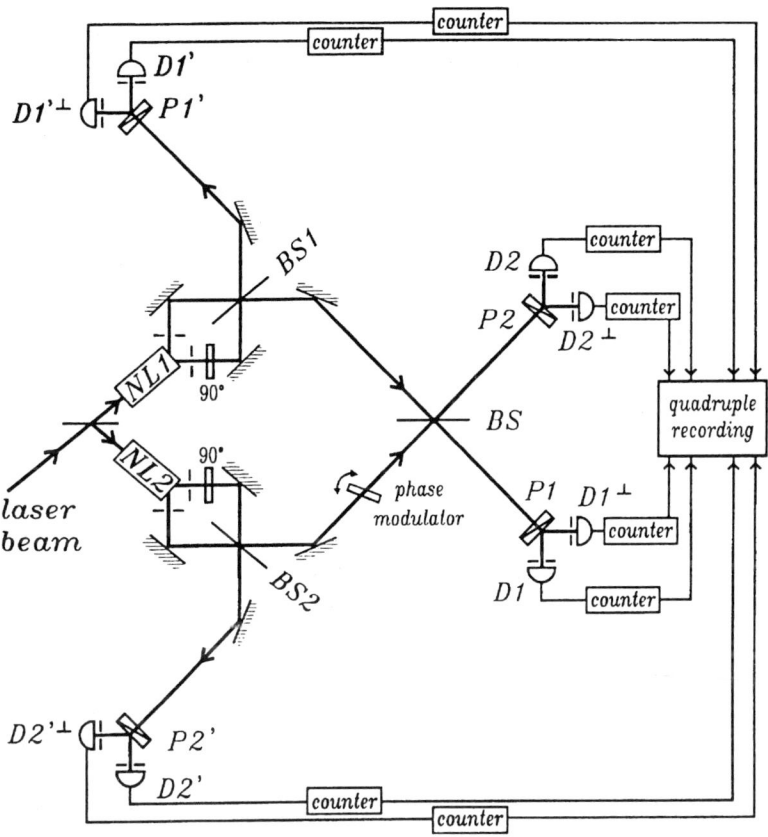

Fig. 6.14. Experimental arrangement for an interferometric test of Bell's inequality using spin-correlated photons (Pavičić [1995]).

where $\mathcal{V}$ is the visibility of the fringes normally obtained by coincidence counting. This probability is given by the ratio of the numbers of coincidence counts,

$$f(\theta_{1'}, \theta_{2'}) = \frac{N(\text{D1}' \cap \text{D2}')}{N\left[(\text{D1}' \cup \text{D1}'^{\perp}) \cap (\text{D2}' \cup \text{D2}'^{\perp})\right]}, \qquad (6.21)$$

divided by 4. For a violation of Bell's inequality, we need

$$\eta(1 + \mathcal{V}\sqrt{2}) > 2, \qquad (6.22)$$

where η is the quantum efficiency of the detectors.

Pavičić [1995] has also proposed a modification that would, in principle, make it possible to lower the required threshold levels for the visibility of the interference fringes and the quantum efficiency of the detectors.

§ 7. Complementarity

Interferometry in the quantum domain is characterized by complementarity: wave *vs* particle, certainty in photon number *vs* certainty of phase, visibility of interference fringes *vs* certainty of the photon path. The paradox of the undular and corpuscular aspects of light, which flow from the quantum description, has led to many experiments to study complementarity.

In § 3.2, we discussed some experiments on interferometry with single-photon input states by Grangier, Roger and Aspect [1986]. Although the quality of the interference fringes produced by single-photon states is impressive, the most striking aspect of the experiment is the fact that the apparatus could be transformed easily to exhibit either wave-like or particle-like behavior by a single photon. At the same time, it does not follow that two distinct experiments are required to reveal complementary features of the photon. Wootters and Zurek [1979] employed an information-theoretic approach to show how, in a double-slit experiment, one could obtain some information on the path taken by the photon (particle-like behavior) while retaining an interference pattern with some degree of clarity (wave-like behavior). Measurements are not restricted therefore to either one or the other of these complementary quantities, and some information on both can always be obtained, subject to the limits set by complementarity.

7.1. QUANTUM-NONDEMOLITION MEASUREMENTS

Heisenberg's principle states that the uncertainty in the number of quanta n in a beam of light and the uncertainty in its phase ϕ are linked through the relation (Heitler [1954])

$$\Delta n \Delta \phi \geq \tfrac{1}{2}. \tag{7.1}$$

It follows from this relation that, if we know the exact number of photons in a beam, we have no knowledge of the phase.

However, in principle, experiments based on photon-number quantum non-demolition measurements are possible (Milburn and Walls [1983], Yamamoto, Imoto and Machida [1986], Braginsky [1989]), in which the photon number of the light field can be measured in such a way that, following the measurement, the number of light quanta remains unchanged. Several schemes have been proposed for this purpose (Roch, Roger, Grangier, Courty and Reynaud [1992]).

One method is based on the phase shift of an electron wave produced by a light beam through the Aharonov–Bohm effect (Chiao [1970], Lee, Yin, Gustafson and Chiao [1992]). Another uses the phase shift in a probe beam resulting from the index change produced through the Kerr effect by a signal beam (Imoto, Haus and Yamamoto [1985], Kitagawa and Yamamoto [1986]). Yet other proposals use Rydberg atoms to give indirect information on the number of photons in a microwave cavity (Haroche, Brune and Raimond [1992], Walther [1992]).

Since a quantum-nondemolition measurement allows the determination of the presence of a single photon without annihilating it, complementarity requires a disturbance to the interference fringes. A theoretical treatment, by Sanders and Milburn [1989], of a photon-number quantum-nondemolition measurement in one arm of a Mach–Zehnder interferometer, with single-photon inputs into one port of the interferometer, demonstrates that the interference fringes are progressively reduced in visibility as greater certainty of the path of the photon through the interferometer is obtained. The presence of the photon is detected by the phase shift of a probe field that interacts with the photon *via* a nonlinear Kerr medium. Greater certainty of the path of the photon requires a reduction of the phase fluctuations in the probe field. This reduction requires a corresponding increase in the amplitude fluctuations of the probe field which feed, in turn, into the phase fluctuations of the field within the interferometer and destroy the interference fringes.

7.2. DELAYED-CHOICE EXPERIMENTS

An interesting question raised by von Weiszäcker [1931] and by Wheeler [1978] is whether the result of Young's double-slit experiment would be changed if the decision to observe either interference, or the path of the photon, was made after the photon had passed through the slits.

Wheeler's proposal envisaged a Mach–Zehnder interferometer illuminated by a light pulse with photo detectors placed in the two outputs. A decision would be made "whether to put in the second beam splitter, or take it out, at the very last minute". This would make it possible to decide whether the photon had come by one route, or by both routes, after it had already completed its journey.

A delayed-choice experiment along these lines was carried out by Hellmuth, Walther, Zajonc and Schleich [1987] with an interferometer incorporating 5 m long single-mode fibers in the two paths. The light source was a mode-locked krypton-ion laser emitting pulses with a duration of 150 ps at a repetition rate of 81 MHz. An acousto-optic switch was used to select one pulse out of 8000, thereby ensuring that the time between pulses was much longer than the transit time of the light through the interferometer. An optical attenuator reduced the average number of photons per pulse to less than 0.2. A combination of a Pockels cell and a polarizing prism was used as a switch in one arm to interrupt the light after it had passed the first beam splitter. Data were recorded as the mode of operation was switched between normal and delayed-choice for successive light pulses.

The results obtained showed no observable difference between the normal and delayed-choice modes of operation, in agreement with the predictions of quantum mechanics. However, since the picosecond pulse is in a coherent state, the second-order correlation function $g^{(2)}(0)$ is nonzero, and perfect path information cannot be obtained.

Another delayed-choice experiment, performed by Baldzuhn, Mohler and Martienssen [1989], used photon pairs produced by parametric down-conversion (see § 2.5.2). One photon served as a trigger to switch between registration of "which-path" information and phase information. In this case also, the result was independent of whether the switching took place before, or after, the photon passed the first beam splitter of the interferometer.

7.3. THE QUANTUM ERASER

Another consequence of the uncertainty principle is that any attempt to identify the path of a photon leads to an irreversible change in its momentum, which in turn washes out any interference effects (Bohr [1983]). However, measurements which do not involve a reduction of the state vector can be reversible in some sense.

Normally, whenever "which-path" information is available, the paths in an interferometer are no longer indistinguishable, and interference effects cannot be observed. However, interference effects may reappear if the distinguishing information can somehow be "erased" by correlating the results of the measurements with the results of properly chosen measurements on the physical system. This procedure is the basis of what is now commonly known as the "quantum eraser" (Scully and Druhl [1982], Scully, Englert and Walther [1991], Zajonc, Wang, Zou and Mandel [1991]).

One demonstration of a quantum eraser (Kwiat, Steinberg and Chiao [1992]) used the interferometer shown in fig. 7.1. A half-wave plate inserted in one of the paths before the beam splitter was used to rotate the plane of polarization

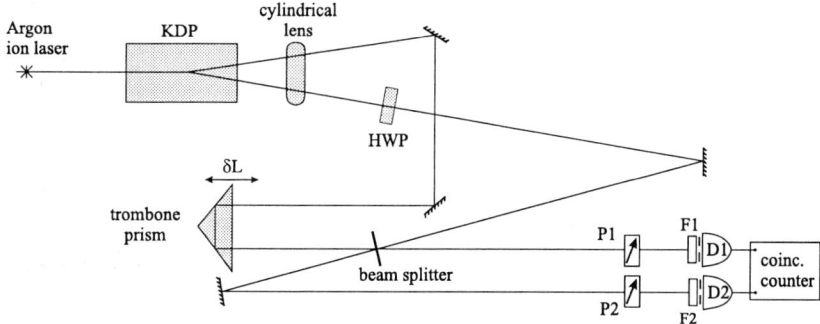

Fig. 7.1. Two-photon interferometer used as a quantum eraser (Kwiat, Steinberg and Chiao [1992]).

of one of the beams. When the polarization of this beam was orthogonal to that of the other beam, the coincidence null disappeared, since it became possible to identify the paths taken by each of the photons. However, this information could be erased by inserting two polarizers just in front of the photo detectors, after the photons had left the beam splitter.

In particular, if the initial polarization of the down-converted photons was horizontal, and the half-wave plate rotated one polarization to vertical, polarizers at 45° before each detector restored the original coincidence null. Interference could not be restored with a single polarizer in front of one detector, since "which-path" information was available from the photon reaching the other photodetector. In addition, as shown in fig. 7.2, if one polarizer was set at 45° and the other at −45°, an interference peak was observed instead of a dip.

The quantum-eraser concept could also be realized with the interferometer shown in fig. 6.7 (Ou, Wang, Zou and Mandel [1990]). In this case, removal of the beam splitter BS_B, which at first sight should not affect the results, destroyed the interference. The explanation is that since the signal and idler photons are produced simultaneously, it then became possible from the output of the photo detector D_B to decide whether the corresponding signal photon came from NL_1 or NL_2. Insertion of BS_B mixed the idlers and erased the information on the paths taken by the photons (Zajonc, Wang, Zou and Mandel [1991]).

The experimental arrangement shown in fig. 6.10 (Zou, Wang and Mandel [1991]) could also be modified to demonstrate this concept by using a half-wave plate between the two crystals to rotate the polarization of the idler photons from NL_1, so that it was orthogonal to the polarization of the idlers from NL_2. In this arrangement, interference could be recovered by using a polarizer in front of D_i and correlating the counts of the two detectors.

With fast detectors and a rapidly switchable polarizer, it should even be

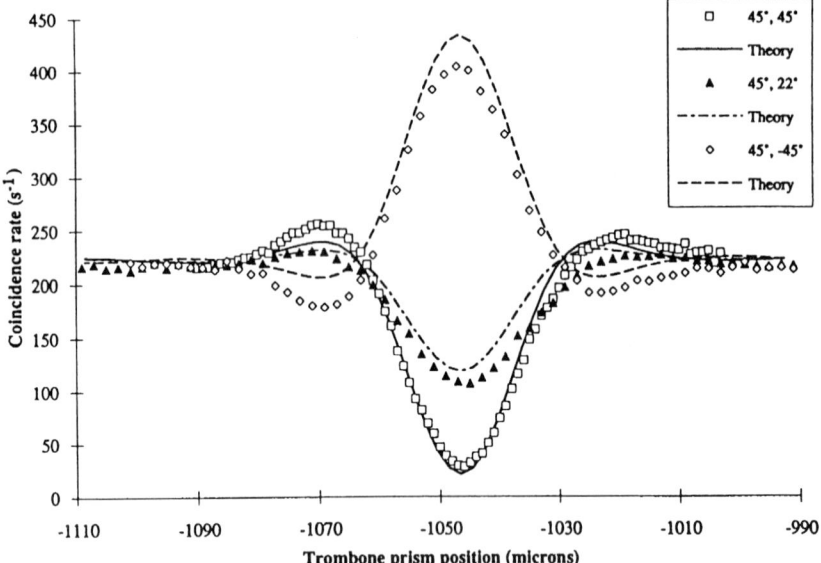

Fig. 7.2. Experimental data obtained with a two-photon interferometer with the polarizer in one path set at 45° and the polarizer in the other path set at various angles (Kwiat, Steinberg and Chiao [1992]).

possible to choose the orientation of the polarizer after the signal photon is detected, making possible a delayed-choice decision to observe particle-like behavior ("which-path" information) or wave-like behavior (interference) (Kwiat, Steinberg and Chiao [1994]).

In all these cases, it appears that the state vector reflects not only what is known about the photon, but also whatever information is available in principle. The additional measurements needed to obtain this information, either on the source or on the path of the detected photon, need not actually be carried out; it is enough for them to be possible, in principle, for the interference effects to be destroyed.

7.4. SINGLE-PHOTON TUNNELING

If two right-angle prisms are placed with their hypotenuse faces opposite each other, but separated by an air gap, a beam of light incident on the interface at an angle greater than the critical angle is totally internally reflected. However, if the air gap is reduced to a fraction of a wavelength, some of the light is transmitted.

The fact that light tunnels through such a gap by evanescent coupling confirms the wave-like behavior of light.

On the other hand, if the same experiment is repeated with single-photon states, nonclassical effects are observed (Ghose, Home and Agarwal [1991]). With this arrangement, as we have seen earlier, quantum mechanics predicts that photons will be detected in perfect anticoincidence in the transmitted and reflected beams. This prediction has been verified experimentally by Mizobuchi and Ohtake [1992]. Accordingly, we have a situation where single-photon states display wave-like properties (tunneling) as well as particle-like properties (anticoincidence).

7.4.1. Tunneling time

The phenomenon of tunneling is actually a fundamental consequence of quantum mechanics, which states that all quantum particles, in principle, can tunnel through normally forbidden regions of space. However, the question of how much time it takes for a particle to tunnel through a barrier is quite controversial (Büttiker and Landauer [1982], Hauge and Støvneng [1989], Fertig [1990]). Interferometric experiments have made it possible to study this aspect of photon tunneling.

7.4.2. Dispersion cancellation

It follows from the uncertainty principle that, to make measurements of transit times with high resolution, it is necessary to make the energy uncertainty or spectral bandwidth quite large. With such large spectral bandwidths, any dispersive effects can result in significant broadening of a pulse, and a consequent decrease in time resolution (Franson [1992]). This problem can be avoided by making measurements with correlated photon pairs. It is then possible to take advantage of quantum-mechanical effects to obtain an effective cancellation of dispersion (Steinberg, Kwiat and Chiao [1992a,b, 1993]).

With photon pairs produced in an entangled state, the frequencies of the individual photons are not defined sharply, but the sum of their frequencies is fixed. If we use an interferometer similar to that described by Hong, Ou and Mandel [1987] (see fig. 5.4), with a dispersive medium (say, a glass plate) in one beam, one photon of each pair travels through the dispersive medium while its conjugate travels through a path containing only air. However, after the photons are recombined at the beam splitter, it becomes impossible to determine which one of them travelled through the glass plate. This indistinguishability leads to a cancellation of first-order dispersion effects, so that there is no broadening of the coincidence minimum. As a result, the shift in the position of the minimum in the rate of coincidences can be used to make high-resolution measurements of the propagation delay produced by the glass plate.

7.4.3. Measurements of tunneling time

The experimental arrangement used for measurements of tunneling time is shown in fig. 7.3 (Steinberg, Kwiat and Chiao [1993]). The tunnel barrier was a multilayer dielectric mirror consisting of 11 alternating layers of low- and high-index material, each a quarter of a wavelength thick at the wavelength used (700 nm in air), coated on one half of the surface of a high-quality optical flat. In such a periodic structure, the multiple reflections interfere so as to exponentially damp the incident wave, resulting in the equivalent of a photonic bandgap (Yablonovitch [1993]) at which more than 99% of the incident light is reflected.

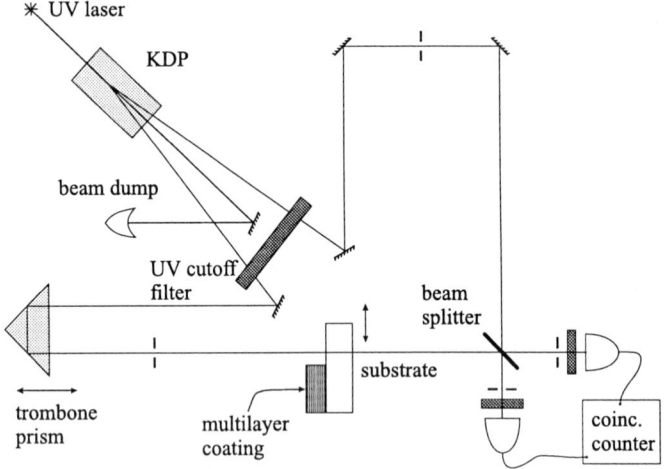

Fig. 7.3. Apparatus used for measurements of the single-photon tunneling time (Steinberg, Kwiat and Chiao [1993]).

To make measurements of the tunneling time, the multilayer structure was moved periodically into and out of the beam, while the optical path difference between the beams was varied slowly by translating the reflecting prism. A Gaussian curve was then fitted to each of the two dips in the rate of coincidences, and the distance between their centers was calculated. Figure 7.4

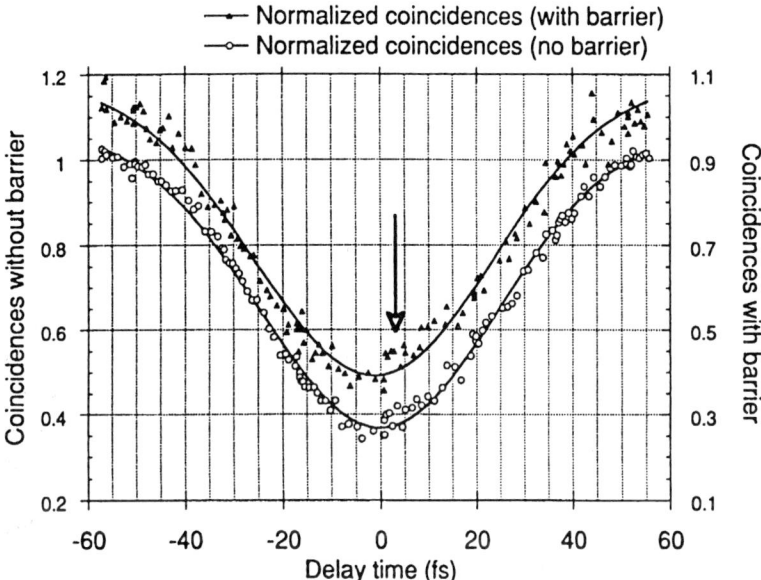

Fig. 7.4. Variation of the rate of coincidence counts with the delay time, with and without the tunnel barrier in the optical path. With the barrier, the minimum occurs approximately 2 fs earlier than without the barrier (Steinberg, Kwiat and Chiao [1993]).

shows a typical set of data. The average of several such measurements showed that the peak arrived 1.47±0.21 fs earlier when the multilayer was in the path (Steinberg, Kwiat and Chiao [1993]). In an extension of this experiment, Steinberg and Chiao [1995] determined the delay times for the transmission of photons through a dielectric mirror as a function of the angle of incidence. These measurements made it possible to study the energy dependence of the tunneling time.

The interpretation of the apparently superluminal velocities observed has been discussed by Landauer [1993]. One explanation is that the whole transmitted wave packet comes from the leading edge of the much larger incident wave packet.

7.5. INTERACTION-FREE MEASUREMENTS

An interesting application of complementarity discussed by Elitzur and Vaidman [1993] and by Vaidman [1994] is in interaction-free measurements. At issue is the determination of whether or not a perfectly efficient detector occupies a certain region of space, without actually triggering this detector. To dramatize the

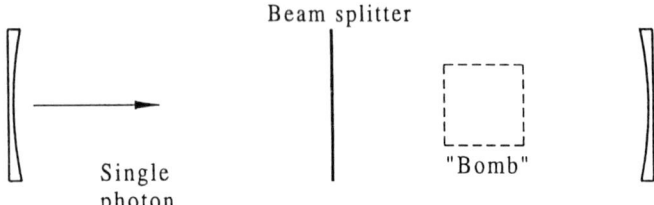

Fig. 7.5. Scheme for interaction-free measurements (Kwiat, Weinfurter, Herzog, Zeilinger and Kasevich [1995]).

situation, the detector is pictured as a bomb which has unit detection efficiency and is triggered by the absorption of a single photon; one must determine whether the bomb is present, or not, without allowing a single photon to be absorbed.

A Mach–Zehnder interferometer can be set up so that all photons exit through a specified output port if the detector, or bomb, is not located in one arm of the interferometer. The presence of the bomb in this arm destroys the interference required for all photons to exit only through the specified port. It follows that the presence of the bomb can be detected through the observation of a photon exiting from the other port. However, for an interferometer constructed with 50:50 beam splitters, the probability of triggering the bomb is 50%, while the probability of knowing unambiguously that the bomb is present, without triggering the bomb, is only 25%. Accordingly, while this scheme permits, in principle, interaction-free measurements of the presence of the bomb, it is far from ideal. The basic strategy for interaction-free measurements is, therefore, to exploit the wave-like behavior of light to increase the probability of establishing the presence of an absorber, while reducing, or eliminating, the probability of a photon traversing the path in which the absorber lies.

Kwiat, Weinfurter, Herzog, Zeilinger and Kasevich [1995] have proposed an interaction-free measurement scheme which, assuming a loss-free system, can raise this ratio for interaction-free measurements to as close to unity as desired. As shown in fig. 7.5, a beam splitter is placed in an optical cavity; the single photon is generated on one side of the beam splitter (say, the left) and the detector (the bomb) is placed on the other side. For a reflectivity

$$R = \cos^2\left(\frac{\pi}{2N}\right), \tag{7.2}$$

the photon will be found in the right side of the cavity, after N time cycles, with probability $\cos^2(\pi/2N)$ if the bomb is not present; however, if the bomb is present, the wave function of the photon is continually projected back on to the left half of the cavity, with the probability of finding the photon in the left half of the cavity, after N cycles, tending to unity as $N \to \infty$.

§ 8. Quantum Limits to Interferometry

8.1. NUMBER–PHASE UNCERTAINTY RELATION

Dirac [1927], in his formulation of quantum electrodynamics, quantized the field by treating the photon number n and phase ϕ as canonically conjugate quantities, rather than the variables X_1 and X_2 of eq. (2.17). However, quantization of the phase variable is not straightforward, due to the periodicity of the phase and the lower bound for the spectrum of the photon number operator (Susskind and Glogower [1964], Paul [1974]). Despite this difficulty, the number–phase uncertainty relation defined by eq. (7.1) has proved useful for characterizing the limited precision of phase measurements with laser light sources (Serber and Townes [1960], Friedburg [1960]). The inequality defined by eq. (7.1) quantifies the trade-off between reducing photon-number fluctuations in the source and reducing the phase noise, and the coherent state of light can be regarded as a minimum-uncertainty state in the strong-field limit (Carruthers and Nieto [1965, 1968]). The difficulty with the periodicity of the phase can be alleviated by replacing the phase operator by noncommuting operators corresponding to $\cos\phi$ and $\sin\phi$ (Louisell [1963], Susskind and Glogower [1964]), leading to a modified version of the uncertainty relation (7.1) involving Δn, $\Delta\cos\phi$ and $\Delta\sin\phi$.

Gerhardt, Welling and Frölich [1973] and Gerhardt, Buchler and Litfin [1974] attempted to measure directly the phase fluctuations of a microscopic radiation field, in order to check the number–phase uncertainty relation. They sent coherent light from a laser into a Mach–Zehnder interferometer and attenuated the light in one path to a mean photon number between 3 and 12. The resultant increase in phase fluctuations induced a random phase shift in the beam in that arm. This field was then amplified by a Q-switched laser with a gain factor of 10^{10}. The field in the second arm of the interferometer served as the reference field for homodyne detection of the output. In order to minimize external disturbances, the phase deviation between two pulses separated by less than a microsecond was measured. The results of these experiments did not agree with the uncertainty relations of Carruthers and Nieto [1965], but Nieto [1977] showed that they agreed better with measurements of a phase-difference operator, rather than with measurements of absolute phase.

It must be kept in mind that although direct measurements of phase, and, therefore, of number–phase uncertainty relations, cannot be performed, indirect measurements of phase are possible. Alternate versions of phase operators can be constructed for particular phase-sensitive measurements (Barnett and

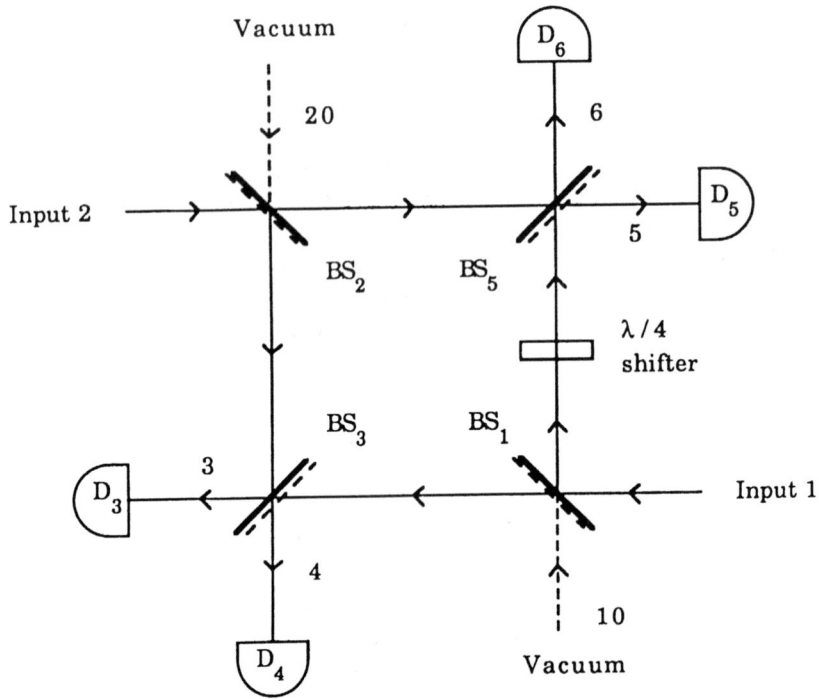

Fig. 8.1. Interferometer used to measure the phase difference between two beams of light (Noh, Fougères and Mandel [1991]).

Pegg [1986], Noh, Fougères and Mandel [1991, 1992a,b]) or inferred from measurements of quasiprobability distributions such as the Wigner function, as in the experiments performed by Smithey, Beck, Cooper and Raymer [1993].

The optical system used by Noh, Fougères and Mandel [1991, 1992a,b] is shown in fig. 8.1. In this arrangement, the two input fields are mixed by four beam splitters, and four photo detectors are used to count the photons emerging from four output ports. A 90° phase shifter is inserted in one arm of the interferometer.

For this measurement scheme, the cosine and sine operators can be taken to be

$$\widehat{C}_M = (\hat{n}_4 - \hat{n}_3)\left[(\hat{n}_4 - \hat{n}_3)^2 + (\hat{n}_6 - \hat{n}_5)^2\right]^{-1/2},$$
$$\widehat{S}_M = (\hat{n}_6 - \hat{n}_5)\left[(\hat{n}_4 - \hat{n}_3)^2 + (\hat{n}_6 - \hat{n}_5)^2\right]^{-1/2},$$
(8.1)

where n_3, n_4, n_5 and n_6 correspond to the photon counts registered by the detectors D_3, D_4, D_5 and D_6, respectively. The operators describe the measurement statistics well in the limit that the photon-number fluctuations at each detector are small compared to the mean photon number.

A theoretical analysis then shows that for input fields with $\langle \hat{n}_1 \rangle, \langle \hat{n}_2 \rangle \ll 1$, the dispersions of $\hat{C}_M$ and $\hat{S}_M$ obey the inequalities

$$\frac{\langle (\Delta \hat{C}_M)^2 \rangle^{1/2}}{|\langle \hat{C}_M \rangle|} \geq 1, \qquad \frac{\langle (\Delta \hat{S}_M)^2 \rangle^{1/2}}{|\langle \hat{C}_M \rangle|} \geq 1, \tag{8.2}$$

so that both the cosine and the sine of the phase difference are ill-defined. This result is confirmed experimentally. The probability distribution $\mathcal{P}(\phi_2 - \phi_1)$ of the phase difference can then be derived by imposing a phase shift ϕ_s on the field at input port 2, and repeating the measurements for a range of values of ϕ_s from $-\pi$ to π.

8.2. THE STANDARD QUANTUM LIMIT

The quantum limit in interferometry is usually obtained from an argument which balances the error due to photon-counting statistics against the disturbances of the end mirrors produced by fluctuations in radiation pressure (Edelstein, Hough, Pugh and Martin [1978], Forward [1978]). According to this argument, since the number of photons which pass through the interferometer in the measurement time τ is

$$n = \frac{P\tau}{\hbar\omega}, \tag{8.3}$$

where P is the laser power, fluctuations in the laser power produce an uncertainty in n given by the relation

$$\Delta n \approx n^{-1/2}. \tag{8.4}$$

The existence of this quantum limit is now well established, but the argument leading to it has been open to question, since it relies on the assumption that the power fluctuations in the two arms are uncorrelated.

A more rigorous analysis (Caves [1980]) reveals two different, but equivalent, points of view on the origin of the fluctuations. The first attributes them to the fact that each photon incident on the beam splitter is scattered independently, thereby producing binomial distributions of photons in the two arms which are precisely anticorrelated. The second ascribes them to vacuum (zero-point) fluctuations in the field entering the interferometer from the other input port. This field acts in antiphase on the laser fields in the two arms. It follows, therefore, that the photon-counting error is an intrinsic property of the interferometer.

The standard quantum limit (SQL) in interferometry is obtained by inserting eq. (8.4) into eq. (7.1) to obtain the corresponding uncertainty in the measured values of the phase:

$$\Delta\phi \geq \frac{1}{2\sqrt{\bar{n}}}, \tag{8.5}$$

where $\bar{n}$ is the mean photon number. This limit poses problems in measuring extremely small displacements and is critical in such applications as the detection of gravitational waves. One way to overcome the SQL is by injecting squeezed states into one or both ports of the interferometer (Caves [1981], Bondurant and Shapiro [1984]).

8.3. INTERFEROMETRY BELOW THE SQL

A schematic of a Michelson interferometer designed to detect gravitational radiation (Caves [1980]) is shown in fig. 8.2. This interferometer uses a delay line in each arm, and although only two reflections on each mirror are shown, a larger number can be used to increase the effective length of each arm (Billing, Maischberger, Rüdiger, Schilling, Schnupp and Winkler [1979]). A change in the difference of the lengths of the optical paths in the two arms due to a gravitational wave results in a change in the phase difference between the beams, which can be measured by the intensity change at the detector.

The SQL can be overcome in such an interferometer by using squeezed light. A limit,

$$\Delta\phi \approx \frac{1}{\bar{n}}, \tag{8.6}$$

is achievable in principle by feeding suitably constructed squeezed states into both input ports of the interferometer. In actual experiments with a polarization interferometer, an increase in the signal-to-noise ratio of 2 dB, relative to the shot-noise limit, has been achieved using squeezed light generated by an optical parametric amplifier (Grangier, Slusher, Yurke and LaPorta [1987]). The maximum improvement in sensitivity can be obtained by preparing the light entering each of the two input ports of the interferometer in a state that consists of exactly j photons (Yurke, McCall and Klauder [1986], Holland and Burnett

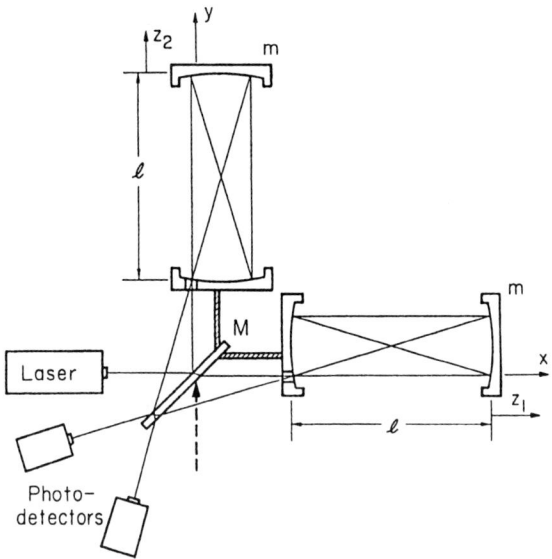

Fig. 8.2. Schematic of a Michelson interferometer for detecting gravitational waves (Caves [1980]).

[1993], Sanders and Milburn [1995]). States conforming to this requirement can be generated, for example, with two-mode four-wave mixers (Yurke, McCall and Klauder [1986]). The difference in the photo counts at the two output ports can then be processed to obtain the phase difference. However, to achieve maximum sensitivity, the deviation of the phase difference ϕ from zero must be less than $1/\bar{n}$. This requirement can be met by the use of a feedback loop that holds ϕ at zero.

8.4. INTERFEROMETERS USING ACTIVE ELEMENTS

An analysis based on the theory of Lie groups shows that conventional interferometers using only passive elements can be characterized by an SU(2)-group symmetry. An alternative class of interferometers, characterized by an SU(1, 1)-group symmetry (Yurke, McCall and Klauder [1986]), exploits the fact that the output of an active element, such as a four-wave mixer or degenerate parametric amplifier, depends on the relative phases of the pump and the incoming signal. In these interferometers, beam splitters are replaced by such active elements, and no light is fed into the input ports. They offer the possibility of attaining a phase sensitivity approaching $1/\bar{n}$ with fewer optical elements.

In one arrangement, the beam splitters of a conventional interferometer are replaced by four-wave mixers. When the two optical paths are equal, no light is delivered to the photodetectors, since the pairs of pump photons converted into pairs of four-wave output photons at the first four-wave mixer are converted back into pump photons at the second four-wave mixer. An alternative arrangement uses two degenerate parametric amplifiers. The output is then sensitive to the difference between the phases accumulated by the signal and pump beams.

§ 9. Conclusions

The dawn of the twentieth century saw the introduction of a duality to the behavior of light. Depending on the experiment chosen, light appeared to behave either as a wave or as a collection of particles. The quantum theory of light evolved in response to the need to reconcile these two contradictory aspects.

Starting from the early work of Dirac [1927] on the quantization of light, which treated photon number and phase as complementary quantities, and the use of coherent states of light to make the transition from classical to quantum descriptions of radiation, this progression led, eventually, to a single cohesive theory of quantum optics which has successfully described and predicted a diverse range of phenomena.

However, despite the success of quantum theory in explaining nonclassical effects, the mystery of the quantum world is deep enough to provoke many tests of the theory itself. Optical interferometry has played an increasingly important part in these tests, ranging from early experiments which verified Dirac's famous dictum to recent experiments involving Bell's inequality. In all these cases, optical interferometry has provided evidence supporting quantum theory and refuting alternative theories.

The mysterious aspects of quantum theory arise essentially because of the complementarity of wave and particle behavior. In the case of light, the complementarity of photon number and phase is still provoking research. The question of "What is phase?" has compelled researchers to conduct experiments to probe the limits to phase measurements, and theoreticians to ponder a definition of phase that is consistent with the axioms of quantum theory.

The duality inherent in the quantum description of light provides two alternate pictures for describing interference phenomena in an intuitive way: the "semi-classical picture", which treats the quantized light field as being composed of classical waves which can interact with quantized matter, and the "photon picture", which treats the light quanta as fundamental. Each picture has its advantages and limitations, depending on the experiment considered, but the two pictures overlap in discussions of weak semiclassical fields where the number of photons is small, and individual photons can be detected.

In most situations, the "photon picture" gives a good description of the interference phenomena observed, provided that Dirac's statement that "a photon interferes only with itself..." is borne in mind. There are conceptual difficulties in applying this dictum to some of the higher-order interference effects described in this review, but a way out is to extend it to read "each pair of photons interferes only with itself".

An alternative explanation is to regard optical interference as due to the existence of indistinguishable paths; on this basis, the effects observed are explained as a manifestation of the correspondence between the mutual coherence and the indistinguishability of two light beams (Mandel [1991]). We can then apply Feynman's rules for interference, after taking into account the fact that outcomes which are distinguishable, even in principle, do not interfere. In all other cases, the coherent addition of the probability amplitudes associated with each path, and the evaluation of the squared modulus of this sum, yields the probability of detection of a photon. This far-reaching principle appears to provide an intuitive understanding of all interference effects.

References

Adam, A., L. Janossy and P. Varga, 1955a, Acta Phys. Hung. **4**, 301.
Adam, A., L. Janossy and P. Varga, 1955b, Ann. Phys. (N.Y.) **16**, 408.
Agarwal, G.S., and P. Hariharan, 1993, Opt. Commun. **103**, 111.
Alford, W.P., and A. Gold, 1958, Am. J. Phys. **26**, 481.
Arecchi, F.T., A. Berné and P. Burlamacchi, 1966, Phys. Rev. Lett. **16**, 32.
Aspect, A., J. Dalibard and G. Roger, 1982, Phys. Rev. Lett. **49**, 1804.
Aspect, A., and P. Grangier, 1987, Hyperfine Interactions **37**, 3.
Aspect, A., P. Grangier and G. Roger, 1981, Phys. Rev. Lett. **47**, 460.
Baldzuhn, J., E. Mohler and W. Martienssen, 1989, Z. Phys. B **77**, 347.
Barnett, S.M., and D.T. Pegg, 1986, J. Phys. A **19**, 3849.
Bell, J.S., 1965, Physics **1**, 195.
Berry, M.V., 1984, Proc. R. Soc. (London) A **392**, 45.
Berry, M.V., 1987, J. Mod. Opt. **34**, 1401.
Bhandari, R., and J. Samuel, 1988, Phys. Rev. Lett. **60**, 1211.
Billing, H., K. Maischberger, A. Rüdiger, R. Schilling, L. Schnupp and W. Winkler, 1979, J. Phys. E **12**, 1043.
Bohm, D., 1951, Quantum Theory (Prentice Hall, Englewood Cliffs, NJ) p. 614.
Bohr, N., 1983, in: Quantum Theory and Measurement, eds J.A. Wheeler and W.H. Zurek (Princeton University Press, Princeton, NJ) p. 32.
Bondurant, R.S., and J.H. Shapiro, 1984, Phys. Rev. D **30**, 2458.

Braginsky, V.B., 1989, in: 3rd Int. Symp. on Foundation of Quantum Mechanics in the Light of New Technology, eds S. Kobayashi, H. Ezawa, Y. Murayama and S. Nomura (The Physical Society of Japan, Tokyo) p. 135.
Brendel, J., W. Dultz and W. Martienssen, 1995, Phys. Rev. A **52**, 2551.
Brendel, J., E. Mohler and W. Martienssen, 1991, Phys. Rev. Lett. **66**, 1142.
Brendel, J., E. Mohler and W. Martienssen, 1992, Europhys. Lett. **20**, 575.
Brendel, J., S. Schütrumpf, R. Lange, W. Martienssen and M.O. Scully, 1988, Europhys. Lett. **5**, 223.
Burnham, D.C., and D.L. Weinberg, 1970, Phys. Rev. Lett. **25**, 84.
Büttiker, M., and R. Landauer, 1982, Phys. Rev. Lett. **49**, 1739.
Campos, R.A., B.E.A. Saleh and M.C. Teich, 1989, Phys. Rev. A **40**, 1371.
Carruthers, P., and M.M. Nieto, 1965, Phys. Rev. Lett. **14**, 387.
Carruthers, P., and M.M. Nieto, 1968, Rev. Mod. Phys. **40**, 411.
Casabella, P.A., and T. Gonsiorowski, 1980, Am. J. Phys. **48**, 393.
Caves, C.M., 1980, Phys. Rev. Lett. **45**, 75.
Caves, C.M., 1981, Phys. Rev. D **23**, 1693.
Caves, C.M., and B.L. Schumaker, 1985, Phys. Rev. D **31**, 3068.
Chiao, R.Y., 1970, Phys. Lett. A **33**, 177.
Chiao, R.Y., A. Antaramian, K.M. Ganga, H. Jiao, S.R. Wilkinson and H. Nathel, 1988, Phys. Rev. Lett. **60**, 1214.
Chyba, T.H., L.J. Wang, L. Mandel and R. Simon, 1988, Opt. Lett. **13**, 562.
Clauser, J.F., 1974, Phys. Rev. D **9**, 853.
Clauser, J.F., M.A. Horne, A. Shimony and R.A. Holt, 1969, Phys. Rev. Lett. **23**, 880.
Clauser, J.F., and A. Shimony, 1978, Rep. Prog. Phys. **41**, 1881.
Croca, J.R., A. Garuccio, V.L. Lepore and R.N. Moreira, 1990, Found. Phys. Lett. **3**, 557.
de Broglie, L., 1969, The Current Interpretation of Wave Mechanics (Elsevier, Amsterdam).
Dempster, A.J., and H.F. Batho, 1927, Phys. Rev. **30**, 644.
Diedrich, F., and H. Walther, 1987, Phys. Rev. Lett. **58**, 203.
Dirac, P.A.M., 1927, Proc. R. Soc. (London) A **114**, 243.
Dirac, P.A.M., 1958, The Principles of Quantum Mechanics, 4th Ed. (Oxford University Press, Oxford) p. 9.
Eberhard, P.H., 1993, Phys. Rev. A **47**, R747.
Edelstein, W.A., J. Hough, J.R. Pugh and W. Martin, 1978, J. Phys. E **11**, 710.
Einstein, A., B. Podolsky and N. Rosen, 1935, Phys. Rev. **47**, 777.
Elitzur, A.C., and L. Vaidman, 1993, Found. Phys. **23**, 987.
Fearn, H., and R. Loudon, 1987, Opt. Commun. **64**, 485.
Fearn, H., and R. Loudon, 1989, J. Opt. Soc. Am. B **6**, 917.
Ferguson, J.B., and R.H. Morris, 1978, Appl. Opt. **17**, 2924.
Fertig, H.A., 1990, Phys. Rev. Lett. **65**, 2321.
Feynman, R.P., R.B. Leighton and M. Sands, 1963, Lectures on Physics, Vol. 3 (Addison-Wesley, London) ch. 3.
Forrester, A.T., R.A. Gudmundsen and P.O. Johnson, 1955, Phys. Rev. **99**, 1691.
Forward, R.L., 1978, Phys. Rev. D **17**, 379.
Franson, J.D., 1989, Phys. Rev. Lett. **62**, 2205.
Franson, J.D., 1991a, Phys. Rev. Lett. **67**, 290.
Franson, J.D., 1991b, Phys. Rev. A **44**, 4552.
Franson, J.D., 1992, Phys. Rev. A **45**, 3126.
Freedman, S.J., and J.F. Clauser, 1972, Phys. Rev. Lett. **28**, 938.
Friberg, S., C.K. Hong and L. Mandel, 1985, Phys. Rev. Lett. **54**, 2011.

Friedburg, H., 1960, in: Quantum Electronics, ed. C.H. Townes, (Columbia University Press, New York) p. 228.
Gans, R., and P. Miguez, 1917, Ann. Phys. **52**, 291.
Gerhardt, H., U. Buchler and G. Litfin, 1974, Phys. Lett. A **49**, 119.
Gerhardt, H., H. Welling and D. Frölich, 1973, Appl. Phys. **2**, 91.
Ghose, P., D. Home and G.S. Agarwal, 1991, Phys. Lett. A **153**, 403.
Ghosh, R., C.K. Hong, Z.Y. Ou and L. Mandel, 1986, Phys. Rev. A **34**, 3962.
Ghosh, R., and L. Mandel, 1987, Phys. Rev. Lett. **59**, 1903.
Glauber, R.J., 1963a, Phys. Rev. **130**, 2529.
Glauber, R.J., 1963b, Phys. Rev. **131**, 2766.
Grangier, P., A. Aspect and J. Vigué, 1985, Phys. Rev. Lett. **54**, 418.
Grangier, P., G. Roger and A. Aspect, 1986, Europhys. Lett. **1**, 173.
Grangier, P., R.E. Slusher, B. Yurke and A. LaPorta, 1987, Phys. Rev. Lett. **59**, 2153.
Grayson, T.P., J.R. Torgerson and G.A. Barbosa, 1994, Phys. Rev. A **49**, 626.
Hanbury Brown, R., 1974, The Intensity Interferometer (Taylor and Francis, London).
Hanbury Brown, R., and R.Q. Twiss, 1956, Nature (London) **177**, 27.
Hardy, L., 1992a, Phys. Rev. Lett. **68**, 2981.
Hardy, L., 1992b, Phys. Lett. A **167**, 17.
Hardy, L., 1993, Phys. Rev. Lett. **71**, 1665.
Hariharan, P., N. Brown, I. Fujima and B.C. Sanders, 1993, J. Mod. Opt. **40**, 1477.
Hariharan, P., N. Brown, I. Fujima and B.C. Sanders, 1995, J. Mod. Opt. **42**, 565.
Hariharan, P., N. Brown and B.C. Sanders, 1993, J. Mod. Opt. **40**, 113.
Hariharan, P., and M. Roy, 1992, J. Mod. Opt. **39**, 1811.
Hariharan, P., M. Roy, P.A. Robinson and J.W. O'Byrne, 1993, J. Mod. Opt. **40**, 871.
Haroche, S., M. Brune and J.M. Raimond, 1992, in: 13th Int. Conf. on Atomic Physics, AIP Conf. Proc. **275**, 261.
Harris, S.E., M.K. Oshman and R.L. Byer, 1967, Phys. Rev. Lett. **18**, 732.
Hauge, E.H., and J.A. Støvneng, 1989, Rev. Mod. Phys. **61**, 917.
Heitler, W., 1954, The Quantum Theory of Radiation, 3rd Ed. (Oxford University Press, London).
Hellmuth, T., H. Walther, A. Zajonc and W. Schleich, 1987, Phys. Rev. A **35**, 2532.
Herzog, T., J.G. Rarity, H. Weinfurter and A. Zeilinger, 1994, Phys. Rev. Lett. **72**, 629.
Holland, M.J., and K. Burnett, 1993, Phys. Rev. Lett. **71**, 1355.
Holland, P.R., and J.P. Vigier, 1991, Phys. Rev. Lett. **67**, 402.
Hong, C.K., and L. Mandel, 1985, Phys. Rev. A **31**, 2409.
Hong, C.K., and L. Mandel, 1986, Phys. Rev. Lett. **56**, 58.
Hong, C.K., Z.Y. Ou and L. Mandel, 1987, Phys. Rev. Lett. **59**, 2044.
Horne, M.A., A. Shimony and A. Zeilinger, 1989, Phys. Rev. Lett. **62**, 2209.
Imoto, N., H.A. Haus and Y. Yamamoto, 1985, Phys. Rev. A **32**, 2287.
Janossy, L., and Z. Naray, 1957, Acta Phys. Acad. Sci. Hung. **7**, 403.
Javan, A., E.A. Ballik and W.L. Bond, 1962, J. Opt. Soc. Am. **52**, 96.
Jerrard, H.G., 1954, J. Opt. Soc. Am. **44**, 634.
Jordan, T.F., 1994, Phys. Rev. A **50**, 62.
Jordan, T.F., and F. Ghielmetti, 1964, Phys. Rev. Lett. **12**, 607.
Kimble, H.J., and D.F. Walls, 1987, J. Opt. Soc. Am. B **4**, 1450.
Kitagawa, M., and Y. Yamamoto, 1986, Phys. Rev. A **34**, 3974.
Klyshko, D.N., 1967, JETP Lett. **6**, 23.
Kocher, C.A., and E.D. Commins, 1967, Phys. Rev. Lett. **18**, 575.
Kwiat, P.G., and R.Y. Chiao, 1991, Phys. Rev. Lett. **66**, 588.
Kwiat, P.G., P. Eberhard, A.M. Steinberg and R.Y. Chiao, 1994, Phys. Rev. A **49**, 3209.

Kwiat, P.G., A.M. Steinberg and R.Y. Chiao, 1992, Phys. Rev. A **45**, 7729.
Kwiat, P.G., A.M. Steinberg and R.Y. Chiao, 1993, Phys. Rev. A **47**, R2472.
Kwiat, P.G., A.M. Steinberg and R.Y. Chiao, 1994, Phys. Rev. A **49**, 61.
Kwiat, P.G., A.M. Steinberg, R.Y. Chiao, P. Eberhard and M. Petroff, 1993, Phys. Rev. A **48**, R867.
Kwiat, P.G., W.A. Vareka, C.K. Hong, H. Nathel and R.Y. Chiao, 1990, Phys. Rev. A **41**, 2910.
Kwiat, P.G., H. Weinfurter, T. Herzog, A. Zeilinger and M.A. Kasevich, 1995, Phys. Rev. Lett. **74**, 4763.
Landauer, R., 1993, Nature **365**, 692.
Larchuk, T.S., R.A. Campos, J.G. Rarity, P.R. Tapster, E. Jakeman, B.E.A. Saleh and M.C. Teich, 1993, Phys. Rev. Lett. **70**, 1603.
Lee, B., E. Yin, T.K. Gustafson and R.Y. Chiao, 1992, Phys. Rev. A **45**, 4319.
Louisell, W.H., 1963, Phys. Lett. **7**, 60.
Maeda, M.W., P. Kumar and J.H. Shapiro, 1987, Opt. Lett. **12**, 161.
Magyar, G., and L. Mandel, 1963, Nature **198**, 255.
Mandel, L., 1962, J. Opt. Soc. Am. **52**, 1335.
Mandel, L., 1964, Phys. Rev. **134**, A10.
Mandel, L., 1976, in: Progress in Optics, Vol. 13, ed. E. Wolf (North-Holland, Amsterdam) p. 27.
Mandel, L., 1982, Phys. Lett. A **89**, 325.
Mandel, L., 1983, Phys. Rev. A **28**, 929.
Mandel, L., 1991, Opt. Lett. **16**, 1882.
Mandel, L., and E. Wolf, 1965, Rev. Mod. Phys. **37**, 231.
Milburn, G.J., 1984, J. Phys. A **17**, 737.
Milburn, G.J., and D.F. Walls, 1983, Phys. Rev. A **28**, 2065.
Mizobuchi, Y., and Y. Ohtake, 1992, Phys. Lett. A **168**, 1.
Morris, R.H., J.B. Ferguson and J.S. Warniak, 1975, Appl. Opt. **14**, 2808.
Newton, T.D., and E.P. Wigner, 1949, Rev. Mod. Phys. **21**, 400.
Nieto, M.M., 1977, Phys. Lett. A **60**, 401.
Noh, J.W., A. Fougères and L. Mandel, 1991, Phys. Rev. Lett. **67**, 1426.
Noh, J.W., A. Fougères and L. Mandel, 1992a, Phys. Rev. A **45**, 424.
Noh, J.W., A. Fougères and L. Mandel, 1992b, Phys. Rev. A **46**, 2840.
Ou, Z.Y., 1988, Phys. Rev. A **37**, 1607.
Ou, Z.Y., C.K. Hong and L. Mandel, 1987, Opt. Commun. **63**, 118.
Ou, Z.Y., and L. Mandel, 1988a, Phys. Rev. Lett. **61**, 50.
Ou, Z.Y., and L. Mandel, 1988b, Phys. Rev. Lett. **61**, 54.
Ou, Z.Y., L.J. Wang, X.Y. Zou and L. Mandel, 1990, Phys. Rev. A **41**, 566.
Ou, Z.Y., X.Y. Zou, L.J. Wang and L. Mandel, 1990a, Phys. Rev. Lett. **65**, 321.
Ou, Z.Y., X.Y. Zou, L.J. Wang and L. Mandel, 1990b, Phys. Rev. A **42**, 2957.
Pancharatnam, S., 1956, Proc. Indian Acad. Sci. A **44**, 247. Reprinted: 1975, in: Collected Works of S. Pancharatnam, (Oxford University Press, Oxford) p. 77.
Paul, H., 1974, Fortschr. Phys. **22**, 657.
Paul, H., 1986, Rev. Mod. Phys. **58**, 209.
Pavičić, M., 1994, Phys. Rev. A **50**, 3486.
Pavičić, M., 1995, J. Opt. Soc. Am. B **12**, 821.
Pavičić, M., and J. Summhammer, 1994, Phys. Rev. Lett. **73**, 3191.
Pfleegor, R.L., and L. Mandel, 1967a, Phys. Lett. A **24**, 766.
Pfleegor, R.L., and L. Mandel, 1967b, Phys. Rev. **159**, 1084.
Pfleegor, R.L., and L. Mandel, 1968, J. Opt. Soc. Am. **58**, 946.
Pipkin, F.M., 1978, in: Advances in Atomic and Molecular Physics, Vol. 14, eds D.R. Bates and B. Bederson (Academic Press, New York) p. 281.

Ramaseshan, S., and R. Nityananda, 1986, Curr. Sci. (India) **55**, 1225.
Rarity, J.G., and P.R. Tapster, 1989, J. Opt. Soc. Am. B **6**, 1221.
Rarity, J.G., and P.R. Tapster, 1990, Phys. Rev. Lett. **64**, 2495.
Rarity, J.G., and P.R. Tapster, 1992, Phys. Rev. A, **45**, 2054.
Rarity, J.G., P.R. Tapster, E. Jakeman, T. Larchuk, R.A. Campos, M.C. Teich and B.E.A. Saleh, 1990, Phys. Rev. Lett. **65**, 1348.
Roch, J.F., G. Roger, P. Grangier, J. Courty and S. Reynaud, 1992, Appl. Phys. B **55**, 291.
Sanders, B.C., and G.J. Milburn, 1989, Phys. Rev. A **39**, 694.
Sanders, B.C., and G.J. Milburn, 1995, Phys. Rev. Lett. **75**, 2944.
Santos, E., 1992, Phys. Rev. A **46**, 3646.
Scully, M.O., and K. Druhl, 1982, Phys. Rev. A **25**, 2208.
Scully, M.O., B.-G. Englert and H. Walther, 1991, Nature **351**, 111.
Serber, R., and C.H. Townes, 1960, in: Quantum Electronics, ed. C.H. Townes (Columbia University Press, New York) p. 233.
Shelby, R.M., M.D. Levenson, S.H. Perlmutter, R.G. de Voe and D.F. Walls, 1986, Phys. Rev. Lett. **57**, 691.
Shih, Y.H., and C.O. Alley, 1988, Phys. Rev. Lett. **61**, 2921.
Shih, Y.H., A.V. Sergienko and M.H. Rubin, 1993, Phys. Rev. A **47**, 1288.
Simon, R., H.J. Kimble and E.C.G. Sudarshan, 1988, Phys. Rev. Lett. **61**, 19.
Slusher, R.E., L.W. Hollberg, B. Yurke, J.C. Mertz and J.F. Valley, 1985, Phys. Rev. Lett. **55**, 2409.
Smithey, D.T., M. Beck, J. Cooper and M.G. Raymer, 1993, Phys. Rev. A **48**, 3159.
Steinberg, A.M., and R.Y. Chiao, 1995, Phys. Rev. A **51**, 3525.
Steinberg, A.M., P.G. Kwiat and R.Y. Chiao, 1992a, Phys. Rev. Lett. **68**, 2421.
Steinberg, A.M., P.G. Kwiat and R.Y. Chiao, 1992b, Phys. Rev. A **45**, 6659.
Steinberg, A.M., P.G. Kwiat and R.Y. Chiao, 1993, Phys Rev. Lett. **71**, 708.
Sudarshan, E.C.G., 1963, Phys. Rev. Lett. **10**, 277.
Susskind, L., and J. Glogower, 1964, Physics **1**, 49.
Taylor, G.I., 1909, Proc. Cambridge Philos. Soc. **15**, 114.
Tiwari, S.C., 1992, J. Mod. Opt. **39**, 1097.
Tomita, A., and R.Y. Chiao, 1986, Phys. Rev. Lett. **57**, 937.
Torgerson, J.R., D. Branning and L. Mandel, 1995, Appl. Phys. B **60**, 267.
Torgerson, J.R., D. Branning, C.H. Monken and L. Mandel, 1995, Phys. Lett. A **204**, 323.
Vaidman, L., 1994, Quantum Opt. **6**, 119.
Vain'shtein, L.A., V.N. Melekhin, S.A. Mishin and E.R. Podolyak, 1981, Sov. Phys. JETP **54**, 1054.
Vigué, J., J.A. Beswick and M. Broyer, 1983, J. Phys. (Paris) **44**, 1225.
Vigué, J., P. Grangier, G. Roger and A. Aspect, 1981, J. Phys. (Paris) **42**, L531.
von Weiszäcker, C.F., 1931, Z. Phys. **70**, 114.
Walls, D.F., 1977, Am. J. Phys. **45**, 952.
Walls, D.F., 1983, Nature **306**, 141.
Walther, H., 1992, in: 13th Int. Conf. on Atomic Physics, AIP Conf. Proc. **275**, 287.
Wang, L.J., X.Y. Zou and L. Mandel, 1991, Phys. Rev. Lett. **66**, 1111.
Wheeler, J.A., 1978, in: Mathematical Foundations of Quantum Theory, ed. A.R. Marlow (Academic Press, New York) p. 9.
Wolf, E., 1955, Proc. R. Soc. (London) **230**, 246.
Wootters, W.K., and W.H. Zurek, 1979, Phys. Rev. D **19**, 473.
Wu, L., H.J. Kimble, J.L. Hall and H. Wu, 1986, Phys. Rev. Lett. **57**, 2520.
Yablonovitch, E., 1993, J. Opt. Soc. Am. B **10**, 283.
Yamamoto, Y., N. Imoto and S. Machida, 1986, in: 2nd Int. Symp. on Foundations of Quantum

Mechanics in the Light of New Technology, eds S. Kobayashi, H. Ezawa, Y. Murayama and S. Nomura (The Physical Society of Japan, Tokyo) p. 265.
Yuen, H.P., 1976, Phys. Rev. A **13**, 2226.
Yurke, B., S.L. McCall and J.R. Klauder, 1986, Phys. Rev. A **33**, 4033.
Zajonc, A.G., L.J. Wang, X.Y. Zou and L. Mandel, 1991, Nature **353**, 507.
Zeeman, P., 1925, Physica (The Hague) **5**, 325.
Zou, X.Y., T.P. Grayson, G.A. Barbosa and L. Mandel, 1993, Phys. Rev. A **47**, 2293.
Zou, X.Y., T.P. Grayson and L. Mandel, 1992, Phys. Rev. Lett. **69**, 3041.
Zou, X.Y., L.J. Wang and L. Mandel, 1991, Phys. Rev. Lett. **67**, 318.

III

SUPER-RESOLUTION BY DATA INVERSION

BY

MARIO BERTERO

Dipartimento di Fisica, Università di Genova, Via Dodecaneso 33, 16146 Genova, Italy

AND

CHRISTINE DE MOL

*Département de Mathématique, Université Libre de Bruxelles,
Campus Plaine CP 217, Boulevard du Triomphe, 1050 Bruxelles, Belgium*

CONTENTS

	PAGE
§ 1. INTRODUCTION	131
§ 2. RESOLUTION LIMITS AND BANDWIDTH	134
§ 3. LINEAR INVERSION METHODS AND FILTERING	142
§ 4. OUT-OF-BAND EXTRAPOLATION	154
§ 5. CONFOCAL MICROSCOPY	162
§ 6. INVERSE DIFFRACTION AND NEAR-FIELD IMAGING	168
ACKNOWLEDGEMENTS	172
APPENDIX A	173
APPENDIX B	174
REFERENCES	176

§ 1. Introduction

Super-resolution is a widely-used keyword that appears in the literature in several contexts and with quite different significations. Unfortunately, this has contributed to the creation of some confusion about the corresponding concept. In the present tutorial chapter, we attempt to clarify the situation by giving a precise meaning to the word "super-resolution" in some specific and well-defined case-examples.

The conventional concept of resolving power of an optical instrument dates back to the end of the last century with the well-known classical works by Abbe [1873] and Rayleigh [1879]. In § 2, we briefly recall how the classical Rayleigh resolution limit is introduced and defined in terms of the overlap between the images of two point sources. Next we show that, in the framework of modern Fourier optics, the resolving power is characterized instead by specifying the spatial-frequency band associated with the instrument. Indeed, due to diffraction effects, the set of frequencies transmitted by an optical system is confined to some finite region in Fourier space, called the band of the system. A link between the two viewpoints is provided by the sampling theorems for bandlimited functions, which we also recall in § 2.

In the last decades, the development of micro-informatics and of computer-assisted instruments has deeply modified the classical concept of resolving power. Indeed, in many modern optical devices, numerical algorithms can be implemented to process the recorded data, in order to get a better representation of the probed object. Such "reconstructions" of the object may eventually allow resolution of finer details than those visible in the unprocessed image. The corresponding enhancement in resolution is often called "super-resolution", although in the following we will adopt a slightly more restrictive definition of this word. We suppose we have a good model for describing the imaging process in a given instrument. In other words, we assume that we can write down an explicit mathematical relationship between any object under examination and its image produced by the instrument. The determination of the image of a given object according to such a model is called the "direct" imaging problem. Conversely, the recovery or "restoration" of the object from its image is called the "inverse" imaging problem. As is usual in Fourier optics, we assume moreover that the imaging model is linear, so that the corresponding direct and inverse problems are linear as well. This means in particular that we do not consider here any partial-coherence effects or nonlinear inverse problems such as the well-known "phase problem", which consists of retrieving the phase of an object with known modulus.

For the sake of simplicity, we split the inverse imaging problem into two parts and we analyze separately each of the two sub-problems. The first one, which we call "deblurring", consists of recovering the Fourier spectrum of the object inside the band; i.e., for those frequency components which are transmitted by the instrument. Deblurring allows one to correct, e.g., for the effect of aberrations, which are responsible for phase distortion and attenuation inside the band. *The second step, for which we reserve the denomination of "super-resolution", is the attempt at recovering the object spectrum outside the band of the instrument, or in other words, at restoring the object beyond the diffraction limit.* As we will see later, this step involves an extrapolation of the object spectrum outside the band. Such an extrapolation appears to be feasible under some assumptions, such as the analyticity of the object spectrum, which holds, e.g., for an object vanishing outside some finite region in space. Notice that both steps require the processing – more precisely the "inversion" – of the image data in order to get estimates of the object spectrum inside or outside the band.

The aim of the present study is to analyze the resolution enhancement that can be achieved by such data inversion processes and to quantify the corresponding gain in resolving power. The natural framework for investigating this question is provided by the so-called "linear inversion theory", which is a collection of methods, referred to as "regularization" methods, allowing definition of stable solutions of linear inverse imaging problems. In § 3, we recall some important features of such inversion methods, to the extent that they relate to the assessment of resolution limits. The main difficulty encountered in solving inverse problems is their sensitivity to noise in the data, which can be the source of major instabilities in the solutions. The role of regularization is to prevent such instabilities from occurring. However, the price to pay for stability is that we can only recover the solution with a limited resolution. In this framework, the resolution limits arise as a practical limitation imposed by the necessity to control the noise amplification inherent in all inversion procedures. Accordingly, the resolving power appears no longer as a purely intrinsic characteristic of the instrument itself, but rather as a combined property of the hardware (the optical components) and of the implemented inversion software. The concept of "overall impulse response" introduced in § 3.1 allows one to take both aspects into account and to characterize the ultimate resolution capabilities of the instrument.

In § 3.2, we introduce an optimal filtering method, known as the Wiener filter, for the solution of convolution equations. We apply it to the deblurring problem and show that the object spectrum can only be recovered on some "effective" band, which depends on the level of noise in the recorded data through a parameter called the signal-to-noise ratio. This effective band is, in general, smaller than the band of the instrument, so that no super-resolution can be achieved in such a case. To recover the object spectrum outside the effective band, one must take into account a priori information about the localization of the object. This is done in § 3.3, where we analyze some well-known inversion methods in terms of the singular system of the imaging operator. Indeed, because of the object localization, this operator is no longer a pure convolution operator but is instead a compact integral operator. The corresponding overall impulse response allows one to assess the achievable resolution limits. Accordingly, we show that the resolving power depends on the signal-to-noise ratio and, to a lesser extent, on the choice of the data inversion algorithm. We also define the useful concept of "number of degrees of freedom", which generalizes the Shannon number used in information theory.

In the following sections, we consider in more detail three particular problems which can be viewed as paradigms for super-resolution problems. In § 4, we analyze the problem of extrapolating the object spectrum outside the band or the effective band – under the assumption that the object vanishes outside some finite known domain. We show how to assess the amount of achievable super-resolution and we are led to the conclusion that a significant resolution enhancement with respect to the Rayleigh limit is obtained only in the case of a very low space–bandwidth product, or equivalently, of a very small number of degrees of freedom. In § 5, we consider a situation where this condition is fulfilled in practice, namely the case of scanning microscopy. We focus on confocal microscopy and show how the use of data inversion techniques allows one to enhance the resolving power of such microscopes. Finally, in § 6, we consider the problem of inverse diffraction from plane to plane, which consists of back-propagating towards the source plane a field propagating in free space. When the data plane is situated in the far-field region, the problem essentially reduces to that of out-of-band extrapolation considered in § 4. Hence, no significant super-resolution is achievable, except when the space–bandwidth product is very low, which means that the source should be small when compared with the wavelength of the illuminating field. Such a conclusion, however, no longer holds when near-field data are available. We show that in such cases the effective bandwidth is increased significantly due to the presence of evanescent waves, and we assess the improvement as a function of the signal-to-noise ratio and of the distance between the source and data planes. Near-field imaging techniques provide, in fact, an alternative way to overcome the far-field diffraction limit, as demonstrated nowadays, e.g., by the resolving capabilities of near-field scanning microscopes.

§ 2. Resolution Limits and Bandwidth

The resolving power of an optical device can be characterized in various ways. We show that, instead of the classical Rayleigh criterion recalled in § 2.1, it is preferable to analyze and specify the band of the spatial frequencies transmitted by the system and to define the resolution limit on the basis of sampling theorems for bandlimited functions (see § 2.2). We then argue that because of noise, the concept of band must be replaced by the notion of "effective band", and we define "deblurring" as object restoration inside the effective band and "super-resolution" as out-of-band extrapolation of the object spectrum (see § 2.3).

2.1. THE RAYLEIGH RESOLUTION LIMIT

The resolving power of an imaging system is a measure of its ability to separate the images of two neighboring points. According to geometrical optics, in the absence of aberrations, the image of a point source is a perfectly sharp point, and hence the resolving power is unlimited. However, because of diffraction effects, the image is never just a point, but instead a small light patch called the *diffraction pattern*. When two point sources get closer, their diffraction patterns start overlapping progressively, until it is no longer possible to discriminate between one or two point sources. The limit distance between the two sources down to which the discrimination can be done may depend in practice on many factors (including, e.g., the sensitivity of the human eye), which can hardly be quantified. Nevertheless, it appears useful to have some simple and objective criteria to compare the performances of optical systems. Different criteria have been proposed for this purpose, the most famous being the Rayleigh criterion (see the original paper by Rayleigh [1879] or the book by Born and Wolf [1980]). According to the original version of this criterion, the diffraction patterns are considered as just resolved if the central maximum of the first coincides with the first minimum of the other. For the so-called Airy pattern, which is the Fraunhofer (far-field) intensity diffraction pattern at a circular aperture, the distance between the principal maximum and the first zero or dark ring is given by $R = 1.22(\lambda/2\alpha)$, where λ is the wavelength of the light and α is equal to the radius of the diffracting aperture divided by the distance between the aperture and image planes. The distance R is known as the *Rayleigh resolution limit* and $1/R$ as the *resolving power*. The above formula also provides the resolution limit of a microscope if α denotes its numerical aperture (Abbe [1873]), and the angular resolution limit of a telescope if α represents the radius of the objective aperture (see Born and Wolf [1980]).

2.2. BAND OF AN OPTICAL SYSTEM AND SAMPLING THEOREMS

Since the introduction of Fourier methods in optics, it has become more usual to characterize the resolution of an optical system in terms of its *bandwidth*. Diffraction effects are indeed responsible for the existence of a *cut-off frequency*, due to the fact that not all the spatial frequencies of the object are transmitted by the *pupil* of the instrument. The optical system is then said to be *bandlimited* or *diffraction-limited*. More precisely, in the framework of Fourier optics (see, e.g., Born and Wolf [1980], Goodman [1968], Papoulis [1968]), an optical system is viewed as a linear system; i.e., as a black box characterized fully by its impulse response. Hence the equation describing the imaging process is:

$$g(r) = \int dr' \, S(r,r') f(r'). \tag{2.1}$$

The functions $f(r)$ and $g(r)$ are the object (input of the system) and the image (output of the system), respectively; they represent scalar light amplitudes in the case of coherent imaging and light intensities in the case of incoherent imaging. In full generality, the spatial vector coordinate $r = (x, y, z)$ is three-dimensional (3D), since it includes the depth coordinate z along the optical axis. In many instances, however, a two-dimensional (2D) description can be used, in terms of the vector $\rho = (x, y)$ formed by the transverse cartesian coordinates in the object and image planes. The impulse response $S(r, r')$, which is assumed to be known, represents the image at point r of a point source situated at point r'. For space-invariant ("isoplanatic") systems, this impulse response depends only on the difference of the variables $r - r'$ and the image $g(r)$ is then given by the convolution integral

$$g(r) = \int dr' \, S(r-r') f(r'). \tag{2.2}$$

Note that henceforth, unless explicitly specified, the limits of integration are supposed to be infinite. In optics, the function $S(r)$ is usually called the *point spread function* (PSF) and its Fourier transform $\widehat{S}(k)$ is called the *transfer function* (TF) of the optical system. We use the following definition of the Fourier transform:

$$\widehat{S}(k) = \int d\mathbf{r}\, e^{-i\mathbf{k}\cdot\mathbf{r}}\, S(\mathbf{r}). \tag{2.3}$$

In the case of an optical system consisting of well-corrected lenses ("ideal" system) and in the case of coherent illumination, $\widehat{S}(k)$ is the *pupil function* (except for a scaling of coordinates); i.e., the function equal to one inside the pupil and to zero outside. In mathematical terms, $\widehat{S}(k)$ is called the characteristic function of the pupil. In any case, and also for aberrated systems and incoherent illumination, because of diffraction, $\widehat{S}(k)$ vanishes outside some bounded set B called the band of $S(\mathbf{r})$ or the *band of the optical system*. In such cases, the function $S(\mathbf{r})$ is said to be *bandlimited*. Then all images $g(\mathbf{r})$ are also bandlimited with the same band as $S(\mathbf{r})$, as follows from the Fourier convolution theorem:

$$\widehat{g}(k) = \widehat{S}(k)\widehat{f}(k). \tag{2.4}$$

Therefore, the sharpest details in the object, which correspond to high-frequency components lying outside the band, are smoothed off and no longer noticeable in the image, which appears as a blurred version of the object. Rayleigh's criterion can then be reinterpreted in the framework of communication and information theory (see Gabor [1961], Toraldo di Francia [1969]), and related to the sampling theory for bandlimited functions. In particular, in one-dimensional (1D) situations, the Rayleigh resolution distance coincides with the Nyquist sampling distance. To clarify this point, we first consider a 1D ideal coherent imaging system, having as TF the characteristic function of the interval $[-K, +K]$ and hence as PSF $S(x) = \sin(Kx)/(\pi x)$. In such a case, according to Rayleigh's criterion, the resolution distance is $R = \pi/K$. On the other hand, we see from the 1D version of eq. (2.4) that the image has the same spatial *cut-off frequency* K as the PSF. One then says that the function $g(x)$ is bandlimited with bandwidth K (or in short, K-bandlimited). A basic result in information theory – used widely in optics, as discussed by Gabor [1961] – is the following sampling theorem, named after Whittaker [1915] and Shannon [1949] (see Jerri [1977] for a review): any K-bandlimited function can be represented by the sampling expansion

$$g(x) = \sum_{n=-\infty}^{+\infty} g\left(n\frac{\pi}{K}\right) \operatorname{sinc}\left[\frac{K}{\pi}\left(x - n\frac{\pi}{K}\right)\right], \tag{2.5}$$

where

$$\text{sinc}(x) = \frac{\sin(\pi x)}{\pi x}. \tag{2.6}$$

The theorem implies that any K-bandlimited function $g(x)$ can be represented by a sequence of its samples without any loss in information, provided that the samples are taken at equidistant points spaced by the distance π/K, called the *Nyquist distance*. We see that in this simple case the Nyquist and Rayleigh distances coincide.

Let us observe that the set of all the K-bandlimited functions is a linear subspace of the space of all square-integrable functions on the real line. In the mathematical literature, such a subspace is usually called a Paley–Wiener space of functions. Consider now a function $f(x)$ that is not bandlimited and let $f_K(x) = (B_K f)(x)$ be its projection on the subspace of the K-bandlimited functions. In Fourier space, the bandlimiting operator B_K acts by simply multiplying the spectrum $\widehat{f}(k)$ by the characteristic function of the band $[-K, +K]$. Hence by the convolution theorem, we get

$$f_K(x) = (B_K f)(x) = \int_{-\infty}^{+\infty} dy \, \frac{K}{\pi} \, \text{sinc}\left[\frac{K}{\pi}(x-y)\right] f(y). \tag{2.7}$$

For the ideal system considered above, the image $g(x)$ is precisely the projection $f_K(x)$ of the object. We can say that knowing $f_K(x)$ is equivalent to knowing $f(x)$ within the *limit of resolution* $\Delta_x = \pi/K$.

For 2D images, the Rayleigh resolution distance is defined only in the case of a circular band, while sampling expansions can be derived for any geometrical shape of the band. We define the band B of a 2D image, $g(\boldsymbol{\rho}) = g(x, y)$, as the support of its Fourier transform; i.e., the set of all the spatial frequencies $\boldsymbol{\kappa} = \{k_x, k_y\}$ such that $\widehat{g}(\boldsymbol{\kappa}) \neq 0$. We say that the function $g(\boldsymbol{\rho})$ is B-bandlimited if B is a bounded subset of the frequency plane.

When the band B is the rectangle $|k_x| \leq K_x$, $|k_y| \leq K_y$, the sampling expansion (2.5) is replaced by

$$g(x, y) = \sum_{m=-\infty}^{+\infty} \sum_{n=-\infty}^{+\infty} g\left(\frac{n\pi}{K_x}, \frac{m\pi}{K_y}\right) \text{sinc}\left[\frac{K_x}{\pi}\left(x - \frac{n\pi}{K_x}\right)\right] \text{sinc}\left[\frac{K_y}{\pi}\left(y - \frac{m\pi}{K_y}\right)\right]. \tag{2.8}$$

As in dimension one, a function $f(\boldsymbol{\rho})$ that is not bandlimited can be projected on the subspace of the B-bandlimited functions and its projection is denoted by $f_B(\rho)$. When we know that an image is bandlimited to a rectangle, the expansion (2.8) suggests the definition of two limits of resolution, one in the x-variable, $\Delta_x = \pi/K_x$, and one in the y-variable, $\Delta_y = \pi/K_y$.

More generally, one can introduce a direction-dependent resolution limit: given a unit vector $\theta = \{\cos\phi, \sin\phi\}$, if K_θ is the bandwidth of the 1D function $f_{B,\theta}(s) = f_B(s\theta)$, then the limit of resolution in the direction θ is $\Delta_\theta = \pi/K_\theta$. It is easy to see that K_θ is related to the length of the interval obtained by projecting B orthogonally on the direction θ (see Appendix A). This result holds for any shape of the band B. When B is the rectangle considered above, $K_\theta = K_x \cos\phi + K_y \sin\phi$. When B is a disc of radius K, then $K_\theta = K$ and the resolution limit $\Delta_\theta = \pi/K$ does not depend on θ. Let us remark that this limit does not coincide exactly with the resolution distance $R = 1.22(\pi/K)$ prescribed by the Rayleigh criterion and resulting from the position of the first zero of the Fourier transform of the characteristic function of the disc of radius K.

For functions of two variables, it can also be useful to introduce the concept of the *resolution cell,* which in the case of a rectangular band is the fundamental cell of the rectangular lattice characterized by the vectors $\rho_x = (\pi/K_x)\theta_x$ and $\rho_y = (\pi/K_y)\theta_y$; i.e., the rectangle with side lengths π/K_x and π/K_y, θ_x and θ_y being the unit vectors along the x-axis and the y-axis, respectively. Notice that the number of resolution cells per unit area is the number of independent sampling values, per unit area, of the image $f_B(\rho)$. In a sense, it is the number of independent information elements, per unit area, contained in the image $f_B(\rho)$. When B is not a rectangle, it is always possible to find a rectangle of minimum area containing B. However, in such cases, the sampling expansion (2.8) is not in general the most efficient one. It has been proved by Petersen and Middleton [1962] that a function with a bounded band B can be reconstructed in several ways from its samples taken over a periodic lattice, and that the most efficient lattice – in the sense that it requires the minimum number of sampling points per unit area – is not in general rectangular. The fundamental cell of this optimum periodic lattice is the natural extension of the resolution cell defined above. For example, in the case of a circular band with radius K, Petersen and Middleton [1962] showed that the unique optimum sampling lattice is the 120° rhombic lattice with a spacing between adjacent sampling points equal to $2\pi/\sqrt{3}K = 1.16\pi/K$.

The previous analysis can also be extended to deal with functions representing 3D images. This extension is important, for instance, in the case of confocal microscopy, a basic technique for obtaining 3D images of biological objects. Then, the definition of a direction-dependent resolution limit is necessary, since the resolving power of the instrument is different in terms of lateral resolution and of axial resolution (see § 5).

From the previous considerations, it follows that the concept of resolution limit may be somewhat ambiguous in many instances. We believe that a more relevant and unambiguous concept is the band of the optical instrument. A further example to support this statement is the case of imaging by a partially obscured pupil. Indeed, the limit of resolution given by the sampling theorem, the Nyquist distance, coincides with that of the corresponding completely filled pupil. On the other hand, according to the Rayleigh criterion, the resolving power of the partially obscured pupil is greater than that of the completely filled one – see again Born and Wolf [1980]. We assert that neither the Rayleigh nor the Nyquist distance is fully satisfactory for characterizing the resolving power of such a system. Indeed, when analyzing the available bands it is clear that a partially obscured pupil provides less information about the object than the corresponding completely filled pupil.

2.3. DEBLURRING AND SUPER-RESOLUTION

To get a complete understanding of the imaging performances of a given optical system, let us observe that even the concept of band introduced in the previous subsection is not yet sufficient. In general, indeed, the image $g(r)$ is not simply a bandlimited approximation of $f(r)$ because $\widehat{S}(k)$ is not necessarily constant on the band [see eq. (2.4)]. This happens, e.g., in incoherent imaging and in the presence of aberrations. In order to get a bandlimited approximation of $f(r)$ from the image, one should divide $\widehat{g}(k)$ by $\widehat{S}(k)$ for all $k \in \mathcal{B}$. Anticipating the next section, let us already observe that this operation is not feasible for the following reasons:
(i) in general, $\widehat{S}(k)$ tends to zero when k tends to points belonging to the boundary of the band $\mathcal{B}$;
(ii) if the image is detected by an instrument, then the recorded image $g(r)$ is affected by instrumental noise and therefore eq. (2.4) must be replaced by $\widehat{g}(k) = \widehat{S}(k)\widehat{f}(k) + \hat{e}(k)$, where $\hat{e}(k)$ is the Fourier transform of the function $e(r)$ modelling the experimental errors or noise.

If we take these two points into account, we see that the main effect resulting from the division of $\widehat{g}(k)$ by $\widehat{S}(k)$ is expected to be the amplification of the noise in a neighborhood of the boundary of $\mathcal{B}$. This is a well-known difficulty of "inverse filtering" (see, e.g., Frieden [1975], Andrews and Hunt [1977]).

Appropriate methods for the recovery of the object $f(r)$ from its image $g(r)$ through any linear optical system will be discussed in § 3. For the moment, let us simply observe that, in general, it will only be possible to recover a bandlimited approximation of $f(r)$ on a band that is smaller than the band B of the instrument. In § 3, we will define this "effective band", B_{eff}, as the set of frequencies for which the modulus of the TF $\widehat{S}(k)$ is larger than some threshold value depending on the data noise level. In the absence of noise, or when $|\widehat{S}(k)| = 1$ on the band B, the effective band B_{eff} coincides with the band. As already stated in § 1, we use the term *deblurring* to describe any restoration process taking place inside the band of the instrument and, more precisely, inside its effective band. Accordingly, the recovery of a bandlimited approximation of the object on the effective band is just a deblurring process. In general, it is the effective band that determines the "effective" resolution capabilities; i.e., the true performances of the instrument in practical situations.

In some cases, however, we can gain access to spatial-frequency components of the object lying outside this effective band. We use the term *super-resolution* to describe the resulting enhancement in resolution. Super-resolution is achieved when the object spectrum can be somehow extrapolated outside the effective band, so that we can get a bandlimited approximation of the object on a larger band. This clearly requires further assumptions about the object, namely some extra information that can be exploited to restore a piece of its spectrum that is not transmitted by the optical system. The required *out-of-band extrapolation* can be performed, at least in principle, under the assumption that the object has a finite spatial extent (see § 3). Indeed, in such a case, the object spectrum is an entire analytic function and hence is determined uniquely by the part of the spectrum transmitted by the optical system (see Wolter [1961], Harris [1964]). This would allow restoration of the object "beyond the diffraction limit"; i.e., with a better resolution than the Rayleigh or Nyquist distance. This idea for overcoming the Rayleigh limit is quite old, and has been discussed abundantly in the literature (see, e.g., Toraldo di Francia [1955], Wolter [1961], Harris [1964], McCutchen [1967], Rushforth and Harris [1968], Toraldo di Francia [1969]). However, as shown by Viano [1976], this argument does not take into account the instability of analytic continuation in the presence of noise, which in practice prevents an easy recovery of the object spectrum beyond the cut-off frequency. The appropriate framework to address this question and to assess quantitatively the achievable resolution improvement appears to be the regularization theory for inverse problems, which we review in the next section. In such a framework it has been shown by Bertero, De Mol and Viano [1979, 1980] that, because of the ill-posedness of the inverse problem in the presence of noise, only very little super-resolution could be achieved in most practical situations. However, as will be shown in § 4, such a conclusion does not hold for problems characterized by a very small *Shannon number* or, equivalently, *space–bandwidth product.*

When feasible, to enhance the resolving power of an instrument, an alternative to the above approach based on data processing is of course to modify the instrument itself. One way is to improve the data collection scheme to record data over a wider range and containing more information about the probed object. For example, one can think of measuring image data obtained with different wavelengths. Another possibility is to increase the aperture to make the band larger or to try to modify the PSF to make it narrower. The latter possibility is precisely the idea behind an early attempt at overcoming the Rayleigh limit, by means of so-called *super-resolving pupils* (Toraldo di Francia [1952]). A super-resolving pupil is a pupil that produces a diffraction pattern with a central peak narrower that the initial PSF of the instrument and surrounded by a sufficiently large dark ring. In principle, such a pupil can be realized by means of a suitable coating, changing the TF without modifying the band of the instrument. Hence it is a deblurring technique, which could also be seen as a kind of "apodization". However, let us recall that the scope of apodization is usually to reduce – by changing the TF – the side lobes of the PSF to avoid unpleasant artefacts in the image. It is well known that the usual price to pay for this is an increase of the width of the central peak of the PSF. In a super-resolving pupil, the shrinkage of this central peak is accompanied by an increase of the side-lobes, which may become quite huge. This major drawback is, however, irrelevant in some special circumstances; the side-lobes can just be ignored or chopped when the observed object is small compared with the resolution limit and isolated in an uninteresting background. Another serious limitation is then to keep enough light intensity in the central peak of the PSF. Moreover, the required coating of the pupil should be realized with an extremely high accuracy, since even a small error would completely destroy its super-resolving properties. The design of a super-resolving pupil is a typical example of a *synthesis problem,* and such problems present the same sensitivity to errors as inverse problems do (see § 3). This explains why, as far as we know, no practical super-resolving pupil has ever been manufactured. In any case, synthesis problems are outside the scope of the present work where we focus on resolution enhancement *by means of data inversion.* We assume that the instrument is given and has a specified PSF, and that the wavelength at which it operates as well as the type of the recorded data are fixed. In other words, we work on the basis of a prescribed imaging equation like eq. (2.1), with objects and images also defined on well-specified domains. Then the only way to get super-resolution consists of processing the data to recover the object beyond the diffraction limit. Let us nevertheless notice that super-resolving pupils can be emulated by a suitable post-processing of the image, as it has been shown by De Santis, Gori, Guattari and Palma [1986]. The method can be viewed as a particular linear estimator for object deblurring, and hence will be discussed further in § 3.2.

§ 3. Linear Inversion Methods and Filtering

Because of noise, to perform both object restoration on the band of the instrument (deblurring) and outside the band (super-resolution), we need appropriate data inversion methods. In the present section, we summarize essential features of such methods and we introduce the concept of *overall impulse response*, which is useful for assessing the achievable resolution (§ 3.1). For convolution equations, we use the classical Wiener filter to solve the deblurring problem and to define the *effective bandwidth* of an optical instrument (§ 3.2). To deal with super-resolution problems, we show how to make use of the required assumption about the localization (limited spatial extent) of the object and we introduce the *singular system* of the corresponding imaging operator (§ 3.3). This allows us to define the notion of *number of degrees of freedom* (NDF), which is useful for analyzing super-resolution methods. Both the effective bandwidth and the NDF depend on the instrumental PSF and on a quantity called the *signal-to-noise ratio* (SNR) to be defined in § 3.2.

3.1. THE OVERALL IMPULSE RESPONSE

The *inverse imaging problem* consists of estimating the original object $f(r)$ from a given recorded image $g(r)$ through an optical instrument with known impulse response. When tackling this problem, we must take into account the fact that a measured image is inevitably contaminated by noise. Therefore, we modify the imaging equation (2.1) as follows:

$$g(r) = \int dr' \, S(r, r') f(r') + e(r), \tag{3.1}$$

where $e(r)$ is an unknown function modelling the experimental errors and noise in the data. This equation can be rewritten shortly in operator form as

$$g = Lf + e, \tag{3.2}$$

where L is the *imaging operator* which maps the object on its noise-free image defined by

$$(Lf)(r) = \int dr'\, S(r,r') f(r'). \tag{3.3}$$

We call it L as a reminder that it is a Linear operator describing the effect of the optical system, which is just a Lens in the simplest case. In the following we will also use the adjoint imaging operator, L^*, which can be defined as

$$(L^*g)(r) = \int dr'\, S^*(r',r)\, g(r'), \tag{3.4}$$

denoting by S^* the complex conjugate of S.

Reconstructing the object $f(r)$ from its image is a typical *linear inverse problem*, and we must devise an appropriate method to perform the inversion. We will restrict the discussion to linear estimation and we will denote throughout by $\tilde{f}(r)$ an estimate of the object. Then the most general linear estimate of the object is obtained as a linear superposition of the values of the recorded image. Hence we may write it in the following form:

$$\tilde{f}(r) = \int dr'\, M(r,r')\, g(r'). \tag{3.5}$$

This linear solver or estimator is characterized by the function $M(r,r')$, which we call the *reconstruction kernel* or the *restoration function*. Replacing the imaging equation into eq. (3.5), we get

$$\tilde{f}(r) = \int dr' \int dr''\, M(r,r') S(r',r'') f(r'') + \int dr'\, M(r,r')\, e(r'). \tag{3.6}$$

Reversing the order of integration and putting

$$T(r,r'') = \int dr'\, M(r,r')\, S(r',r''), \tag{3.7}$$

we also obtain

$$\tilde{f}(r) = \int dr''\, T(r,r'') f(r'') + \int dr'\, M(r,r')\, e(r'). \tag{3.8}$$

In the absence of noise, perfect restoration – with unlimited resolution – would correspond to $T(r,r'') = \delta(r - r'')$, where $\delta(r)$ denotes the Dirac distribution. As already mentioned, this cannot be achieved in practice because of the presence of noise. Indeed, the second term in eq. (3.8) – i.e., the noise term – must be kept sufficiently small and, as will be shown later, the necessity of controlling the noise amplification implies a certain widening of the function $T(r, r'')$. In the absence of noise or with a negligible noise term, this function represents the restoration at point r of a point source situated at point r''. Hence it plays the role of an impulse response for the object restoration process. Typically, as a function of the variable r, $T(r, r'')$ presents a peak centered at r'', with a few side-lobes. The width of this peak characterizes the achievable resolution in the restored object at point r''. In fact, this function $T(r, r'')$ allows one to take into account both the initial resolving power of the instrument, when no data processing is performed, and the subsequent improvement due to the use of a restoration algorithm. Indeed, as seen from eq. (3.7), it depends on the PSF $S(r, r')$ and on the reconstruction kernel $M(r, r')$. Therefore, we follow the terminology introduced by Gori and Guattari [1985] and call $T(r, r'')$ the *overall impulse response*, since it characterizes the entity "instrument + implemented inversion algorithm". When this response is space-invariant, we also call it the *overall PSF*.

Many different linear restoration algorithms have been proposed in the literature, corresponding to different choices of the restoration kernel $M(r, r')$. In the following, we present only some of the most common ones, which are sufficient for our purpose of assessing resolution limits. We want to emphasize the fact that there is no "best all-purpose inversion algorithm". The choice of a particular method is often a matter of convenience and is made according to the specificities of the particular inverse problem one must deal with. Nevertheless, inversion algorithms rely on some common ground and present some basic similarities which can be understood from the examples we describe in the following.

3.2. OPTIMAL FILTERING FOR CONVOLUTION EQUATIONS

In the present subsection we assume that the imaging operator is a true convolution, namely that we must deal with the equation

$$g(r) = \int dr' \, S(r - r') f(r') + e(r). \tag{3.9}$$

In Fourier space, this equation becomes

$$\widehat{g}(k) = \widehat{S}(k) \widehat{f}(k) + \hat{e}(k). \tag{3.10}$$

Because of the space-invariance of the system, it is quite natural to consider only space-invariant restoration; i.e., to take an object estimate of the form

$$\tilde{f}(r) = \int dr'\, M(r-r')\, g(r'), \tag{3.11}$$

or in Fourier space,

$$\widehat{\tilde{f}}(k) = \widehat{M}(k)\, \widehat{g}(k). \tag{3.12}$$

The function $\widehat{M}(k)$ and the corresponding restoration method are usually called a *filter*. The restoration process is sometimes also referred to as "image deconvolution" or "image deblurring".

In the absence of noise, we see from eq. (3.10) that perfect restoration of the object spectrum on the band of the system is achieved by means of the *inverse filter* $\widehat{M}(k) = [\widehat{S}(k)]^{-1}$. No out-of-band restoration is thus obtained because this filter can be used only where $\widehat{S}(k)$ does not vanish. We also see that if the inverse filter is applied to the noisy data (3.10), the noise will be amplified considerably for those frequencies where $\widehat{S}(k)$ is very small, and this will cause uncontrolled instabilities in the restored object. Therefore, the formal solution given by the inverse filter cannot be applied straightforwardly in practice and must be replaced by a so-called *regularized algorithm*, for which stability with respect to noise is guaranteed. This is achieved, for example, by the well-known *Wiener filter* (see, e.g., the book by Bell [1962]), which provides the best linear estimate in the least-squares sense of the solution of eq. (3.9) (see also Strand and Westwater [1968], Franklin [1970], Turchin, Kozlov and Malkevich [1971], Cesini, Guattari, Lucarini and Palma [1978]). We briefly recall how such a filter is derived. One must assume that the object f, the image g and the noise e are zero-mean stochastic processes. The power spectrum of the noise is assumed to be given by

$$\langle |\hat{e}(k)|^2 \rangle = \varepsilon^2\, \hat{\rho}_{ee}(k), \tag{3.13}$$

where $\langle \cdot \rangle$ denotes the expectation value or ensemble average. For the so-called "white noise", the spectral density function $\hat{\rho}_{ee}(k)$ is equal to 1. We also assume the power spectrum of the object to be given by

$$\langle |\hat{f}(k)|^2 \rangle = E^2\, \hat{\rho}_{ff}(k), \tag{3.14}$$

and the object to be uncorrelated with the noise; i.e.,

$$\langle \hat{f}(k)\, \hat{e}^*(k) \rangle = 0 \tag{3.15}$$

(in other words, we assume the noise to be additive). Let us look now for the best linear estimate $\tilde{f}$ of the object f in the least-squares sense; i.e., for the estimate that minimizes the quadratic or least-squares error $\langle\|\widetilde{f}-f\|^2\rangle$, where $\|f\|$ denotes the L^2-norm of f, namely

$$\|f\| = \left[\int d\mathbf{r}\,|f(\mathbf{r})|^2\right]^{1/2}. \tag{3.16}$$

Through Parseval's relation, the norm $\|f\|$ is also equal to the L^2-norm $\|\widehat{f}\|$ of $\widehat{f}$ in Fourier space, up to a constant factor. Hence, equivalently, we must find the function $\widehat{M}(\mathbf{k})$ minimizing the quantity

$$\left\langle \int d\mathbf{k}\,|\widehat{M}(\mathbf{k})\,\widehat{g}(\mathbf{k}) - \widehat{f}(\mathbf{k})|^2 \right\rangle. \tag{3.17}$$

Using eq. (3.10) and the assumptions (3.13)–(3.15), we see that this quantity is also given by

$$\int d\mathbf{k}\left[E^2\,\widehat{\rho}_{\mathrm{ff}}(\mathbf{k})\,|\widehat{M}(\mathbf{k})\,\widehat{S}(\mathbf{k}) - 1|^2 + \varepsilon^2\,\widehat{\rho}_{\mathrm{ee}}(\mathbf{k})\,|\widehat{M}(\mathbf{k})|^2\right]. \tag{3.18}$$

It is easily seen that the minimum is provided by

$$\widehat{M}_{opt}(\mathbf{k}) = \frac{\widehat{S}^*(\mathbf{k})}{|\widehat{S}(\mathbf{k})|^2 + \dfrac{\varepsilon^2}{E^2}\dfrac{\widehat{\rho}_{\mathrm{ee}}(\mathbf{k})}{\widehat{\rho}_{\mathrm{ff}}(\mathbf{k})}}, \tag{3.19}$$

which is the optimal Wiener filter.

For white-noise processes, i.e., when $\widehat{\rho}_{\mathrm{ee}}(\mathbf{k}) = \widehat{\rho}_{\mathrm{ff}}(\mathbf{k}) = 1$, we define the *signal-to-noise ratio* (SNR) to be the quantity E/ε. Let us point out that it does not coincide with the usual concept of SNR used in the field of image restoration, where it is defined as the ratio of the power of the blurred image to the power of the noise. Here the SNR refers to the ratio of magnitude of the object to the magnitude of the data noise. Let us also notice that when $\widehat{\rho}_{\mathrm{ff}}(\mathbf{k}) = 1$, i.e., in the case of a white-noise process, the term $E^2\,\widehat{\rho}_{\mathrm{ff}}(\mathbf{k})$ integrated over all frequencies is divergent, corresponding to infinite energy. One should not worry about such divergences, which are easily eliminated [e.g., by considering, instead of the unphysical model of purely white noise, "quasi-white" or "colored" noise, with a flat spectrum extending beyond the support of $\widehat{S}(\mathbf{k})$].

For white- or colored-noise processes such that $\widehat{\rho}_{\mathrm{ee}}(\mathbf{k})/\widehat{\rho}_{\mathrm{ff}}(\mathbf{k}) = 1$ and when introducing the parameter $\alpha = \varepsilon^2/E^2$, the Wiener filter estimate coincides in fact with Tikhonov's regularized solution, which corresponds to the restoration function

$$\widehat{M}_\alpha(k) = \widehat{W}_\alpha(k) \frac{1}{\widehat{S}(k)}, \qquad (3.20)$$

with

$$\widehat{W}_\alpha(k) = \frac{|\widehat{S}(k)|^2}{|\widehat{S}(k)|^2 + \alpha} \qquad (3.21)$$

(see e.g., Tikhonov and Arsenin [1977], Groetsch [1984]). The function $\widehat{W}_\alpha(k)$ can be interpreted as a spectral window that must be used to "apodize" the inverse filter $[\widehat{S}(k)]^{-1}$ in order to prevent noise amplification. Notice that because of the presence of the positive parameter α in the denominator, the Tikhonov filter can be used for all frequencies, even where $\widehat{S}(k)$ vanishes. Indeed, from eqs. (3.20) and (3.21), we see that $\widehat{M}_\alpha(k) = 0$ when $\widehat{S}(k) = 0$. In Tikhonov's method, the parameter α is called the *regularization parameter* and the regularized estimate obtained through eqs. (3.12), (3.20) and (3.21) is viewed as the solution of the variational problem of minimizing the functional

$$\int dk \, |\widehat{S}(k)\widehat{f}(k) - \widehat{g}(k)|^2 + \alpha \int dk \, |\widehat{f}(k)|^2, \qquad (3.22)$$

which through Parseval's relation is also proportional to

$$\Phi_\alpha[f] = \int dr \, |(Lf)(r) - g(r)|^2 + \alpha \int dr \, |f(r)|^2. \qquad (3.23)$$

The second term in these expressions can be interpreted as a penalization functional, the role of which is to stabilize the pure least-squares solution obtained when minimizing the first term alone. Hence Tikhonov's method is a constrained least-squares method. It is also a purely deterministic or functional method as opposed to stochastic methods like the Wiener filter. Various methods for choosing the regularization parameter α have been proposed in the literature (c.f. Tikhonov and Arsenin [1977], Groetsch [1984], Bertero [1989], Davies [1992]). As we have seen above, using a stochastic approach, α is naturally related to the power spectra of the object and of the noise. For white-noise processes, α is simply the inverse of the squared SNR.

The overall impulse response corresponding to eq. (3.20) is clearly space-invariant and is given by the inverse Fourier transform of the filtering window defined by eq. (3.21), which is the *overall transfer function*. This overall TF cannot be equal to one in the presence of noise and vanishes where the transfer function $\widehat{S}(k)$ is zero. Hence we cannot recover the object spectrum outside the band. In other words, super-resolution – in the sense of out-of-band extrapolation as defined in § 2.3 – cannot be achieved by means of the above method. Moreover, the presence of noise prevents perfect deblurring on the band and consequently affects the resolving power of the instrument. As discussed previously, in the absence of noise, the resolving power is determined by the band of the system; i.e., the support of $|\widehat{S}(k)|^2$. To assess the resolving power in the presence of noise, we can proceed as follows. Notice that the filtering window (3.21) is close to 1 for those frequencies for which $|\widehat{S}(k)|^2$ dominates the term α in the denominator. However, it becomes close to zero when $|\widehat{S}(k)|^2$ is much smaller than α, so that the corresponding frequencies are damped and are not used in the restoration of the object. The resolution limits will be determined by an effective band that depends on the noise level and that can be most easily defined in the following way. Assume that the restoration filter (3.20) is such that $\widehat{W}_\alpha(k)$ is the characteristic function of some frequency band $\mathcal{B}_\alpha$:

$$\widehat{W}_\alpha(k) = \begin{cases} 1, & k \in \mathcal{B}_\alpha, \\ 0, & k \notin \mathcal{B}_\alpha. \end{cases} \qquad (3.24)$$

We then define the *effective band* $\mathcal{B}_{\text{eff}}$ as being the band $\mathcal{B}_\alpha$ that minimizes the least-squares error (3.18), allowing for all restoration filters of the form (3.20) with $\widehat{W}_\alpha(k)$ given by eq. (3.24). From the expression of eq. (3.18) in this particular case, it is easily seen that the effective band is the set of frequencies such that

$$|\widehat{S}(k)|^2 > \frac{\varepsilon^2}{E^2} \frac{\widehat{\rho}_{ee}(k)}{\widehat{\rho}_{ff}(k)}, \qquad (3.25)$$

or, in the case of white-noise processes, such that

$$|\widehat{S}(k)| > \frac{\varepsilon}{E}. \qquad (3.26)$$

When the effective band is a simply connected region, we can also define a direction-dependent effective cut-off $K_{\theta,\text{eff}}$ associated to the effective band $\mathcal{B}_{\text{eff}}$, as done in § 2 for the band $\mathcal{B}$. This effective cut-off will determine the effective resolving power in the direction θ in the presence of a given amount of noise.

Criteria other than optimality with respect to noise can be used in the design of linear object estimators, and one can try to realize an overall PSF or TF having desirable properties. For example, if the restoration filter is chosen in such a way that the spectral window (3.21) is replaced by a classical apodizing window like the Hamming or Hanning functions, the overall PSF will have reduced side-lobes, at the price of an increased width of the central peak. An opposite requirement is to make this central peak narrower than that of the PSF $S(r)$ and in fact as narrow as possible, in order to mimic a super-resolving pupil. In an interesting paper, De Santis, Gori, Guattari and Palma [1986] have shown that this can be achieved by emulating numerically a super-resolving pupil. The price to pay here is the presence of large side-lobes and the sensitivity to noise. As already mentioned in § 2.3, the effect of the side-lobes can be neglected in the case where the object has a finite spatial extent and a size of the order of the Rayleigh resolution distance. On the other hand, the amount of achievable resolution is limited by the necessity to control the instability with respect to noise. Although no out-of-band extrapolation is performed in such a way, the method appears to work under an assumption that is similar to one we will make in § 4; viz, the object should have small spatial extent. Then the emulated super-resolving pupil restores a bandlimited object, which obviously has an infinite spatial extent but nevertheless approximates the original space-limited object over its extent. This suggests that the two points of view may somehow be related, and we think that this question deserves further investigation to be fully understood.

3.3. FILTERED SINGULAR-SYSTEM EXPANSIONS FOR COMPACT OPERATORS

In the case of a pure convolution, no access to out-of-band frequencies – and hence no super-resolution according to our definition – is provided by any of the Fourier restoration filters described in the previous subsection. To be able to extrapolate outside the band or the effective band, we clearly need something more. For example, a further constraint on the object, expressing some a priori knowledge about its properties, can often do the job. A classical a priori constraint is to assume that the object has a finite spatial extent; i.e., that it vanishes outside some known finite region, which we will call the *domain of the object*. Then, by a theorem due to Paley and Wiener [1934], its Fourier transform $\hat{f}(k)$ is an entire analytic function. Thanks to the uniqueness of analytic continuation, such a function is (at least in principle) determined entirely by its values on any finite domain, as for example the band of the system. Such a constraint expressing a priori knowledge about the domain of the object can be taken into account by rewriting the imaging equation (3.9) as follows:

$$g(r) = \int dr'\, S(r-r')\, P(r') f(r') + e(r), \tag{3.27}$$

where $P(r)$ is the characteristic function equal to 1 inside and to 0 outside the domain of the object. Notice that the localization of the object can also be expressed through more general and smooth "profile functions" $P(r)$, such as a gaussian function. In such a case, eq. (3.27) can also be used to describe imaging processes where the object is illuminated in a nonuniform way; e.g., by a laser spot with gaussian profile. For such a gaussian profile, the solution of the imaging equation with $e(r) = 0$ is still unique, but uniqueness holds no longer for certain profiles such as bandlimited functions. The imaging problem associated to eq. (3.27) with different profiles has been analyzed in more detail by Bertero, De Mol, Pike and Walker [1984].

For profiles $P(r')$ vanishing outside some finite domain or decreasing sufficiently fast at infinity, the imaging operator L defined by eqs. (3.27) and (3.2) acquires the mathematical property of being compact in an appropriate function space, the most usual choice being a L^2-space of square-integrable objects. The spectral properties of compact operators are quite similar to those of finite-dimensional matrices; namely, they have a discrete spectrum. Compact self-adjoint operators can be expressed in diagonal form by means of their eigensystem, but in general they have an infinite number of eigenvalues. Imaging operators, however, are not necessarily self-adjoint (e.g., because the image and object domains are different or because the optical system is affected by aberrations and hence the PSF is not real). Instead of the eigensystem, one then uses the *singular system* of the imaging operator L. To our knowledge, singular systems were first introduced in optics by Gori, Paolucci and Ronchi [1975], but without referring to them as such and without reference to a more general mathematical framework (for this, see the book by Groetsch [1984] or the review paper by Bertero [1989]). The method was used by Gori, Paolucci and Ronchi [1975] and later by De Santis and Palma [1976] for investigating the number of degrees of freedom of an optical image in the presence of aberrations.

Let us recall that the singular system of L is the set of the triples $\{\sigma_n; u_n, v_n\}, n = 0, 1, \ldots$, which solve the following "shifted eigenvalue problem" (L^* denotes the adjoint operator):

$$Lv_n = \sigma_n u_n ; \qquad L^* u_n = \sigma_n v_n. \qquad (3.28)$$

The singular values σ_n are real and positive by definition, and ordered as follows: $\sigma_0 \geqslant \sigma_1 \geqslant \sigma_2 \geqslant \cdots$. Except in degenerate cases where they are in finite number, the sequence $\{\sigma_n\}$ of the singular values tends to zero when n tends to infinity. The singular functions or vectors $\{u_n\}$, eigenvectors of LL^*, constitute an orthonormal basis in the space of all possible noise-free images, whereas the singular functions or vectors $\{v_n\}$, eigenvectors of L^*L, form an orthonormal basis in the set of all objects if and only if the equation $Lf = 0$ has only the trivial solution $f = 0$. Otherwise they span the subspace orthogonal to the set of invisible (or transparent) objects. An invisible object is an object producing a zero-image, or equivalently, belonging to the so-called null-space of the imaging operator. When the null-space of L is not reduced to zero, the solution of the imaging equation with zero noise is not unique. Indeed, any object can be written as the sum of its visible and invisible parts. The invisible part cannot be retrieved from the data. Hence we are free to assign this component arbitrarily; the most usual choice is to set it equal to zero.

In view of the previous properties, data and solutions can be expanded on the singular functions of L, and restoration kernels as well. The following representations of the operators L and L^* also hold true:

$$Lf = \sum_{n=0}^{\infty} \sigma_n (f, v_n) u_n, \qquad (3.29)$$

$$L^*g = \sum_{n=0}^{\infty} \sigma_n (g, u_n) v_n, \qquad (3.30)$$

where (f, v_n) and (g, u_n) denote the usual scalar products in L^2. With the help of these formulas, one can easily find the object estimate that minimizes the least-squares error or, in the case of Tikhonov's regularization, the functional (3.23). The reconstruction kernel yielding the minimum is given by

$$M(r, r') = \sum_{n=0}^{\infty} \frac{\sigma_n}{\sigma_n^2 + \alpha} v_n(r) u_n(r'), \qquad (3.31)$$

and the corresponding overall impulse response by

$$T(r, r'') = \sum_{n=0}^{\infty} \frac{\sigma_n^2}{\sigma_n^2 + \alpha} v_n(r) v_n(r''). \qquad (3.32)$$

The filtering factor it contains, namely

$$W_{n,\alpha} = \frac{\sigma_n^2}{\sigma_n^2 + \alpha}, \qquad (3.33)$$

is analogous to the spectral window (3.21). The effect of this filter can be discussed along the same lines as for the window (3.21). In particular, one realizes that only the singular values such that $\sigma_n > \sqrt{\alpha}$ yield a significant contribution to the reconstruction kernel (3.31). For this reason, a widely used kernel amounts to sharp truncation to the N first terms of the singular-system expansion; i.e.,

$$M(r,r') = \sum_{n=0}^{N-1} \frac{1}{\sigma_n} v_n(r) u_n(r'), \tag{3.34}$$

corresponding to the following overall impulse response:

$$T(r,r'') = \sum_{n=0}^{N-1} v_n(r) v_n(r''). \tag{3.35}$$

Clearly, the truncation allows one to avoid the unacceptable amplification of the noise term in eq. (3.8) arising from the fact that singular values accumulate to zero. If one keeps only the terms corresponding to singular values larger than the square root of α, then it can be shown that, among all truncated solutions obtained through eq. (3.34), this stopping criterion yields the minimum of the functional (3.23). Moreover, for white-noise processes, the stopping criterion becomes

$$\sigma_n > \frac{\varepsilon}{E}, \tag{3.36}$$

in complete analogy with eq. (3.26). Here ε denotes the standard deviation (i.e., the square-root of the variance) of each noise component (e, u_n) and E the standard deviation of the object components (f, v_n) (for more details, see Bertero, De Mol and Viano [1979]).

As discussed by De Santis and Palma [1976] and by Bertero and De Mol [1981b], the resulting number N of terms in the expansion of the restored object can be considered as its *number of degrees of freedom* (NDF). Indeed, since the expansion is orthogonal, N represents the number of independent "pieces of information" about the object that can be reliably (i.e., stably) retrieved from the noisy data. The number of degrees of freedom, which depends on the operator L and on the value of ε/E, can be viewed as a generalization of the Shannon number, which will be defined in §4 and represents the information content of a bandlimited signal. Notice that instead of sharp truncation of the singular-system expansion, one could also filter the smallest singular values more smoothly, by introducing gently decreasing weighting factors W_n in the expansions (3.34) and (3.35). For example, one can use the classical window shapes of Hanning, Hamming or a triangular filter.

Again the ultimate resolution capabilities of the instrument can be assessed by looking at the overall impulse response $T(r, r'')$. When the imaging equation $Lf = g$ has a unique solution, then in the absence of noise, N tends to $+\infty$ and the singular system provides a resolution of the identity, i.e., $T(r, r'') = \delta(r - r'')$, which yields a perfect restoration. In the presence of noise, however, the expansion must be truncated or filtered to avoid instabilities, and accordingly, $T(r, r'')$ acquires a certain width depending on the signal-to-noise ratio. Notice that when the solution of the imaging equation is not unique, the overall impulse response always has a finite width, even in the absence of noise, because of the existence of invisible objects that can never be retrieved from the data.

The method of singular-system expansions described in the present subsection applies also to the very important case where the data are discrete. In practice, data are always recorded by a finite set of detectors. Hence, instead of a continuous image, one measures a finite-dimensional data vector g. In such cases, the imaging operator is automatically compact, even without any assumption on the domain of the object. All formulas given above still hold provided that one replaces the integrals on the x-variable by discrete sums on the components of the data vector and of the singular vectors u_n. The singular system can then be quite easily computed numerically. For a given imaging operator L, it can be computed once, and then stored. When this is done, the restoration kernel (3.34) provides very fast numerical algorithms for recovering the object. For a thorough discussion of linear inversion from discrete data, we refer the interested reader to the review papers by Bertero, De Mol and Pike [1985, 1988]. In particular, a comparison is made between the linear inverse problem with a continuous data function described by eq. (3.27) and the corresponding problem with discrete data, obtained when only sampled values of this data function are known. It can be shown that provided the sampling points are adequately placed and in sufficient number – roughly of the order of the NDF – then the singular system of the discrete-data problem approximates in some sense the singular system of the corresponding continuous-data problem. Therefore, all the considerations about resolution we make in the present work remain essentially valid in the case of discrete data, which therefore does not require a separate discussion.

Before concluding this section, we want to point out that we have considered here only *linear* restoration methods. Moreover, the only kind of a priori information we use about the object is some knowledge about its localization. As seen, such type of a priori knowledge is easily taken into account in the framework of linear methods. Our choice relies on the fact that it is only using linear inversion methods that one is able to define a resolution limit (through the overall impulse response) which is essentially independent of the particular object under study. However, let us stress the fact that many other inversion methods have been proposed for inverse problems and image restoration. Let us mention iterative methods with and without constraints and statistical methods such as maximum likelihood, maximum entropy, simulated annealing, etc. Most of these methods are non linear, and therefore the achievable resolution is in general strongly dependent on the image or object to be restored. We refer the interested reader to the recent review paper by Biemond, Lagendijk and Mersereau [1990] for iterative methods; to the paper by Meinel [1986] for maximum likelihood and to the paper by Donoho, Johnstone, Hoch and Stern [1992] for maximum entropy. A comparison of several inversion methods in the case of object restoration can be found in the paper by Bertero, Boccacci and Maggio [1995].

§ 4. Out-of-Band Extrapolation

As discussed in § 3.2, by processing the image provided by an optical instrument, it is possible to estimate the Fourier spectrum of the object on the effective band B_{eff} and hence to recover a bandlimited approximation of the object. The problem of estimating $\widehat{f}(k)$ outside the effective band is then a particular case of the general problem of *out-of-band extrapolation,* which can be formulated as follows: given the noisy values of $\widehat{f}(k)$ on a bounded domain B of the frequency space, estimate $\widehat{f}(k)$ outside B from its known noisy values on B. In the applications we consider, B represents either the band of the optical system or the effective band.

We first discuss this problem in the simple case of functions of a single variable x, denoting by k the conjugate variable in Fourier space. Let us assume to know $\widehat{f}(k)$ on the interval $B = [-K, +K]$, with an error $\hat{e}(k)$ which depends on k. Hence the data are given by

$$\widehat{g}(k) = \begin{cases} \widehat{f}(k) + \hat{e}(k), & |k| \leqslant K, \\ 0, & |k| > K. \end{cases} \qquad (4.1)$$

The inverse Fourier transform of $\widehat{g}(k)$ is given by

$$g(x) = \frac{1}{2\pi} \int_{-\infty}^{+\infty} dk\, \widehat{g}(k)\, e^{ikx}, \qquad (4.2)$$

or, using eq. (4.1), by

$$g(x) = \int_{-\infty}^{+\infty} dx' \, \frac{\sin K(x-x')}{\pi(x-x')} f(x') + e(x). \quad (4.3)$$

In other words, $g(x)$ is a noisy bandlimited approximation of $f(x)$. As already observed in § 2.2, knowing $g(x)$ is equivalent to knowing $f(x)$ with a resolution π/K. Equation (4.3) is a special case of the imaging equation (3.9). The problem is how to use this equation in order to estimate $\hat{f}(k)$ on a broader band $[-K', +K']$, and therefore to know $f(x)$ with a better resolution π/K'. If we find a method for doing so, then we say that we have achieved *super-resolution* in the restoration of the object $f(x)$.

The extrapolation of $\hat{f}(k)$ outside $[-K, +K]$ is not possible in general without any additional information about $f(x)$. Clearly, the solution of eq. (4.3) with $e(x) = 0$ is not unique, since the spectrum $\hat{f}(k)$ is perfectly arbitrary outside the band. Uniqueness holds true, however, in the case of a finite-extent object; i.e., if $f(x)$ vanishes outside a finite interval, say $[-X, +X]$. Indeed, by the theorem of Paley and Wiener [1934] already mentioned in §3.3 (see also Papoulis [1968]), the Fourier transform of a finite-extent object is an entire analytic function and therefore the analytic continuation of $\hat{f}(k)$ outside $[-K, +K]$ is unique, provided that it exists. This point is carefully discussed by Wolter [1961], who also points out that small errors on the data can produce completely different analytic continuations. The same problem was discussed by Viano [1976] from the point of view of the regularization theory for ill-posed problems.

The main difficulties of the problem of analytic continuation can be summarized as follows. First, we do not know exactly $\hat{f}(k)$ on $[-K, +K]$, but only $\hat{g}(k)$ as given by eq. (4.1). Since the noise term $\hat{e}(k)$ is not in general an analytic function, then also $\hat{g}(k)$ is not a piece of an analytic function and there is no solution to the problem of analytic continuation of the function $\hat{g}(k)$. Moreover, even if for a very peculiar $\hat{e}(k)$, there would be a solution, then we could find many analytic functions reproducing the data on the band within a prescribed accuracy, but being completely different outside the band. Indeed, it is always possible to find an entire function $\hat{h}(k)$ which is arbitrarily small on $[-K, +K]$ and arbitrarily large outside $[-K, +K]$. This shows the instability of analytic continuation in the presence of noise on the data.

To proceed further, let us write explicitly the corresponding imaging operator to be inverted, taking into account the a priori knowledge about the domain of the object:

$$(Lf)(x) = \int_{-X}^{+X} dx' \frac{\sin K(x-x')}{\pi(x-x')} f(x'), \quad -\infty < x < \infty. \tag{4.4}$$

As shown by Bertero and Pike [1982], this integral operator is a compact operator from the space $L^2(-X,+X)$ of square-integrable functions on $[-X,+X]$ into the space $L^2(-\infty,+\infty)$ of square-integrable functions on the real line. Moreover, the inverse operator exists, and this is just a different way of stating the uniqueness of the out-of-band extrapolation. Since we are now in the case described in § 3.3, we can use the singular system of the operator L. Its singular functions are in fact related to the *prolate spheroidal wave functions* (PSWF) introduced by Slepian and Pollack [1961]. To see this, we introduce the reduced variables $s = x/X$, $s' = x'/X$ and the parameter

$$c = KX, \tag{4.5}$$

which we define as the *space–bandwidth product*. Then the imaging operator becomes

$$(Lf)(s) = \int_{-1}^{+1} ds' \frac{\sin c(s-s')}{\pi(s-s')} f(s'), \quad -\infty < s < +\infty, \tag{4.6}$$

whereas its adjoint is given by

$$(L^*g)(s) = \int_{-\infty}^{+\infty} ds' \frac{\sin c(s-s')}{\pi(s-s')} g(s'), \quad -1 < s < +1. \tag{4.7}$$

It is now easy to compute the operator L^*L, which is precisely the operator

$$(L^*Lf)(s) = \int_{-1}^{+1} ds' \frac{\sin c(s-s')}{\pi(s-s')} f(s'), \quad -1 < s < +1, \tag{4.8}$$

considered by Slepian and Pollack [1961]. Hence the singular system $\{\sigma_n; u_n, v_n\}$ of the operator L is given by

$$\sigma_n = \sqrt{\lambda_n}; \quad u_n(s) = \psi_n(c,s); \quad v_n(s) = \frac{1}{\sqrt{\lambda_n}} \psi_n(c,s), \tag{4.9}$$

for $n = 0, 1, 2, \ldots$, where the $\psi_n(c,s)$ are the PSWF; i.e., the eigenfunctions of the operator (4.8), and the λ_n are the corresponding eigenvalues, which tend to zero when $n \to \infty$. Notice that the functions $u_n(s)$ are defined on the real line, whereas the functions $v_n(s)$ are defined on the interval $[-1, +1]$.

The solution of the out-of-band extrapolation problem can be estimated by means of the methods described in § 3.3. For example, an approximate and stable solution is given by a truncated singular-system expansion, the number of terms being equal to the number of singular values greater than the inverse of the SNR – see eqs. (3.34) and (3.36). This number is also defined in § 3.3 as the number of degrees of freedom. In order to estimate the NDF as a function of c and of the SNR, we need to know the dependence of the prolate eigenvalues λ_n on the space–bandwidth product c. Let us recall that when c is sufficiently larger than one, the eigenvalues λ_n are approximately equal to 1 for $n < 2c/\pi$ and tend to zero very rapidly for $n > 2c/\pi$. The quantity

$$S = \frac{2c}{\pi} = \frac{2KX}{\pi}, \tag{4.10}$$

called the *Shannon number* by Toraldo di Francia [1969], has a very simple meaning: it is the number of sampling points, spaced by the Nyquist distance $R = \pi/K$, interior to the interval $[-X, +X]$. In communication theory, it is interpreted as the information content of the finite-extent object or "signal" transmitted by a bandlimited system or "channel" with cut-off frequency K. From the behavior of the prolate eigenvalues, we can conclude that when the space–bandwidth product is large, the NDF is approximately equal to the Shannon number S and that, accordingly, no significant out-of-band extrapolation can be achieved.

The previous conclusion, however, holds true only when the Shannon number is large; i.e., when the size $2X$ of the object is large when compared with the sampling distance π/K. When S is not much larger than one, the sharp stepwise behavior of the eigenvalues λ_n is no longer observed and the NDF can be significantly larger than S. To demonstrate this, let us assess the number of singular values $N(E/\varepsilon, c)$ satisfying the truncation condition

$$\sqrt{\lambda_n(c)} \geqslant \frac{\varepsilon}{E}. \tag{4.11}$$

This number can be easily determined from the published tables of the prolate eigenvalues (see, for instance, Frieden [1971]). In table 1, we report the NDF $N(E/\varepsilon, c)$ for various values of the SNR and for different values of c corresponding to small Shannon numbers. Let us make a few comments about the interpretation of table 1. To take an example, in the case $c = 5$, we have $S = 3.18$ and therefore the size of the object is approximately three times the

Table 1

Number of degrees of freedom $N(E/\varepsilon, c)$ of the problem of out-of-band extrapolation, as a function of the signal-to-noise ratio E/ε and for various values of the Shannon number $S = 2c/\pi$

$c=1, S=0.64$		$c=2, S=1.27$		$c=5, S=3.18$		$c=10, S=6.37$	
E/ε	$N(E/\varepsilon, c)$	E/ε	$N(E/\varepsilon, c)$	E/ε	$N(E/\varepsilon, c)$	E/ε	$N(E/\varepsilon, c)$
10	2	10	3	10	5	10	9
10^2	3	10^2	4	10^2	7	10^2	10
10^3	4	10^3	5	10^3	8	10^3	12
10^4	5	10^4	6	10^4	9	10^4	13
10^5	5	10^5	7	10^5	10	10^5	15
10^6	6	10^6	8	10^6	11	10^6	16

sampling distance. Then, if $E/\varepsilon = 10^2$, we see that the NDF is equal to 7. Hence the number of parameters which can be estimated stably is approximately twice the number of sampling points associated with the initial bandwidth. This result implies that an improvement in resolution by a factor of 2 can be achieved; i.e., that it is possible to extrapolate $\widehat{f}(k)$ from $[-K, +K]$ into $[-K', +K']$ with $K' \simeq 2K$. More generally, we notice from table 1 that the NDF behaves roughly as follows:

$$N\left(\frac{E}{\varepsilon}, c\right) \cong S + A + B \log_{10}\left(\frac{E}{\varepsilon}\right), \tag{4.12}$$

where A and B are some constants of the order of 1 (in fact, allowing for a slow variation of A and B with c would help to reproduce more accurately the behavior of the NDF). This result implies that *the increase of the NDF with respect to the Shannon number depends only logarithmically on the signal-to-noise ratio.* A similar result holds true for the gain in bandwidth, if we define the bandwidth of the extrapolated image as follows:

$$K' = \frac{\pi}{2X} N\left(\frac{E}{\varepsilon}, c\right) \cong K + \frac{\pi}{2X} A + \frac{\pi}{2X} B \log_{10}\left(\frac{E}{\varepsilon}\right)$$
$$= K\left[1 + \frac{1}{S} A + \frac{1}{S} B \log_{10}\left(\frac{E}{\varepsilon}\right)\right]. \tag{4.13}$$

The corresponding new Nyquist or Rayleigh distance is given by $R' = \pi/K'$; namely,

$$R' = \frac{\pi}{K'} = \frac{S}{S + A + B \log_{10}(E/\varepsilon)} R. \tag{4.14}$$

The main conclusions which can be derived from the previous analysis are the following:

(i) super-resolution, in the sense of out-of-band extrapolation, is feasible only when the known size of the object is not too large when compared with the resolution limit of the imaging system;

(ii) the amount of achievable super-resolution depends on the space–bandwidth product $c = KX$ and on the signal-to-noise ratio, even if the latter dependence is rather weak (since it is only logarithmic).

To actually perform the out-of-band extrapolation of the object spectrum, one needs a numerical method. As discussed above, one possibility is to use a truncated singular-system expansion, but this method is not very efficient from the computational point of view. It also presents the drawback of producing restorations which may not be satisfactory at the edges of the object domain. Indeed, the PSWF become quite large at the endpoints ± 1, for large values of the index n, and therefore the truncated singular-system expansion cannot reproduce accurately a continuous object vanishing at those endpoints.

A simple iterative method that is easier to implement has been proposed by Gerchberg [1974]. In the case where the object domain is the interval $[-1,+1]$ and the band is the interval $[-c,+c]$, the procedure works as follows. We start from the noisy data function $\widehat{g}(k)$ given by eq. (4.1) with $K = c$, or equivalently from its Fourier transform $g(s)$. The initial object estimate $f_1(s)$ is obtained by truncating $g(s)$ to the object domain $[-1,+1]$. Then the Fourier transform $\widehat{f}_1(k)$ of $f_1(s)$ is computed and the values of this function in the interval $[-c,+c]$ are replaced by the known values of $\widehat{g}(k)$. The inverse Fourier transform of the resulting composite function, truncated to the interval $[-1,+1]$, forms the new object estimate $f_2(s)$, and so on. If we introduce the function

$$\text{rect}(x) = \begin{cases} 1, & |x| \leqslant 1, \\ 0, & |x| > 1, \end{cases} \tag{4.15}$$

and if we recall that $\widehat{g}(k) = 0$ for $|k| > c$, this iterative procedure can be written as follows:

$$\begin{cases} f_1(s) = \text{rect}(s)\, g(s), \\ \widehat{g}_{j+1}(k) = \widehat{g}(k) + [1 - \text{rect}(k/c)]\, \widehat{f}_j(k), & j = 1, 2, \ldots \\ f_{j+1}(s) = \text{rect}(s)\, g_{j+1}(s), \end{cases} \tag{4.16}$$

In the absence of noise, the convergence of this method to the solution of the equation $Lf = g$, with L given by eq. (4.6), has been proved by Papoulis [1975] and by De Santis and Gori [1975]. A generalization of the Gerchberg method to the case of an image produced by an optical instrument with an arbitrary non-negative TF has been proposed by Gori [1975]. This method applies, for instance, to both coherent and incoherent imaging through ideal systems. It was later recognized by Abbiss, De Mol and Dhadwal [1983] that the Gerchberg method – as well as its generalization due to Gori – is a particular form of the method of Landweber [1951] for the solution of first-kind Fredholm integral equations, such as are the imaging equations of the form $Lf = g$ considered here. Indeed, we show in Appendix B that the iterative procedure (4.16) coincides with the following one:

$$\begin{cases} f_0 = 0, \\ f_{j+1} = L^*g + [I - L^*L]f_j, \end{cases} \qquad (4.17)$$

where I denotes the identity operator and L^* and L^*L are, respectively, the operators (4.7) and (4.8). Equation (4.17) defines precisely the Landweber iterative method applied to the integral equation $Lf = g$. In Appendix B, we also derive from eq. (4.17) a result due to De Santis and Gori [1975] and Gori [1975]: the j-th estimate f_j can be obtained by applying a suitable filter to the solution of the integral equation. In fact, the number of iterations plays the role of a regularization parameter and therefore, for any given noisy function to be extrapolated, there exists an optimum number of iterations that depends on the function and on the noise. The application of this iterative method to the case of regularized solutions was investigated by Cesini, Guattari, Lucarini and Palma [1978] and by Abbiss, De Mol and Dhadwal [1983]. The repeated Fourier transforms used by the Gerchberg method are usually implemented numerically by means of the FFT algorithm, which makes the method work faster, even though in some cases, the required number of iterations can be quite large.

The previous analysis of the out-of-band extrapolation problem can be extended in the following two directions:

(i) *The use of other constraints, in addition to that on the domain of the solution.* One of the most frequent is the positivity of the solution, which can be imposed at each step of the Gerchberg procedure by simply setting to zero all negative values. The constraint of positivity was considered by Lent and Tuy [1981], who also proved a convergence result. More general classes of constraints were considered by Youla and Webb [1982] (for a tutorial, see Youla [1987]).

(ii) *The extrapolation of functions of two and three variables.* The analysis of the 1D case based on singular-system expansions can be extended to 2D and 3D problems by means of the generalized spheroidal wave functions introduced by Slepian [1964]. Estimates of the amount of super-resolution have been derived by Bertero and Pike [1982] for functions of two variables, assuming that both the band of the image and the domain of the object are either squares or discs. To our knowledge, no estimates have been obtained for more general shapes and for functions of three variables. On the other hand, the Gerchberg method extends quite naturally to the case of two and three variables, and is then still equivalent to the Landweber iterative method. In the expansion (B.11), one must substitute the PSWF by the generalized prolate spheroidal wave functions, making also the corresponding substitution for the eigenvalues.

An example of the application of the Gerchberg method to the problem of limited-angle tomography, with the additional constraint of positivity, is given in the paper by Lent and Tuy [1981] cited above. In that case, the shape of the available band is an angular sector determined by the directions accessible to measurements. In three dimensions, out-of-band extrapolation can also be applied to inverse scattering problems at fixed energy in the Born approximation, for which measurements only provide access to the so-called Ewald sphere (Wolf and Habashy [1993], Habashy and Wolf [1994]).

Let us still observe that, for didactic purposes, we have in fact split the problem of super-resolving the image provided by an imaging system into two steps:

(i) The first step, mainly considered in § 3.2, consists of the solution of an image deblurring problem; the result is an estimate of the Fourier transform of the object on the effective band of the imaging system;

(ii) The second step consists of the extrapolation of the Fourier transform of the object out of the effective band of the imaging system.

This separation is useful for clarifying the role of the finite domain of the object in the second step. It is important to mention, however, that one can find methods mixing together the two steps. The simplest one is the constrained Landweber method applied directly to the imaging equation. This possibility is mentioned, for instance, by Schafer, Mersereau and Richards [1981], although in their paper, the method is written in the form originally due to van Cittert [1931], and the proof of convergence is not correct. In such a method, it is easy to introduce at each iteration step, the constraint on the domain of the object as well as other constraints such as positivity. Another method, which probably requires further investigation, is the constrained conjugate-gradients method (see Lagendijk, Biemond and Boekee [1988]).

Statistical methods can be used as well for this problem, such as maximum entropy (see, e.g., the paper by Frieden [1975]) and maximum likelihood (see, e.g., Richardson [1972], Lucy [1974], Shepp and Vardi [1982]). It has been shown by Donoho, Johnstone, Hoch and Stern [1992] that the method of maximum entropy has a super-resolving effect in the case of so-called *nearly-black objects*; i.e., of objects which are essentially zero in the vast majority of samples. This result seems to be in agreement with the analysis of super-resolution presented in this section.

§ 5. Confocal Microscopy

The results of the analysis made in the previous section indicate that a considerable resolution enhancement is to be expected if the theory can be put into practice in some optical devices. As we have seen, to get a significant amount of super-resolution, the size of the object must be of the same order as or smaller than the resolution distance of the imaging system. This condition is rarely satisfied in usual imaging systems. One exception we can cite is the problem of resolving a double star that cannot be resolved by the available telescope, in the case where this object is isolated in a dark background.

Another situation where the above requirement is satisfied can be found in scanning microscopy, as noticed by Bertero and Pike [1982], who suggested the possibility of using inversion techniques, based on singular-system expansions, to enhance the resolution of such microscopes. In a scanning device, for each scanning position, only a very small portion of the object is illuminated at a time, so that the resulting space–bandwidth product is very low. Bertero and Pike [1982] treat the case of uniform illumination of the object through some small diaphragm. More general illumination profiles have been considered by Bertero, De Mol, Pike and Walker [1984], resulting in an imaging equation of the type (3.27) for each scanning position. Because of the great practical relevance of this technique, we will focus here on the case of *confocal scanning microscopy* (Sheppard and Choudhury [1977], Brakenhoff, Blom and Barends [1979]; for a survey, see the book by Wilson and Sheppard [1984]).

In a confocal microscope, focused laser illumination allows one to achieve low space–bandwidth products as well as high SNR values, resulting in an enhanced resolving power with respect to ordinary microscopes. Each portion of the specimen is illuminated by means of a sharp laser spot focused by a first lens, the *illumination lens*. The image is formed by means of a second lens, the *imaging* or *collector lens*, whose focal plane also coincides with the plane containing the specimen. This plane is called the *confocal plane*, while the common focus of the two lenses is called the *confocal point*. Finally, the central part of this image is detected by placing a pinhole in front of the detector (e.g., a photomultiplier). A scan through the specimen is then performed by translating it with respect to the confocal system. The situation described above corresponds to the so-called transmission mode of the confocal microscope. In the reflection mode, the same lens is used as both the illumination and the imaging lenses. Coherent imaging can be used, as well as incoherent imaging in the case of fluorescence microscopy. Fluorescence microscopy is used widely because it allows one to also achieve 3D imaging, with a good resolution in the axial direction as well (Wijnaendts-van-Resandt, Marsman, Kaplan, Davoust, Stelzer and Stricker [1985], Brakenhoff, van der Voort, van Spronsen and Nanninga [1986]).

Let us first consider the case of 2D images and planar objects. In such a case, the imaging equations corresponding to the transmission and to the reflection modes have essentially the same structure. For each fixed scanning position, the imaging process can be modelled by an equation similar to eq. (3.27), but with 2D transverse coordinates in the object and image planes; namely, the equation:

$$g(\sigma) = \int d\rho' \, S_2(\sigma - \rho') \, S_1(\rho') f(\rho'), \tag{5.1}$$

which expresses that the object $f(\rho')$ is first multiplied by the illumination profile, which is the PSF $S_1(\rho)$ of the illumination lens, and then convolved with the PSF $S_2(\rho)$ of the imaging lens.

The images formed at the different scanning positions ρ are obtained by translating the object; i.e., they are given by

$$g(\sigma; \rho) = \int d\rho' \, S_2(\sigma - \rho') \, S_1(\rho') f(\rho' + \rho). \tag{5.2}$$

For each scanning position ρ, only one value of the image is recorded by means of the detector, namely its central value on the optical axis (corresponding to $\sigma = 0$) or, more exactly, a central value obtained by integration over the finite pinhole. Then, if $P(\sigma)$ denotes the characteristic function of the pinhole, the output of the detector is proportional to:

$$G(\rho) = \int d\sigma \, P(\sigma) \, g(\sigma; \rho). \tag{5.3}$$

Substituting eq. (5.2) into eq. (5.3), by a change of the integration order and by a change of variable, one finds:

$$G(\rho) = \int d\rho' \, H(\rho - \rho') f(\rho'), \tag{5.4}$$

where

$$H(\rho) = S_1(-\rho) \int d\sigma \, P(\sigma) \, S_2(\sigma + \rho). \tag{5.5}$$

In a typical confocal microscope, no object restoration is performed and the recorded image $G(\rho)$ is taken as the estimate of the object. We see that it is simply the convolution of the object by the PSF $H(\rho)$. This PSF is also bandlimited, but its band is broader than the band of $S_1(\rho)$ or of $S_2(\rho)$. This leads to an improvement in resolution with respect to an ordinary microscope using uniform illumination of the object and only one imaging lens with PSF $S_2(\rho)$. For example, in an ideal 1D coherent-illumination case [when $S_1(x)$ and $S_2(x)$ are both sinc-functions with bandwidth K] and for a very small pinhole [i.e., $P(s) = \delta(s)$], $H(x)$ is a sinc^2-function with bandwidth $2K$. Since $\widehat{H}(k)$ has a triangular shape, the resulting bandwidth of the instrument is approximately twice the bandwidth of each single lens (the effective bandwidth in the presence of noise being slightly smaller since $\widehat{H}(k)$ goes to zero at the edges of the band). Unfortunately, one cannot use a very small pinhole, because one needs sensible values of the signal-to-noise ratio. The effect of the finite size of the pinhole is an additional reduction of the effective bandwidth of the instrument. It follows that, in practice, the effective gain in resolving power provided by confocal imaging is close to a factor of 1.4 (Cox, Sheppard and Wilson [1982], Brakenhoff, Blom and Barends [1979]).

A means to further enhance the resolving power of a confocal microscope has been suggested by Bertero and Pike [1982], and later investigated in a series of subsequent papers (Bertero, De Mol, Pike and Walker [1984], Bertero, Brianzi and Pike [1987], Bertero, De Mol and Pike [1987], Bertero, Boccacci, Defrise, De Mol and Pike [1989], Bertero, Boccacci, Davies and Pike [1991]). The original idea – a more recent modification will be discussed in a moment – was to replace the single on-axis detector of the conventional confocal microscope by an array of detectors in order to measure, at each scanning position, the complete diffraction image. In other words, one should record, for each fixed scanning position ρ, the image $g(\sigma; \rho)$, given by eq. (5.2), and solve this equation in order to estimate $f(\rho)$. We notice that, up to a change of origin, the imaging equation is the same for each scanning position, so that one can use the same linear estimator at every point.

As already mentioned, the integral operator appearing in eq. (5.1) has the same structure as the imaging operator of eq. (3.27), and in fact it is not difficult to prove that it is a compact operator whenever $S_1(\rho)$ and $S_2(\rho)$ are bandlimited PSF (with or without aberrations). In 1D ideal coherent imaging, when both $S_1(x)$ and $S_2(x)$ are sinc-functions, a nice feature happens, namely that the singular system of the imaging operator defined by eq. (5.1) can be determined analytically, as was shown by Gori and Guattari [1985]. The properties of the singular systems of the imaging operators corresponding to various other situations (coherent and incoherent illumination, one-dimensional and bi-dimensional systems) have been investigated numerically in the series of papers by Bertero, Pike and coworkers mentioned above. The investigation of the 1D coherent and incoherent cases was motivated essentially by the fact that, due to the difficulty of the general problem, they provide sufficiently simple inverse problems, whose solution facilitates the understanding of the structure of the singular system corresponding to more general cases.

The method we used in the papers mentioned above for estimating the object $f(\rho' + \rho)$ in eq. (5.2), on the optical axis (i.e., for $\rho' = 0$) and for each scanning position ρ, is a truncated singular-system expansion. The result is expressed by the following estimator

$$\tilde{f}(\rho) = \int d\sigma \, M(\sigma) \, g(\sigma; \rho), \tag{5.6}$$

where [see eq. (3.34)]

$$M(\sigma) = \sum_{n=0}^{N-1} \frac{1}{\sigma_n} v_n(0) \, u_n(\sigma). \tag{5.7}$$

The result of these investigations is that the effective band of this modified confocal microscope essentially coincides with the band of the PSF $H(\rho)$, as given by eq. (5.5) with $P(\sigma) = \delta(\sigma)$, and is therefore broader than the effective band of $II(\rho)$. Indeed, in general, $\tilde{H}(k)$ tends to zero rather rapidly at the boundary of the band and this causes a substantial loss in useful frequencies. In other words, the method described above is a way to utilize efficiently the full theoretical band of the confocal system. It amounts to extending the object spectrum from the effective band of $H(\rho)$ to its full band. Hence, the modified confocal microscope is super-resolving compared to the usual confocal microscope where no data inversion is performed.

Experimental confirmations of the theoretical results discussed above were obtained by Walker [1983], Young, Davies, Pike and Walker [1989] and Young, Davies, Pike, Walker and Bertero [1989], using both coherent and incoherent confocal microscopes, with low numerical aperture. However, the use of multiple detectors presents the disadvantage that the detectors or detector elements must be calibrated. In addition, for the coherent case, signals relating to the complex amplitude are required, thus necessitating the use of interferometry or some other method of phase measurement. However, looking at eq. (5.6), we remark that the only operations required for implementing the inversion method are multiplication of the image $g(\sigma, \rho)$ by the "mask" function $M(\sigma)$, and integration of the result over the image plane. These operations can in fact be implemented by means of optical processors, without needing an array of detectors. The use of such optical processors has been proposed by Walker, Pike, Davies, Young, Brakenhoff and Bertero [1993] for both the coherent and the incoherent cases. Properties of the optical masks $M(\sigma)$ corresponding to various situations were investigated theoretically by Bertero, Boccacci, Davies, Malfanti, Pike and Walker [1992]. Preliminary experimental results were obtained by Grochmalicki, Pike, Walker, Bertero, Boccacci and Davies [1993], again in good agreement with the theory.

A more general linear estimator than eq. (5.6) has been considered by Defrise and De Mol [1992]. It is given by

$$\tilde{f}(\rho) = \int d\sigma \int d\rho' \, M(\sigma; \rho - \rho') \, g(\sigma; \rho'), \tag{5.8}$$

and is assumed to be shift-invariant with respect to the scanning position, but not with respect to the detector position, since the integration on σ may be limited to a finite domain. With the help of such an estimator, one can derive some interesting results for the ideal 1D coherent case where $S_1(x) = S_2(x) = (K/\pi) \operatorname{sinc}(Kx/\pi)$. For instance, in the absence of noise, one can devise different choices of the restoration function leading to the following overall impulse response:

$$T(x, x') = \frac{\sin[2K(x - x')]}{\pi(x - x')}, \tag{5.9}$$

or, equivalently, to the overall transfer function $\widehat{T}(k) = 1$ on the band $[-2K, +2K]$ and $\widehat{T}(k) = 0$ outside. This corresponds to a resolving power effectively enhanced by a factor 2 with respect to an ordinary microscope with PSF $S_1(x)$. A first possibility for such restoration relies on the 1D exact inversion formula for the imaging equation (5.1) derived by Bertero, De Mol and Pike [1987]:

$$M(s; x) = \frac{4\pi}{K} \cos(Ks) \, \delta(x), \tag{5.10}$$

where s is the 1D coordinate corresponding to σ. In this case, we obtain an estimator having the same structure as the estimator (5.6). However, in addition to eq. (5.10), and because of the redundancy of the data (5.2), a whole family of reconstruction kernels can be constructed, all yielding eq. (5.9) as the overall impulse response (Defrise and De Mol [1992]), including the following one:

$$M(s; x) = \frac{\pi}{K} \delta(x - \tfrac{1}{2}s), \tag{5.11}$$

which was first proposed by Sheppard [1988].

Let us now consider the problem of producing 3D images of thick specimens. The usual case is that of a fluorescent object characterized by the distribution function of the fluorescent material, denoted by $f(r)$, where r is the 3D vector coordinate. We also use the notation $r = (\rho, z)$, where z denotes the depth coordinate along the optical axis and ρ is the transverse coordinate. In general, for 3D imaging, the confocal microscope is operated in the transmission mode and circularly polarized excitation light is used. Then, if we denote by $W(r) = W(\rho, z)$ the time-averaged electrical energy distribution in the focal region of the lens, the imaging equation (5.1) must be replaced by the following one (Bertero, Boccacci, Brakenhoff, Malfanti and van der Voort [1990]):

$$g(\sigma) = \int d\rho' \, dz' \, W(\sigma - \rho', z') \, W(\rho', z') f(\rho', z'). \tag{5.12}$$

For simplicity, we have neglected the effect of the difference between the primary and the emission wavelengths. A more refined imaging model is discussed by van der Voort and Brakenhoff [1990]. The function $g(\sigma)$ is proportional to the intensity distribution in the image plane. For lenses with a high numerical aperture, the Debye approximation is not adequate and one must use the more accurate approximation of Richards and Wolf [1959] for the computation of $W(\rho, z)$.

A 3D scanning is performed and the images at different scanning positions $r = (\rho, z)$ are given by

$$g(\sigma; \rho, z) = \int d\rho' \, dz' \, W(\sigma - \rho', z') \, W(\rho', z') f(\rho' + \rho, z' + z). \tag{5.13}$$

Therefore, if one takes into account the effect of the pinhole as in the 2D case, one obtains the following expression for the recorded 3D image:

$$G(r) = \int dr' \, H(r - r') f(r'), \tag{5.14}$$

where

$$H(r) = W(-r) \int d\sigma \, P(\sigma) \, W(\sigma + \rho, z), \tag{5.15}$$

which is an extension of eq. (5.5). The function $H(r)$ is bandlimited, but its band is not a sphere. If the optical system has a circular symmetry around the optical axis z, the band of $H(r)$ has a circular symmetry around the k_z-axis, but normally its extension in the direction of the k_z-axis is roughly one third of its extension in the transverse directions and, consequently, axial resolution is poorer than lateral resolution (see van der Voort and Brakenhoff [1990]).

Methods of image deconvolution similar to those discussed in § 3.2 have been applied to some 3D images produced by a confocal microscope by Bertero, Boccacci, Brakenhoff, Malfanti and van der Voort [1990], demonstrating the possibility of obtaining an improvement of the image quality, especially in the axial direction. In the same paper, one can also find an estimation of the resolution improvement obtained by detecting, for each scanning position ρ, the complete image $g(\sigma;\rho)$ and by solving the integral equation (5.13) to estimate the object $f(r)$. It was shown that this method leads to an improvement of the effective bandwidth in all directions.

§ 6. Inverse Diffraction and Near-Field Imaging

In this section, we analyze another possibility for improving the resolution limits, which consists in recording – whenever possible – near-field data. The conclusions derived from the previous examples, which deal only with far-field data, no longer hold true in such a case. As we shall see, near-field imaging techniques allow considerable enhancement of the resolving power compared to far-field imaging with the same wavelength. Strictly speaking, this is not true super-resolution in the sense of out-of-band extrapolation as defined above. Indeed, the near-field data contain information – conveyed by the so-called evanescent waves – about spatial frequencies of the object which are no longer present in the far-field region.

A good theoretical laboratory for investigating the resolution enhancement arising from the effect of evanescent waves is provided by the so-called *inverse diffraction problem*, which consists in back-propagating (towards the sources) a scalar field propagating in free space according to the Helmholtz equation. The simplest geometry to formulate this problem is the case of *inverse diffraction from plane to plane*, where the field propagates in the half-space $z \geq 0$ and where one must recover the field on the boundary plane $z = 0$ from its values on the plane $z = a > 0$. This problem was first considered by Sherman [1967] and Shewell and Wolf [1968] (see also the discussion in the book by Nieto-Vesperinas [1991]).

Let us consider a scalar monochromatic field propagating in the half-space $z \geq 0$, the complex field amplitude $u(r) = u(x,y,z)$ being a solution of the Helmholtz equation

$$\Delta u + K^2 u = 0, \qquad (6.1)$$

satisfying the Sommerfeld radiation condition at infinity

$$\lim_{r\to\infty}[r(\frac{\partial u}{\partial r}-iKu)]=0, \quad |\theta|<\pi/2, \tag{6.2}$$

where $r = |r|$, θ is the polar angle and $K = 2\pi/\lambda$ is the wavenumber.

If the field amplitude is known on the source plane $z = 0$ and given by $u(x,y,0) = f(x,y)$, then the field $u(x,y,a) = g_a(x,y)$ in the plane $z = a$, $a > 0$, is uniquely determined and given by

$$g_a = L_a f, \tag{6.3}$$

where the imaging operator L_a acts by convolving the field in the source plane with the forward propagator $S_a^+(\rho)$:

$$(L_a f)(\rho) = \int d\rho'\, S_a^+(\rho-\rho') f(\rho'). \tag{6.4}$$

We have denoted by $\rho = (x,y)$ and $\rho' = (x',y')$ the transverse vector coordinates in the image and source planes, respectively. The propagator $S_a^+(\rho)$ is given by

$$S_a^+(\rho) = \left(\frac{1}{2\pi}\right)^2 \int d\kappa\, e^{i\kappa\cdot\rho}\, e^{ia\, m(\kappa)}, \tag{6.5}$$

with $\kappa = (k_x, k_y)$ and

$$m(\kappa) = \begin{cases} \sqrt{K^2 - |\kappa|^2} & \text{for } |\kappa| \leqslant K \\ i\sqrt{|\kappa|^2 - K^2} & \text{for } |\kappa| > K. \end{cases} \tag{6.6}$$

The real part of $m(\kappa)$ corresponds to the so-called *homogeneous waves*, propagating without attenuation, whereas the imaginary part of $m(\kappa)$ corresponds to the *inhomogeneous* or *evanescent waves*, whose amplitude decreases exponentially with the distance a between the source and data planes.

In Fourier space, the convolution (6.3) becomes simply

$$\widehat{g}_a(\kappa) = e^{ia\, m(\kappa)}\, \widehat{f}(\kappa). \tag{6.7}$$

Hence we see that backward propagation is described by

$$\widehat{f}(\kappa) = e^{-ia\, m(\kappa)}\, \widehat{g}_a(\kappa). \tag{6.8}$$

Because of the exponential damping, the restoration of the part of the object spectrum corresponding to evanescent waves is unstable. When the data plane goes to the far-field region, the evanescent waves are no longer present in the data. Only the spatial frequencies $|\kappa| \leqslant K$ propagate to the far-field and then the corresponding propagator $S_a^+(\rho)$ is just the Fourier transform of the disc of radius K, except for a phase factor. Hence, the inverse diffraction problem from far-field data is equivalent to inverse imaging through a circular pupil, in the presence of aberrations because of the phase factor $e^{ia\,m(\kappa)}$. The corresponding resolution limit is given by the Rayleigh distance $R = 1.22\,(\lambda/2)$.

In the absence of further information about the unknown object, which is the field in the source plane, all we can recover from the far-field data g_a with $a \gg \lambda$ is the following bandlimited approximation of $f(\rho)$:

$$f_B(\rho) = \left(\frac{1}{2\pi}\right)^2 \int_{|\kappa|\leqslant K} d\kappa\; e^{i\kappa\cdot\rho}\, e^{-ia\,m(\kappa)}\, \widehat{g}_a(\kappa). \tag{6.9}$$

This is equivalent to the inversion formula derived by Shewell and Wolf [1968]. On the band $|\kappa| \leqslant K$, the inverse of the imaging operator is simply given by its adjoint L_a^*, which is the convolution operator whose kernel is the backward propagator $S_a^-(\rho) = (S_a^+)^*(-\rho)$. In such a case no super-resolution is achieved in the inversion process.

For any value of a, the possibility of going beyond the Rayleigh limit through the use of regularization theory has been explored by Bertero and De Mol [1981a], using Fourier techniques and constraints on the spectrum of the object. The case of finite-extent objects, vanishing outside an a priori known domain, has been considered by Bertero, De Mol, Gori and Ronchi [1983]. As follows from eq. (6.9), in the case of far-field data, the problem of inverse diffraction is exactly equivalent to the out-of-band extrapolation discussed in § 4, and the increase in resolution with respect to the bandlimited approximation given by eq. (6.9) can be assessed with the methods already described. Let us recall the main conclusion that the Rayleigh limit can be significantly improved only for objects of small spatial extent, corresponding to a small space–bandwidth product. Since the cut-off frequency is given here by $K = 2\pi/\lambda$, this means that the linear spatial dimensions of the object must be of the order of or smaller than the wavelength λ of the field. Such objects are sometimes referred to as *subwavelength sources*.

In the case in which one has access to near-field data, the use of the information conveyed by the evanescent waves before they become damped allows one to increase the effective band available for the restoration. According to formula (3.26), the effective band is the set of spatial frequencies such that

$$|\widehat{S_a^+}(\kappa)| = \exp\left(-a\sqrt{|\kappa|^2 - K^2}\right) > \frac{\varepsilon}{E}, \tag{6.10}$$

Table 2
Resolution ratio R/R_{eff} for inverse diffraction from near-field data, as a function of the signal-to-noise ratio E/ε and for various values of the distance a between the source and data planes

$a = \lambda/20$		$a = \lambda/10$		$a = \lambda/2$		$a = \lambda$	
E/ε	R/R_{eff}	E/ε	R/R_{eff}	E/ε	R/R_{eff}	E/ε	R/R_{eff}
10	7.40	10	3.80	10	1.24	10	1.06
10^2	14.69	10^2	7.40	10^2	1.77	10^2	1.24
10^3	22.01	10^3	11.04	10^3	2.42	10^3	1.49
10^4	29.33	10^4	14.69	10^4	3.10	10^4	1.77
10^5	36.66	10^5	18.35	10^5	3.80	10^5	2.09
10^6	43.99	10^6	22.01	10^6	4.51	10^6	2.42

where E/ε is the signal-to-noise ratio and $|\widehat{S_a^+}(\kappa)|$ is the modulus of the Fourier transform of the forward propagator. The corresponding effective cut-off frequency K_{eff} is defined by

$$a\sqrt{K_{\text{eff}}^2 - K^2} = \ln\left(\frac{E}{\varepsilon}\right), \tag{6.11}$$

or else

$$K_{\text{eff}} = K\left[1 + \frac{1}{(Ka)^2}\ln^2\left(\frac{E}{\varepsilon}\right)\right]^{1/2}. \tag{6.12}$$

In table 2, we report some numerical values for the resolution improvement; i.e., for the ratio $R/R_{\text{eff}} = K_{\text{eff}}/K$, where R is the far-field resolution distance. Notice that in this problem we have again a logarithmic dependence on the SNR.

A further gain in resolution is to be expected from the knowledge of the domain of the object. As in the case of far-field data, a significant additional gain is obtained only for subwavelength sources. The resolution improvement can then be estimated through a numerical computation of the singular values of the imaging operator and of the corresponding number of degrees of freedom. Some numerical results were reported by Bertero, De Mol, Gori and Ronchi [1983] for the one-dimensional problem of a field amplitude invariant with respect to one of the lateral variables and with domain in a slit of width $2X$ in the boundary plane.

The conclusions of the previous analysis provide a theoretical model for understanding why and to what extent near-field imaging devices allow one to supersede the classical far-field resolution limit of half a wavelength. Whereas the idea of a super-resolving scanning near-field microscope was already present in a theoretical paper by Synge [1928], the first experimental realization of such a device was achieved by Ash and Nicholls [1972] in the microwave range. Nowadays, near-field imaging techniques are developing rapidly and are implemented in various ways. Let us mention, for example, near-field acoustic holography (NAH – see Williams and Maynard [1980]) and scanning near-field optical microscopy (SNOM – see, e.g., Betzig and Trautman [1992], Pohl and Courjon [1993], and the references therein). However, in this latter case, it is not a simple matter to derive a good theoretical model for the imaging process and to write a closed-form imaging equation like eq. (6.3). As an example of recent attempts in that direction, let us cite a paper by Bozhevolnyi, Berntsen and Bozhevolnaya [1994].

The problem of inverse diffraction can be formulated in other geometries, corresponding to backward propagation from one closed surface to another. The corresponding uniqueness problem for the solution is discussed extensively by Hoenders [1978]. For simple surfaces such as spheres and cylinders, expansions in terms of Bessel or Hankel functions can be written for the solution. The corresponding imaging operator is then compact, so that the regularization techniques described in § 3.3 can be used (see the paper by Bertero, De Mol and Viano [1980] for a discussion of the case of inverse diffraction from cylinder to cylinder).

The results derived in the scalar case can also be generalized to the full vector case, described by the Maxwell equations (see Hoenders [1978]). Such a treatment is certainly required for near-field imaging, where polarization effects cannot be neglected. Similar conclusions about resolution limits should also hold true in the vector case, in analogy with those derived above from the simpler model of scalar propagation.

Acknowledgements

The authors would like to thank Franco Gori and Michel Defrise for a critical reading of the manuscript. They are particularly grateful to Prof. Gori for his suggestions and his invaluable help in clarifying some delicate points.

Christine De Mol is 'Maître de recherches' with the Belgian National Fund for Scientific Research.

Appendix A

Let $S(r)$ be the PSF of an imaging system. It is a function of two variables for a system providing 2D images and a function of three variables for a system providing 3D images (for example, a confocal microscope).

We define the *PSF in the direction* $\boldsymbol{\theta}$ as the function of one variable given by

$$S_\theta(s) = S(s\boldsymbol{\theta}), \tag{A.1}$$

where $\boldsymbol{\theta}$ is a vector of unit length. The width of this function is related to the resolution achievable in the direction $\boldsymbol{\theta}$.

In this Appendix, we derive the relationship between the TF associated with $S_\theta(s)$; i.e., $\widehat{S}_\theta(k)$, and the TF associated with $S(r)$; i.e., $\widehat{S}(\boldsymbol{k})$. To this purpose we must recall some definitions related to the Radon transform, which is basic in computerized tomography – see, e.g., the book by Natterer [1986].

We define the *projection of* $\widehat{S}(\boldsymbol{k})$ *in the direction* $\boldsymbol{\theta}$ as the function of one variable defined by

$$(R_\theta \widehat{S})(k) = \int_{\boldsymbol{\xi}\cdot\boldsymbol{\theta}=0} d\boldsymbol{\xi}\ \widehat{S}(k\boldsymbol{\theta}+\boldsymbol{\xi}). \tag{A.2}$$

In the case of a function of two variables, $(R_\theta \widehat{S})(k)$ is the integral of $\widehat{S}(\boldsymbol{k})$ over the straight line perpendicular to $\boldsymbol{\theta}$ with signed distance k from the origin. In the case of a function of three variables, $(R_\theta \widehat{S})(k)$ is the integral of $\widehat{S}(\boldsymbol{k})$ over the plane perpendicular to $\boldsymbol{\theta}$ with signed distance k from the origin.

A basic theorem of the theory of Radon transform is the so-called *projection theorem* or *Fourier slice theorem* (see Natterer [1986]) which provides the needed relationship between the transfer functions. For completeness, we reproduce here the proof, which is quite simple.

If we take the inverse Fourier transform of $R_\theta \widehat{S}$, we get

$$\frac{1}{2\pi} \int_{-\infty}^{+\infty} dk\ (R_\theta \widehat{S})(k)\ e^{isk} = \frac{1}{2\pi} \int_{-\infty}^{+\infty} dk\ e^{isk} \int_{\boldsymbol{\xi}\cdot\boldsymbol{\theta}=0} d\boldsymbol{\xi}\ \widehat{S}(k\boldsymbol{\theta}+\boldsymbol{\xi}). \tag{A.3}$$

Therefore if we introduce the variable $\boldsymbol{k} = k\boldsymbol{\theta} + \boldsymbol{\xi}$, we have

$$d\boldsymbol{k} = dk\ d\boldsymbol{\xi}\ ; \qquad sk = s\boldsymbol{\theta} \cdot \boldsymbol{k}, \tag{A.4}$$

so that

$$\frac{1}{2\pi} \int_{-\infty}^{+\infty} dk\ (R_\theta \widehat{S})(k)\ e^{isk} = \frac{1}{2\pi} \int d\boldsymbol{k}\ \widehat{S}(\boldsymbol{k})\ e^{is\boldsymbol{\theta}\cdot\boldsymbol{k}} = (2\pi)^{n-1} S(s\boldsymbol{\theta}), \tag{A.5}$$

where n is the number of variables. We conclude that *the inverse Fourier transform of the projection of* $\widehat{S}(\boldsymbol{k})$ *in the direction* $\boldsymbol{\theta}$ *is precisely the PSF in* the direction θ multiplied by $(2\pi)^{n-1}$. We can also state the result in a direct form: *the Fourier transform of $S_\theta(s)$ is the projection of $\widehat{S}(k)$ in the direction θ multiplied by $(2\pi)^{1-n}$.*

In this way one can associate with each direction θ a TF and a band, as well as a bandwidth or an effective bandwidth.

Appendix B

In this Appendix we prove that the Gerchberg method [eq. (4.16)] coincides with the Landweber method [eq. (4.17)]. To this purpose, it is convenient to consider the functions f_j in eq. (4.17) as defined everywhere on $(-\infty, +\infty)$, by putting $f_j(s) = 0$ outside $[-1, +1]$. Then the operator L^* and L^*L, given respectively by eqs. (4.7) and (4.8), can be written as follows:

$$(L^*g)(s) = \text{rect}(s) \int_{-\infty}^{+\infty} ds' \, \frac{\sin c(s-s')}{\pi(s-s')} g(s'), \tag{B.1}$$

and

$$(L^*Lf_j)(s) = \text{rect}(s) \int_{-\infty}^{+\infty} ds' \, \frac{\sin c(s-s')}{\pi(s-s')} f_j(s'), \tag{B.2}$$

where $\text{rect}(s)$ is the function defined by eq. (4.15).

Now, if the Fourier transform of $g(s)$ is zero outside $[-c, +c]$, we have:

$$g(s) = \int_{-\infty}^{+\infty} ds' \, \frac{\sin c(s-s')}{\pi(s-s')} g(s'), \tag{B.3}$$

and therefore, from eq. (B.1), we get for such a bandlimited function

$$(L^*g)(s) = \text{rect}(s) \, g(s). \tag{B.4}$$

Noticing that from eq. (4.17) we have $f_1 = L^*g$, we conclude from eq. (B.4) that the initial estimate of the procedure (4.17) coincides with the initial estimate of the procedure (4.16). Moreover, since $f_j(s) = \text{rect}(s) f_j(s)$, from eqs. (4.17), (B.4) and (B.2), it results that $f_{j+1}(s)$ can be written as follows:

$$f_{j+1}(s) = \text{rect}(s) \, g_{j+1}(s), \tag{B.5}$$

where

$$g_{j+1}(s) = g(s) + f_j(s) - \int_{-\infty}^{+\infty} ds' \, \frac{\sin c(s-s')}{\pi(s-s')} f_j(s'). \tag{B.6}$$

By taking the Fourier transform of both sides of eq. (B.6) we obtain the second of eqs. (4.16).

In order to obtain the expression of f_j in terms of the singular values and singular functions of the operator L, we use eq. (4.17) and the following representations of the operators L^* and L^*L, which can be derived from eqs. (3.29) and (3.30):

$$L^*g = \sum_{n=0}^{\infty} \sigma_n \, (g, u_n)_\infty \, v_n, \tag{B.7}$$

where $(g, u_n)_\infty$ is the scalar product in $L^2(-\infty, +\infty)$, and

$$L^*Lf_j = \sum_{n=0}^{\infty} \sigma_n^2 \, (f_j, v_n)_1 \, v_n, \tag{B.8}$$

where $(f_j, v_n)_1$ is the scalar product in $L^2(-1, +1)$. The singular values σ_n and singular functions u_n, v_n are given in eq. (4.9). Now, from eq. (4.17) we obtain, for any n,

$$(f_{j+1}, v_n)_1 = \sigma_n \, (g, u_n)_\infty + (1 - \sigma_n^2) \, (f_j, v_n)_1. \tag{B.9}$$

It is easy to show by induction that

$$(f_j, v_n)_1 = \sigma_n \left\{ 1 + (1 - \sigma_n^2) + (1 - \sigma_n^2)^2 + \cdots + (1 - \sigma_n^2)^{j-1} \right\} (g, u_n)_\infty$$

$$= \left[1 - (1 - \sigma_n^2)^j \right] \frac{(g, u_n)_\infty}{\sigma_n}. \tag{B.10}$$

Therefore one obtains for the jth iterate the following filtered expansion on the PSWF

$$f_j = \sum_{n=0}^{\infty} W_{n,j} \, \frac{1}{\sigma_n} \, (g, u_n)_\infty \, v_n, \tag{B.11}$$

the filter being given by

$$W_{n,j} = 1 - (1 - \sigma_n^2)^j. \tag{B.12}$$

Since σ_n^2 is always less than one (even if it can be very close to one), we have $0 < W_{n,j} < 1$. In fact, when σ_n^2 is close to one, then $W_{n,j} \simeq 1$. On the other hand, when σ_n^2 is much smaller than 1, then $W_{n,j} \simeq j\sigma_n^2$ and therefore $W_{n,j}$ is also very small if $j\sigma_n^2 \ll 1$. Comparing this with eq. (3.33), we see that the inverse of the number of iterations j plays the same role here as the regularization parameter α in the Tikhonov method.

References

Abbe, E., 1873, Archiv. Microsk. Anat. Entwicklungsmech. **9**, 413.
Abbiss, J.B., C. De Mol and H.S. Dhadwal, 1983, Opt. Acta **30**, 107.
Andrews, H.C., and B.R. Hunt, 1977, Digital Image Restoration (Prentice Hall, Englewood Cliffs, NJ).
Ash, E.A., and G. Nicholls, 1972, Nature **237**, 510.
Bell, D.A., 1962, Information Theory and its Engineering Applications (Pitman, New York).
Bertero, M., 1989, Linear Inverse and Ill-Posed Problems, in: Advances in Electronics and Electron Physics, Vol. 75, ed. P.W. Hawkes (Academic Press, New York) pp. 1–120.
Bertero, M., P. Boccacci, G.J. Brakenhoff, F. Malfanti and H.T.M. van der Voort, 1990, J. Microsc. **157**, 3.
Bertero, M., P. Boccacci, R.E. Davies, F. Malfanti, E.R. Pike and J.G. Walker, 1992, Inverse Probl. **8**, 1.
Bertero, M., P. Boccacci, R.E. Davies and E.R. Pike, 1991, Inverse Probl. **7**, 655.
Bertero, M., P. Boccacci, M. Defrise, C. De Mol and E.R. Pike, 1989, Inverse Probl. **5**, 441.
Bertero, M., P. Boccacci and F. Maggio, 1995, Int. J. Im. Systems Tech. **6**, 376.
Bertero, M., P. Brianzi and E.R. Pike, 1987, Inverse Probl. **3**, 195.
Bertero, M., and C. De Mol, 1981a, IEEE Trans. Antennas Propag. **AP-29**, 368.
Bertero, M., and C. De Mol, 1981b, Atti Fond. Giorgio Ronchi **36**, 619.
Bertero, M., C. De Mol, F. Gori and L. Ronchi, 1983, Opt. Acta **30**, 1051.
Bertero, M., C. De Mol and E.R. Pike, 1985, Inverse Probl. **1**, 301.
Bertero, M., C. De Mol and E.R. Pike, 1987, J. Opt. Soc. Am. A **4**, 1748.
Bertero, M., C. De Mol and E.R. Pike, 1988, Inverse Probl. **4**, 573.
Bertero, M., C. De Mol, E.R. Pike and J.G. Walker, 1984, Opt. Acta **31**, 923.
Bertero, M., C. De Mol and G.A. Viano, 1979, J. Math. Phys. **20**, 509.
Bertero, M., C. De Mol and G.A. Viano, 1980, The stability of inverse problems, in: Inverse Scattering Problems in Optics, ed. H.P. Baltes (Springer, Berlin) ch. V, pp. 161–214.
Bertero, M., and E.R. Pike, 1982, Opt. Acta **29**, 727, 1599.
Betzig, E., and J.K. Trautman, 1992, Science **257**, 189.
Biemond, J., R.L. Lagendijk and R.M. Mersereau, 1990, Proc. IEEE **78**, 856.
Born, M., and E. Wolf, 1980, Principles of Optics, 6th Ed. (Pergamon Press, Oxford).
Bozhevolnyi, S., S. Berntsen and E. Bozhevolnaya, 1994, J. Opt. Soc. Am. A **11**, 609.
Brakenhoff, G.J., P. Blom and P. Barends, 1979, J. Microsc. **117**, 219.
Brakenhoff, G.J., H.T.M. van der Voort, E.A. van Spronsen and N. Nanninga, 1986, Ann. N.Y. Acad. Sc. **483**, 405.
Cesini, G., G. Guattari, G. Lucarini and C. Palma, 1978, Opt. Acta **25**, 501.
Cox, I.J., C.J.R. Sheppard and T. Wilson, 1982, Optik **60**, 391.
Davies, A.R., 1992, Optimality in regularization, in: Inverse Problems in Scattering and Imaging, eds M. Bertero and E.R. Pike (Adam Hilger, Bristol) pp. 393–410.
Defrise, M., and C. De Mol, 1992, Inverse Probl. **8**, 175.
De Santis, P., and F. Gori, 1975, Opt. Acta **22**, 691.

De Santis, P., F. Gori, G. Guattari and C. Palma, 1986, Opt. Commun. **60**, 13.
De Santis, P., and C. Palma, 1976, Opt. Acta **23**, 743.
Donoho, D.L., I.M. Johnstone, J.C. Hoch and A.S. Stern, 1992, J. R. Statist. Soc. B **54**, 41.
Franklin, J.N., 1970, J. Math. Anal. Applic. **31**, 682.
Frieden, B.R., 1971, Evaluation, design and extrapolation for optical signals, based on the use of prolate functions, in: Progress in Optics, Vol. IX, ed. E. Wolf (North-Holland, Amsterdam) ch. VIII.
Frieden, B.R., 1975, Image enhancement and restoration, in: Picture Processing and Digital Filtering, ed. T.S. Huang (Springer, Berlin) pp. 177–248.
Gabor, D., 1961, Light and information, in: Progress in Optics, Vol. I, ed. E. Wolf (North-Holland, Amsterdam) ch. IV.
Gerchberg, R.W., 1974, Opt. Acta **21**, 709.
Goodman, J.W., 1968, Introduction to Fourier Optics (McGraw-Hill, New York).
Gori, F., 1975, in: Digest of Int. Optical Computing Conf., Washington, DC, April 23–25, 1975 (IEEE Catalog No. 75 CH0941-5C) pp. 137–141.
Gori, F., and G. Guattari, 1985, Inverse Probl. **1**, 67.
Gori, F., S. Paolucci and L. Ronchi, 1975, J. Opt. Soc. Am. **65**, 495.
Grochmalicki, J., E.R. Pike, J.G. Walker, M. Bertero, P. Boccacci and R.E. Davies, 1993, J. Opt. Soc. Am. A **10**, 1074.
Groetsch, C.W., 1984, The Theory of Tikhonov Regularization for Fredholm Equations of the First Kind (Pitman, Boston, MA).
Habashy, T., and E. Wolf, 1994, J. Mod. Optics **41**, 1679.
Harris, J.L., 1964, J. Opt. Soc. Am. **54**, 931.
Hoenders, B.J., 1978, The uniqueness of inverse problems, in: Inverse Source Problems in Optics, ed. H.P. Baltes (Springer, Berlin) pp. 41–82.
Jerri, A.J., 1977, Proc. IEEE **65**, 1565.
Lagendijk, R.L., J. Biemond and D.E. Boekee, 1988, IEEE Trans. Acoust. Speech Signal Process. **ASSP-36**, 1874.
Landweber, L., 1951, Am. J. Math. **73**, 615.
Lent, A., and H. Tuy, 1981, J. Math. Anal. Appl. **83**, 554.
Lucy, L., 1974, Astron. J. **79**, 745.
McCutchen, C.W., 1967, J. Opt. Soc. Am. **57**, 1190.
Meinel, E.S., 1986, J. Opt. Soc. Am. A **3**, 787.
Natterer, F., 1986, The Mathematics of Computerized Tomography (Teubner, Stuttgart).
Nieto-Vesperinas, M., 1991, Scattering and Diffraction in Physical Optics (John Wiley, New York).
Paley, R.E.A.C., and N. Wiener, 1934, Fourier Transforms in the Complex Domain (American Mathematical Society, Providence, RI).
Papoulis, A., 1968, Systems and Transforms with Applications in Optics (McGraw-Hill, New York).
Papoulis, A., 1975, IEEE Trans. Circuits Syst. **CAS-22**, 735.
Petersen, D.P., and D. Middleton, 1962, Inf. Control **5**, 279.
Pohl, D.W., and D. Courjon, eds, 1993, Near-Field Optics (Kluwer, Dordrecht).
Rayleigh, J.W.S., 1879, Phil. Mag. **8**, 261.
Richards, B., and E. Wolf, 1959, Proc. R. Soc. A **253**, 358.
Richardson, W.H., 1972, J. Opt. Soc. Am. **62**, 55.
Rushforth, C.K., and R.W. Harris, 1968, J. Opt. Soc. Am. **58**, 539.
Schafer, R.W., R.M. Mersereau and M.A. Richards, 1981, Proc. IEEE **69**, 432.
Shannon, C.E., 1949, Proc. IRE **37**, 10.
Shepp, L.A., and Y. Vardi, 1982, IEEE Trans. Med. Imaging **MI-1**, 113.
Sheppard, C.J.R., 1988, Optik **80**, 53.

Sheppard, C.J.R., and A. Choudhury, 1977, Opt. Acta **24**, 1051.
Sherman, G.C., 1967, J. Opt. Soc. Am. **57**, 1490.
Shewell, J.R., and E. Wolf, 1968, J. Opt. Soc. Am. **58**, 1596.
Slepian, D., 1964, Bell Syst. Tech. J. **43**, 3009.
Slepian, D., and H.O. Pollack, 1961, Bell Syst. Tech. J. **40**, 43.
Strand, O.N., and E.R. Westwater, 1968, SIAM J. Numer. Anal. **5**, 287.
Synge, E.H., 1928, Phil. Mag. **6**, 356.
Tikhonov, A.N., and V.Y. Arsenin, 1977, Solutions of Ill-posed Problems (Winston/Wiley, Washington, DC).
Toraldo di Francia, G., 1952, Nuovo Cimento Suppl. **9**, 426.
Toraldo di Francia, G., 1955, J. Opt. Soc. Am. **45**, 497.
Toraldo di Francia, G., 1969, J. Opt. Soc. Am. **59**, 799.
Turchin, V.F., V.P. Kozlov and M.S. Malkevich, 1971, Sov. Phys. Usp. **13**, 681.
van Cittert, P.H., 1931, Z. Physik **69**, 298.
van der Voort, H.T.M., and G.J. Brakenhoff, 1990, J. Microsc. **158**, 43.
Viano, G.A., 1976, J. Math. Phys. **17**, 1160.
Walker, J.G., 1983, Opt. Acta **30**, 1197.
Walker, J.G., E.R. Pike, R.E. Davies, M.R. Young, G.J. Brakenhoff and M. Bertero, 1993, J. Opt. Soc. Am. A **10**, 59.
Whittaker, E.T., 1915, Proc. R. Soc. Edinburgh A **35**, 181.
Wijnaendts-van-Resandt, R.W., H.J.B. Marsman, R. Kaplan, J. Davoust, E.H.K. Stelzer and R. Stricker, 1985, J. Microsc. **138**, 29.
Williams, E.G., and J.D. Maynard, 1980, Phys. Rev. Lett. **45**, 554.
Wilson, T., and C.J.R. Sheppard, 1984, Theory and Practice of Scanning Optical Microscopy (Academic Press, London).
Wolf, E., and T. Habashy, 1993, J. Mod. Opt. **40**, 785.
Wolter, H., 1961, On basic analogies and principal differences between optical and electronic information, in: Progress in Optics, Vol. I, ed. E. Wolf (North-Holland, Amsterdam) ch. V.
Youla, D.C., 1987, Mathematical theory of image restoration by the method of convex projections, in: Image Recovery, Theory and Applications, ed. H. Stark (Academic Press, New York) pp. 29–77.
Youla, D.C., and H. Webb, 1982, IEEE Trans. Med. Imaging **MI-1**, 81.
Young, M.R., R.E. Davies, E.R. Pike and J.G. Walker, 1989, Digest of Conf. on Signal Recovery and Synthesis 3 (Optical Society of America, Washington, DC) paper WD4.
Young, M.R., R.E. Davies, E.R. Pike, J.G. Walker and M. Bertero, 1989, Europhys. Lett. **9**, 773.

IV

RADIATIVE TRANSFER: NEW ASPECTS OF THE OLD THEORY

BY

Yu.A. Kravtsov

*Space Research Institute, Russian Academy of Sciences,
Profsoyuznaya Street 84/32, Moscow 117810, Russian Federation*

AND

L.A. Apresyan

*Radio Engineering Institute, Russian Academy of Sciences,
8th March Street, Moscow 125083, Russian Federation*

CONTENTS

	PAGE
§ 1. INTRODUCTION	181
§ 2. RADIATIVE TRANSFER IN FREE SPACE	188
§ 3. PHENOMENOLOGICAL AND STATISTICAL-WAVE DERIVATIONS OF THE RADIATION TRANSFER EQUATION	200
§ 4. NEW ASPECTS OF THE RTE	207
§ 5. NEW APPLICATION FIELDS OF THE RTE	212
§ 6. NEW EFFECTS IN STATISTICAL RADIATIVE TRANSFER	229
§ 7. CONCLUSION	237
REFERENCES	237

§ 1. Introduction

1.1. CLASSICAL RADIATIVE TRANSFER THEORY

The classical theory of radiative transfer evolved almost a century ago in the form of a transport equation for brightness, mainly owing to the efforts of Khvolson [1890] and Schuster [1905]. In transfer theory, the main characteristic is the radiance or brightness, $I(n, R)$, rather than the wave amplitude as in wave theory. The radiance is the flux of energy from a unit area of a source or a luminous (real or imaginary) surface into a unit solid angle in the direction indicated by a unit vector n. It characterizes the angular distribution of radiation; that is, it represents an angular spectrum. This spectrum has a local character, for the radiance varies from one point R to another.

The radiance is an energy variable, and it therefore obeys the energy balance equation that is derived from phenomenological considerations and takes the form of a radiative transport equation (RTE). According to this equation a variation of radiance in the direction n, that is, a change of energy flux in this direction is defined by

(i) a decrement of the energy flux due to the absorption and scattering processes,

(ii) an increment of the energy flux in the direction n as a result of scattering from other directions n', and

(iii) an increment of the energy flux due to sources of energy.

In general, the balance equation for energy fluxes may be represented in block form as follows:

$$\begin{aligned}
&\left\{\begin{array}{l}\text{variation of radiance } I(n, R), \text{ i.e.,}\\ \text{variation of energy flux in direction } n\end{array}\right\} \\
&= -\left\{\begin{array}{l}\text{decrement of energy flux due to absorption and}\\ \text{scattering from direction } n \text{ to } n'\end{array}\right\} \\
&\quad + \left\{\begin{array}{l}\text{increment of energy flux in direction } n\\ \text{due to scattering from other directions } n'\end{array}\right\} \\
&\quad + \{\text{increment of flux due to sources}\}.
\end{aligned} \quad (1.1)$$

All the classic radiation transport theory reduces essentially to the analysis of the balance equation (1.1). From the point of view of wave theory, such an approach is obviously not exhaustive, because it operates only with the law of energy conservation and tells nothing about the wave amplitude. Nevertheless, the transfer equation is closed with respect to the radiance and in this sense constitutes a self-consistent theory.

Thus the entire classical transfer theory is based on simple and intuitively appealing considerations of energy balance. Unlike other disciplines of physics which are known for their complete physical foundations, transfer theory has long had no reliable substantiation in wave theory and has been developed as if independent from the latter. A transition from wave fields to an energy balance equation is usually not treated explicitly. In the best case, simple and obviously insufficient considerations are invoked of the type of assumption on incoherence of wave beams. Even the best monographs, including books by Sobolev [1956, 1975], Chandrasekhar [1960], Bekefi [1966], Zheleznyakov [1970] and Born and Wolf [1980] introduce the main concepts of transfer theory in a purely phenomenological manner, actually as in the early works on transfer theory, without any attempts to endow these equations with a more rigorous statistical and wave meaning.

This situation can be explained in part by the simplicity and physical transparency of the geometrical optics concepts used by transfer theory. The concept of a "ray tube" and considerations of "incoherence of wave beams", which from the very outset allow one to operate with wave intensities, seem to be rather convincing and at first glance do not require any additional explanations. These seemingly evident heuristic considerations appear to be illusive in close view. This is especially true for the propagation of laser light and radio waves through a turbid atmosphere or scattering media where interference and diffraction phenomena cannot be neglected any longer and the applicability of the ray representations becomes questionable.

Indeed, the very terms of transfer theory imply the possibility to operate with plane or quasi-plane waves, which in the general case of an inhomogeneous medium is valid only in the geometrical optics limit of vanishingly small wavelengths $\lambda \to 0$. At finite wavelengths ($\lambda \neq 0$) characteristic of modern optical measurements and radio engineering applications, the concept of "independent and incoherent beams" is frequently inapplicable.

For scattering media, the phenomenological transfer theory has indirectly implied some averaging over an infinitesimally small (i.e., small when compared with the wavelength) volume of the medium and also tacitly implied independence of successive scattering events. As to scattering, the procedure has been reduced to a simple substitution of the term "mean energy" for the term "energy". The hypothesis about independent scattering events has been presumed to be evident for media with "sufficiently random" distribution of scatterers. In scattering with notable variation of frequency, say in Raman scattering or in scattering at movable inhomogeneities, the neglect of interference in successive scattering events may still be justified by the mutual incoherence of the incident and scattered waves. In these circumstances, the applicability limits of the transport equation turn out to be markedly wider. However, when scattering proceeds without alterations of frequency, such an incoherent summation of intensities is no longer valid.

A clear understanding of the fundamentals and applicability conditions of radiative transfer theory can be achieved only by invoking the statistical wave formalism of the propagation of waves in random media. Two recent decades have seen substantial reconsideration of radiative transfer theory on the base of the statistical wave formalism. The first attempts of this revision were directed toward the substantiation of the condition of incoherent addition of wave beams, which underlies the energy balance in the radiative transport equation. However, much more has come to the surface than one might have expected; namely, that the familiar energy balance conceals an equation for the coherence function of the wave field.

1.2. RTE AND BASIC EQUATIONS OF STATISTICAL-WAVE THEORY

The status of transfer theory is no longer viewed as uncertain as it was twenty years ago. The difference between transfer theory and the rigorous statistical approach has been learned to be characteristic for one dimensional (1-D) problems, and in all probability for 2-D problems. In the final analysis it is due to the phenomenon of strong Anderson localization. Up until the 1970s this phenomenon was studied only in quantum solid state physics in connection with the dynamics of electrons in doped systems, and no one thought about possible manifestations of Anderson localization in classical physics; e.g., in optics and acoustics.

In 1-D and 2-D random systems, owing to strong localization, the eigenfunctions are localized and vanish at infinity, whereas phenomenological transfer theory is based on traveling quasi-plane waves. In 3-D problems, strong localization does not manifest itself up until a certain level of disorder in the system. In contrast to the case of electrons in a solid, in optics this level is not easy to realize: the Anderson transition takes place when the wavelength becomes comparable to the scattering length, $\lambda \approx L_{sc}$ (Ioffe–Regel condition).

On the other hand, ordinary transfer theory is applicable in the region of weak disorder, $\lambda \leqslant L_{sc}$, characteristic of most classical wave problems.

The traditional treatment of the radiative transport equation as a simple condition of energy balance was changed radically only when a direct connection was established between the radiance and the coherence function of the wave field. For an equilibrium thermal radiation, such a relation can be deduced readily from the general theory of thermal fluctuations developed by Rytov [1953]. This work emphasized the statistical meaning of radiance treated as the spatial spectrum of thermal radiation and analyzed some consequences of this fact.

However, the equilibrium thermal radiation is homogeneous and does not transport energy from point to point. For inhomogeneous radiation, a relation of radiance to the coherence function was established by Dolin [1964a, 1968], who demonstrated, in particular, that in the small angle approximation, the transport equation for radiance is equivalent to a parabolic (truncated wave) equation for the coherence function. These pioneering papers have revealed the statistical meaning of the radiance as a spectrum of wave field fluctuations. However, the concept of a spectrum is defined unambiguously only for statistically uniform fluctuations, whereas for nonuniform fields, several generalizations exist.

For nonuniform fields, all versions of generalized spectra retain the familiar properties of a common spectrum only if these fields are quasi-uniform; i.e., if their difference from the uniform fields is not very large. Treating the radiance as a generalization of the spectrum on nonuniform fields, the classical transfer theory actually becomes the theory of quasi-uniform random wave fields. With this approach, the transfer equation follows from stochastic wave equations as a "quasi-uniform" limiting case.

The concept of the spectrum can be generalized in many ways, thus making the transfer theory ambiguous: giving different generalizations of the spectrum one may use the wave equations to obtain somewhat different transfer equations.

Despite a more-or-less evident connection between the coherence function and radiance, the way from the first principles of statistical wave theory to the radiative transport equation was not straightforward. From the first advances in the development of multiple scattering theory in the 1960s to the successive substantiation of transfer theory in the 1970s and 1980s there was a stretch of 15–20 years. Nevertheless, a reliable bridge between wave theory and the phenomenological theory of radiative transfer had been built at last.

Important elements of this bridge, whose significance has not been recognized until recent time, have been two plain, indirectly used hypotheses – on the weakness of scattering and on the statistical quasi-homogeneity of a primary

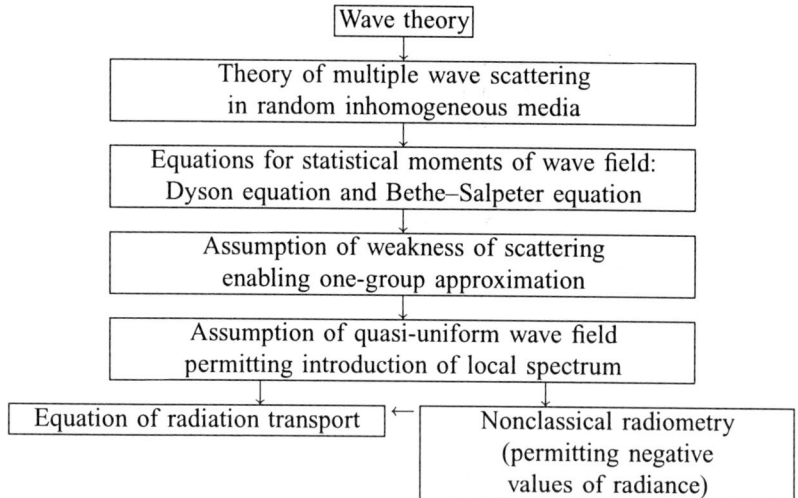

Fig. 1.1. Connection of radiative transfer theory with the fundamentals of wave theory.

wave field. The hypothesis on the relative weakness of a scattering process has led Finkelberg and Barabanenkov to the *one-group approximation* (Finkelberg [1967], Barabanenkov and Finkelberg [1968]), which radically simplified the stochastic equation for the coherence function of a wave field. The hypothesis on quasi-homogeneity has enabled a meaningful transition from the coherence function to radiance. Various forms of this hypothesis have been introduced in wave theory from the late 1960s (Walther [1968], Barabanenkov, Vinogradov, Kravtsov and Tatarskii [1972], Ovchinnikov and Tatarskii [1972], Apresyan [1973, 1975], Wolf and Carter [1977], Wolf [1978]; see also Friberg [1993]). A logical path from the foundations of wave theory to the radiative transfer theory is presented schematically in fig. 1.1.

Thus, after considerable effort it has become clear that radiation transfer theory in general is equivalent to the statistical theory of quasi-uniform wave fields. This is the principal, but not unique, result of the reconsideration of the fundamentals of radiative transfer theory. Other possibilities include

(i) evaluation of the coherence function of the wave field from the radiance,
(ii) incorporation of the coherent component in transfer theory,
(iii) elucidation of applicability conditions for the radiative transport equation, and
(iv) consideration of coherent channels in multiple scattering.

This last point is of highest interest because it associates the radiation transfer equation with some new coherent wave effects considered in § 6.

For a long time, experts in radiometry using the equation of radiative transport have deemed that cooperative, or alternatively, coherent effects are beyond the framework of transfer theory and can only limit the size of this framework (Rozenberg [1970]). Whereas the latter has proved unconditionally true, the former is true only in part. Indeed, the transport equation is based on the incoherent (intensity) summation of waves propagating in different directions. In a real scattering medium there always exist coherent pairs of channels responsible for the effect of backscatter enhancement. This effect concerns a maximum in the radiation pattern of a wave scattered backwards by a medium. These pairs of coherent channels, first reported by Watson [1969], seemed to have a purely wave nature and so, in principle, could not be described by transport theory. However, a procedure has been proposed recently that describes the backscattering effect in terms of transport theory. Roughly speaking, one should double the contribution due to the terms responsible for multiple scattering. Thus, transport theory has proved to be a rather flexible formalism to take into account even some coherent effects in multiple scattering.

When the place of transfer theory with respect to statistical wave theory was established, its range of capabilities became clear as well. While it enables one to describe the diffraction of the quasi-homogeneous part of a wave field, it says nothing about the behavior of the other part of the spectrum, which is normally small, but does present in actual media. Further on, transfer theory fails to describe the effects of spatial dispersion related to the finite size of the effective inhomogeneities of the medium and other phenomena arising in strongly scattering media. Specifically, the effects of strong (Anderson) localization, which owe their existence entirely to interference phenomena, do not fit in the radiative transport formalism. To describe such phenomena, one must resort to the initial Bethe–Salpeter equation in a more general formulation without reduction to the form of an ordinary transport equation for radiance.

1.3. DIFFRACTION CONTENT OF RADIATIVE TRANSPORT EQUATION. NONCLASSICAL RADIOMETRY

Despite the physical transparency of the links between the fundamentals of wave theory and radiative transport theory presented in fig. 1.1, some sequels of wave theory may nonplus a novice to the field. The main sequel of this kind is negative radiance under certain conditions; for example, in a bold computation of the angular spectrum $I(n, R)$ for a field passed through an aperture. From the standpoint of traditional radiometry, which treats radiance as energy density, negative values of radiance are nonsense. From the point of view of wave theory, they are a natural consequence of basic principles.

Let a coherent wave pass through an aperture in an opaque screen. If we associate with this wave an angular spectrum $I(n, R)$ in a formal manner, then this spectrum need not necessarily be positive. The reason is that a field passed through an aperture does not meet the quasi-homogeneity condition. On the other hand, radiance becomes positive far from the screen, where the field from the aperture acquires the properties of a quasi-uniform field.

Admission of positive and negative values for radiance brought about a separate branch of transfer theory, called "nonclassical radiometry" in the books and review by Apresyan and Kravtsov [1983, 1984, 1996]. In the scope of the transport equation, nonclassical radiometry includes those diffraction objects which would have been unthinkable in the framework of classical radiometry. Quite appropriately, this radiometry may be also called "diffraction" transfer theory to contrast the classic geometrical optics transfer theory. Accordingly, in fig. 1.1 nonclassical radiometry appears as a separate box leading to the transport equation from the side of negative radiance, $I(n, R) < 0$.

This extended treatment of radiance does not exhaust the diffraction content of transfer theory. There are many other examples illustrating the fact that results of transfer theory contain diffraction effects. Three of them follow.

(i) In the small angle approximation, the transport equation proves to be identical to the diffraction parabolic equation for the coherence function (Dolin [1964a, 1968]).

(ii) Consideration of weak scattering in the transport equation by perturbation theory leads to exactly the same results as wave scattering theory in the first Born approximation.

(iii) The radiance $I(n, R)$ obeys the wave uncertainty relation, which cannot be explained in the framework of radiometric representations.

Summing up, nonclassical radiometry enables us to describe and solve some diffraction problems. This is the radical change of the role of the transport equation in physics. It ceases to be a simple analog of Boltzmann's formula because it follows from the basic principles of wave theory and has a diffraction content.

Of course, we do not recommend that one attempt to solve diffraction problems by solving radiative transport equations, the more so that the capabilities of the transport equation are limited by simplifying idealizations made in its derivation. However, the very possibility of the transport equation to be a carrier of diffraction information is wonderful.

1.4. PURPOSE AND CONTENT OF THE REVIEW

The major aim of this review is to present the statistical-wave foundation of radiative transfer theory. With this purpose, we consider in § 2 the relation between the radiometric radiance and the coherence function of the wave field using plane sources and free wave fields as an illustration. This example admits a rigorous wave consideration of the problem and allows us to estimate the applicability limits of transfer theory. § 2 hinges around the concept of the spectrum of a statistically quasi-uniform field. We introduce this concept using the Wigner function as a more rational vehicle.

§ 3 is devoted to the derivation of the radiative transfer equation, first on the basis of phenomenological energy-balance equations, and then directly from the first principles; in this case, from stochastic wave field equations. Radiative transfer equations evolve as a direct sequel of the Bethe–Salpeter equation for the wave-field coherence function. The statistical wave derivation of the radiative transfer equation indicates that transfer theory reduces in fact to the theory of statistically weakly nonuniform fields.

The wave derivation of the radiative transfer equation reveals the diffraction content of radiative transfer theory. § 4 presents three vivid proofs of the diffraction nature of the radiative transfer equation: incorporation of the coherent field into this equation (§ 4.1), identity between the radiative transfer equation and the parabolic equation for the coherence function in the small angle approximation (§ 4.2), and a possibility to introduce the correlation scattering cross section (§ 4.3). Some considerations concerning the limiting resolving power of radiometric measurements are outlined in § 4.4.

§ 5 is devoted to new application areas of transfer theory. These are applications of the radiative transfer equation to analysis of weak localization phenomena (§ 5.1), the transfer to the nonlinear forms of the radiative transfer equation for analysis of strong localization (§ 5.2), description of radiative transfer in media with fluctuating macroparameters (§ 5.3), the phenomena of "warming" and "cooling" of a scattering medium (§ 5.4), and critical phenomena in thermal radiation from rough surfaces (§ 5.5).

Finally, the use of transfer theory for analysis of new correlation effects is discussed in § 6.

§ 2. Radiative Transfer in Free Space

2.1. RADIOMETRY AND COHERENCE

In order to elucidate the relation between radiometry and coherence theory, we consider a simple problem of radiation of plane sources in free space. Our analysis will be limited to the main features of this problem which are essential to an understanding of the general pattern. A detailed presentation of multiple results for the case of plane sources may be found in reviews by Wolf [1978], Baltes [1978], and Apresyan and Kravtsov [1984]. We also recommend a good collection of papers edited by Friberg [1993], which contains about 70 papers devoted to this problem published during the last three decades. For a more substantial list of references, the interested reader is referred to the recent book by Apresyan and Kravtsov [1996].

2.1.1. Radiometric description of radiation in free space

We will first recapitulate very briefly the fundamental results of radiometry in optics and in the wave theory for radiation in free space. The radiometric description of free radiation is given by the spectral radiance, $I_\omega = I_\omega(\mathbf{n}, \mathbf{r})$, which is defined at each point $\mathbf{r}$ of space for all directions $\mathbf{n}$ and frequencies ω (for simplicity, we assume that the radiation pattern is stationary; i.e., time invariable). In radiometry, a wave field is treated as a set of incoherent ray beams which give independent energy contributions. Thus, if we place an arbitrarily oriented area with normal $\hat{z}$ at a point $\mathbf{r}$, these beams will pierce it in all possible directions. If the radiance I_ω is known, the mean energy density $\overline{W}$ may be expressed by the usual relationship:

$$\overline{W} = \overline{W}(\mathbf{r}) = \frac{1}{c} \int I_\omega(\mathbf{n}, \mathbf{r}) \, d\Omega_n \, d\omega, \tag{2.1}$$

where c is the speed of light. A similar expression may also be written for the mean energy flux density.

We restrict out treatment to the case of monochromatic radiation and examine the corresponding case of the spectral density:

$$W_\omega = \frac{d\overline{W}}{d\omega} = \frac{1}{c} \int I_\omega \, d\Omega_n. \tag{2.2}$$

Here, for brevity, we omitted the arguments.

Elementary considerations based on energy conservation in a ray tube lead us to the fundamental radiometric equation for the radiance, $I = I_\omega$. In a homogeneous medium, the solid angle at which the source is visible from the point of observation decreases in proportion to the ray tube broadening with the distance from the source. Therefore, the radiance remains constant along the ray. We may put this condition for a stationary field in the form of a transport equation:

$$\frac{dI}{ds} = 0. \tag{2.3}$$

Here, $d/ds = \mathbf{n}\nabla$ is the derivative along the ray.

Thus, classical radiometry offers, on the one hand, relationships (2.1) and (2.2) which relate the mean energy density to the radiance and, on the other hand, the transport equation (2.3) which determines the behavior of the radiance. We consider the radiance I to be a nonnegative (energy) quantity; otherwise, it can be arbitrary. The overbar on $\overline{W}$ denotes averaging over certain space scales $L \gg \lambda$ and over a time interval $T \gg \tau$, where λ is the characteristic wavelength and τ is the characteristic period of the radiation (the latter statement is somewhat vague, since it is generally characteristic of the phenomenological theory; otherwise it is fairly evident).

Below we compare the radiometric relations (2.2) and (2.3) with the rigorous wave theory, where the main quantity is the random complex amplitude of the wave field rather than the radiance.

2.1.2. Radiance and the correlation function of the wave field

The classical transfer theory is based on geometrical optics (ray) concepts. The latter are valid only asymptotically, in the limit of a vanishingly small wavelength $\lambda > 0$. However, the condition $\lambda \to 0$ is not decisive to introduce the radiance, defined as a quantity proportional to the mean intensity of plane waves which collectively constitute the wave field.

Indeed, from the standpoint of statistical theory, a set of intensities of plane waves defines a *spatial spectrum* of the wave field. The wave field is assumed to be statistically uniform so that the concept of the spectrum can be defined unambiguously. The relation of the radiance to the spectral density of radiation enables one to express the radiance directly in terms of the correlation function of radiation. For equilibrium thermal radiation, Rytov developed the respective formalism in 1953. Later, Levin and Rytov [1967] derived a relation of the radiance to space–time correlation functions of thermal radiation.

We illustrate the relation between the radiance and the spatial spectrum by way of a simple example of a random scalar wave field satisfying the wave equation:

$$\left(\Delta + k_0^2\right) u = 0. \tag{2.4}$$

Here, $k_0 = 2\pi/\lambda$ is the wave number, λ is the wavelength, and the time factor $\exp(-i\omega t)$ is omitted. For a plane wave $u = \exp(i\mathbf{k}\mathbf{r})$, in view of eq. (2.4), the wave vector $\mathbf{k}$ satisfies the dispersion equation:

$$k^2 - k_0^2 = 0. \tag{2.5}$$

This equation suggests that, for a statistically uniform field, the spectrum must also be localized at the dispersion surface $k = k_0$; that is, may be written in the form $I_k = B_n \delta(k - k_0)/k_0^2$, where $\mathbf{n}$ is a unit vector, so that B_n acquires the sense of the *angular spectrum* of the field. Substituting this expression in the Wiener–Khinchin formula,

$$\Gamma \equiv \langle u(\mathbf{r}_1) u^*(\mathbf{r}_2) \rangle = \int I_k \exp[i\mathbf{k}(\mathbf{r}_1 - \mathbf{r}_2)] \, d^3k, \tag{2.6}$$

that relates the spectrum I_k with the correlation function Γ, yields:

$$\Gamma = \int B_n \exp[ik_0 \mathbf{n} (\mathbf{r}_1 - \mathbf{r}_2)] \, d\Omega_n. \tag{2.7}$$

Now, for the mean intensity of the field, we obtain:

$$\langle |u|^2 \rangle = \Gamma|_{\mathbf{r}_1 = \mathbf{r}_2} = \int B_n \, d\Omega_n. \tag{2.8}$$

Accurate to within a multiplicative factor, this expression coincides with the similar radiometric expression (2.1) for the intensity of radiation if we identify the energy density W with $\langle |u|^2 \rangle$ and the angular spectrum B_n with the radiance I_ω.

2.1.3. Radiance as the spectrum of quasi-uniform fluctuations

As shown above, for a statistically uniform field, the radiance is a spatial, or, to be more specific, angular spectrum of fluctuations. However, the extension of the spectrum to real fields which do not possess the property of statistical uniformity, faces certain difficulties associated with the fact that the spectrum of nonuniform fluctuations may be introduced in various ways. The question of ambiguity occurs not only in radiative transfer theory, but also in many problems associated with the description of random (or non-random but complicated) signals and fields, for example, in signal processing theory (Cohen [1989]).

Consider a random field $u(x)$ that depends on time and a spatial argument so that $x = (r, t)$. If this field is uniform in x with zero mean $\langle u(x) \rangle = 0$, then the spectrum J_K is defined as the Fourier transform of the correlation function $\Gamma(\rho) = \langle u(x) u^*(x + \rho) \rangle$:

$$J_K = \int \Gamma(\rho) \exp(iK\rho) \, d^4\rho, \tag{2.9}$$

where $K = (k, \omega)$ is the (four-dimensional) wave spectrum, and $K\rho = kr - \omega t$. Denoting the Fourier transform of $u(x)$ by $u(K)$, we obtain

$$\langle u(K_1) u^*(K_2) \rangle = J_K \delta(K_1 - K_2). \tag{2.10}$$

Here, the quantity J_K has the sense of the intensity of plane waves with the wave vector K: $J_K \propto \langle |u(K)|^2 \rangle$.

The problem is how to introduce the local spectrum $J_K(R)$ of the statistically nonuniform field $u(x)$ that would characterize the intensity of fluctuations with wave vector K in the neighborhood of a point R rather than in the entire space.

From physical considerations it should be clear that the local spectrum $J_K(R)$ must be given by some linear transformation $\widehat{Q}$ of the second moment Γ, which, in the case of uniform fluctuations, when Γ_{12} depends only upon the difference $\rho = x_1 - x_2$, would coincide with eq. (2.9). In the general form, such a transformation may be written as

$$J_K(R) = \widehat{Q}\Gamma \equiv \int Q(R, K; R', \rho') \langle u(R' + \tfrac{1}{2}\rho') u^*(R' - \tfrac{1}{2}\rho') \rangle \, d^4R' d^4\rho'. \tag{2.11}$$

Here, the function $Q = Q(R, K; R', \rho')$ must satisfy the condition

$$\int Q(R, K; R', \rho') \, d^4R' = (2\pi)^{-4} \exp(-iK\rho'), \tag{2.12}$$

which converts eq. (2.11) to the ordinary spectrum in the case of a statistically uniform field. In addition, it would be natural to require that the function Q depend only upon the difference $R - R'$: $Q = Q(K, R - R', \rho')$, to reflect the homogeneity and stationarity of $\widehat{Q}$. Indeed, in this case, to the second moment $\Gamma' = \Gamma(x_1 + a, x_2 + a)$, obtained from Γ by a translation of the origin by a, there corresponds the spectrum $J'_K(R) = J_K(R + a)$ obtained from the spectrum $J_K(R)$ by the same shift in the argument R.

The reader may find different definitions of the spectra of statistically nonuniform fluctuations corresponding to different functions Q. Walther [1973] used the spectrum $\langle u(x)e^{iKx}u^*(K)\rangle$. Page [1952] and Lampard [1954] proposed their own versions for nonstationary random processes – the instantaneous Page–Lampard spectrum. A definition used widely is the dynamic spectrum based on a finite segment of the process (for this approach, the reader is referred to the book by Rytov, Kravtsov and Tatarskii [1989b]). All these definitions are covered by the general formalism (2.11)–(2.12).

We note other interesting attempts of introducing the so-called physical spectrum (Marc [1970]) related directly to the method of detection of physical fields; specifically, to the rate of count of photons (Eberly and Wodkiewicz [1977]), or to the blackening of the photographic plate in optical measurement (Bartelt, Brenner and Lohmann [1980]). By refusing to strictly follow a condition like eq. (2.12), one may choose the function Q such that the physical spectrum is a nonnegative quantity. Such attempts are justified when particular experiments need be described, but they are hardly applicable for the construction of a general theory of quasi-uniform field since they lead to a definition of the spectrum that characterizes not only the field but also the method of measurements.

The more popular version of the local fluctuation spectrum is the *Wigner function*, which for the space–time fields under consideration may be defined as:

$$W(K,R) = \int \langle u(R+\tfrac{1}{2}\rho) u^*(R-\tfrac{1}{2}\rho)\rangle e^{-iK\rho} \frac{d^4\rho}{(2\pi)^4}$$
$$\equiv \int \Gamma(R,\rho) e^{iK\rho} \frac{d^4\rho}{(2\pi)^4}, \qquad (2.13)$$

where $K\rho = k\rho - \omega\tau$. This definition can be applied readily to the case of spatial fields or temporal signals.

2.1.4. Wigner function as a local spectral density: advantages and disadvantages

The Wigner function (2.13) was first introduced in quantum mechanics (Wigner [1932]) in a somewhat different context; i.e., as a quasi-probability, and has been used by many authors (see, e.g., Mori, Oppenheim and Ross [1962], Balescu [1963], Tatarskii [1983]). At present, in addition to quantum mechanics and statistical optics, Wigner's function is used widely in radar signal processing, in pattern recognition systems, and in many other applications (see, e.g., the review of Cohen [1989]).

Apart from its relative simplicity, Wigner's function has many other advantages when used as a local spectrum of quasi-uniform fields in statistical optics; namely,

- it is directly related to the mean square energy characteristics of radiation which are measured in common optical experiments,
- it naturally includes the case of partially coherent fields,
- it is convenient for a rigorous mathematical description in terms of wave equations,
- it ensures conversion to the case of the ordinary spectrum for a statistically uniform field,
- for free radiation satisfying the d'Alembert wave equation, it strictly obeys the transfer equation in free space and remains constant along the ray (although, in the general case of nonuniform fields, the concept of a ray loses its informal physical sense), and
- it remains formally useful in the case of nonergodic fields when statistical means do not coincide with the mean over the segment of one realization.

All these advantages outweigh by far the disadvantages of this function; namely,
- Wigner's function can assume negative values (for a quasi-uniform field, this function is positive), and
- the function $W(K,R)$ may be nonzero where the field $u(R)$ vanishes.

The last property evolves from the meaning of the non-local argument R as a center of gravity of the observation points x_1 and x_2. We will discuss this property in more detail below (§ 2.1.6).

2.1.5. Quasi-uniform fields and their spectra

We call the coherence function quasi-uniform if the derivatives with respect to the coordinates of the center of gravity $R = (x_1 + x_2)/2$ are far smaller than the derivatives with respect to the difference argument $\rho = x_1 - x_2$; that is, if

$$|\partial_R \Gamma| \ll |\partial_\rho \Gamma|. \tag{2.14}$$

We note that this definition of quasi-uniformity does not require factorization of Γ as do the functions R and ρ and as has been assumed in the literature (see, e.g., Carter and Wolf [1977]). Simple considerations demonstrate that, in real situations, the quasi-uniformity condition (2.14) cannot be valid for all values of R and ρ because real fields are bounded in space and time. The condition (2.14) should be understood as follows.

In the general case, the coherence function Γ may be represented as a sum of the quasi-uniform part $\Gamma^{(1)}$ and the nonuniform part $\Gamma^{(2)}$:

$$\Gamma = \Gamma^{(1)} + \Gamma^{(2)}, \tag{2.15}$$

where $\Gamma^{(1)}$ obeys eq. (2.14) whereas $\Gamma^{(2)}$ does not. The nonuniform part $\Gamma^{(2)}$ becomes essential near the interfaces, at the surfaces of inhomogeneities, in the presence of regular field nonuniformities near focal points, and at the time the field is switched on and off (nonstationarity). Conversely, far from the interfaces and in the absence of regular nonuniformities, the quasi-uniform part $\Gamma^{(1)}$ provides the predominant contribution.

Even when both terms in eq. (2.15) are of the same order of magnitude, in contrast to $\Gamma^{(1)}$, $\Gamma^{(2)}$ is frequently a rapidly oscillating function of the center-of-gravity coordinate R. This behavior permits us to get rid of $\Gamma^{(2)}$ by smoothing in R over distances of several initial characteristic wavelengths.

It is important that, using a certain asymptotic procedure, one can obtain closed equations for the quasi-uniform part $\Gamma^{(1)}$, so that in a certain sense $\Gamma^{(1)}$ is a quantity independent of $\Gamma^{(2)}$. We will not examine specific procedures of splitting Γ into $\Gamma^{(1)}$ and $\Gamma^{(2)}$. When discussing the quasi-uniform coherence function Γ below, we will keep in mind the possibility to isolate the quasi-uniform part $\Gamma^{(1)}$, thus digressing from the nonuniform part $\Gamma^{(2)}$. Thus, the forthcoming exposition will be based on discarding the nonuniform term $\Gamma^{(2)}$ in eq. (2.15) and on the approximation of Γ by $\Gamma^{(1)}$, keeping in mind, of course, that actually every coherence function contains a nonuniform part, which may be small or rapidly oscillating.

The requirement of quasi-uniformity (2.14) may be rewritten as:

$$|\partial_R \Gamma| \sim \mu |\partial_\rho \Gamma|, \qquad (2.16)$$

where

$$\mu \approx \frac{L_\rho}{L_R} \ll 1 \qquad (2.17)$$

is a small auxiliary parameter equal to the ratio of the scales L_ρ and L_R characterizing the variation rates of Γ in the arguments ρ and R, respectively. Since ρ is the difference variable, we will require that inequality (2.14) holds locally, only for moderate values of ρ. In order of magnitude estimations, we shall assume simply that $\rho = 0$.

The quantity L_R entering in eq. (2.17) has the meaning of the scale of the statistical nonuniformity of the field. For a statistically uniform field, $\Gamma(R, \rho) = \Gamma(\rho)$, and L_R is infinite. The scale L_ρ is usually within the field correlation radius, but need not coincide with this radius. For example, for a plane wave $u = A \exp(iKx)$ with random amplitude A, we have $\Gamma = \langle |A|^2 \rangle \exp(iK\rho)$. Then L_ρ is of the order of the characteristic wavelength or period of the field, $L_\rho \approx K^{-1}$, whereas the correlation radius, that depends on the behavior of the magnitude of Γ, is infinite in this case.

The conditions (2.14) and (2.17) imply that Γ varies slowly along the center-of-gravity coordinate $R = (x_1 + x_2)/2$ as compared with the fast variations in the difference variable $\rho = x_1 - x_2$. For simplicity we assume that the nonuniformity and nonstationarity are described by one small parameter μ, so that eq. (2.16) means the identical relative smallness of the spatial and temporal statistical nonuniformity of the coherence function Γ. In the limit as $\mu \to 0$, we arrive at the condition $\partial_R \Gamma = 0$; i.e., we obtain the coherence function Γ that depends only on the difference variable. For this function, the spectrum (2.13) is a nonnegative quantity. Therefore, it is natural to expect that, for small but nonzero values of μ (i.e., for a quasi-uniform field) the spectrum (2.13) will also be nonnegative.

For brevity we will refer to the limit transition when $\mu \to 0$ as the *quasi-uniform limit*. In this approximation, one might expect that all definitions of the spectra of nonuniform fluctuations which satisfy the conditions (2.14) and (2.17) will be asymptotically equivalent; i.e., will tend to a common limit – the spectrum of statistically uniform fluctuations. In the quasi-uniform limit, this behavior allows us to treat spectrum (2.13) as a local energy characteristic of the field, viewing values of K as values of the local wave vector near point R. Actually, the transition from the spectrum of uniform fluctuations to the spectrum of quasi-uniform fluctuations is in many aspects analogous to the geometrical optics transition from a plane wave to a quasi-plane wave that is also characterized by some local value of the wave vector K. Therefore, in the case of eq. (2.13) we may speak of the local value of the wave vector, implying the usual requirements of slow dependence of R; i.e., only for the quasi-uniform field.

2.1.6. *Local quasi-uniform coherence function in the geometrical optics approximation*

We now demonstrate that, in the absence of multipath propagation, when only one ray passes through a point, an ordinary geometrical optics asymptotic representation of the field satisfies the quasi-uniformity conditions (2.14) and (2.17). Indeed, geometrical optics takes the field u in the form:

$$u(x) = A(\mu_1 x) \exp\left[\frac{i\Psi(\mu_1 x)}{\mu_1}\right], \tag{2.18}$$

where A is the amplitude, Ψ is the phase, and μ_1 is a small parameter of the type of eq. (2.17) comparable to the ratio of the wavelength to the characteristic length over which the properties of the medium or wave vary appreciably: $\mu_1 \approx \lambda/L_R$. The coherence function for this field has the form

$$\Gamma = A\left(\mu_1\left(R + \tfrac{1}{2}\rho\right)\right) A^*\left(\mu_1\left(R - \tfrac{1}{2}\rho\right)\right)$$
$$\times \exp\left\{\frac{i}{\mu_1}\left[\Psi\left(\mu_1\left(R + \tfrac{1}{2}\rho\right)\right) - \Psi\left(\mu_1\left(R - \tfrac{1}{2}\rho\right)\right)\right]\right\}.$$

Assuming that the gradients of the amplitude and phase are quantities of one order of magnitude, $\nabla A/A \approx \nabla \Psi \approx \mu_1$, and letting $\mu_1 \to 0$ we obtain:

$$\partial_R \Gamma |_{\rho=0} \approx \mu_1 \, \partial_\rho \Gamma |_{\rho=0} \ll \partial_\rho \Gamma |_{\rho=0}. \tag{2.19}$$

In this case, the small geometrical optics parameter μ_1 plays the role of the small quasi-uniformity parameter (2.17).

This reasoning leads us to the important conclusion that the geometrical optics asymptotic of the coherence function Γ may be viewed as a particular case of the quasi-uniform limit. Therefore, in a theory of the quasi-uniform field below we will also cover the results which can be obtained for the second moment Γ using the method of geometrical optics for the field u. In other words, for the coherence function Γ, the quasi-uniform field theory is a certain generalization of the ordinary method of geometrical optics.

In the limiting case of coherent radiation, for relatively small ρ, the quasi-uniformity (*local* quasi-uniformity) does not mean the *global* quasi-uniformity; that is, the fulfillment of the quasi-uniformity conditions (2.14) for all ρ. To illustrate this statement we consider the coherence function $\Gamma(R,\rho)$ for a monochromatic geometrical optics field in the form of two mutually coherent beams. One can expect that the respective coherence function may be localized at small ρ and $R \approx r_1$ or $R \approx r_2$ near each beam, where $r_{1,2}$ are two points on the beams. However, in the case of coherent beams, the coherence function Γ acquires an additional term with values $R \approx (r_1 + r_2)/2$ and $\rho = r_1 - r_2$. If we forget for the moment that the argument R is the radius vector of the center of gravity between two observation points, then this addend may be associated with an imaginary interference beam laying between r_1 and r_2. Of course, a measuring instrument put in the way of this interference beam could not detect it. Interference rays occur only in interpretations of interference experiments where the field has been detected at distant points (in this case, at points r_1 and r_2) and they reflect the nonlocal character of the auxiliary argument R.

The addend correspondent to point $R \approx (r_1 + r_2)/2$ is, generally speaking, nonuniform and oscillates rapidly in R, so that its description would go beyond the scope of the quasi-uniform field. This type of addend occurs naturally in descriptions of interference phenomena.

In multipath propagation, the field is represented as a sum of terms similar to eq. (2.17). In this case, the coherence function will contain rapidly oscillating interference terms which do not obey the quasi-stationarity conditions. Description of such terms is beyond the applicability limits of the theory of the quasi-uniform field; however, they are often of low significance, because they vanish upon space or time averaging over lengths exceeding the characteristic scales of interference fringes.

In the forthcoming discussion we intend to use the following simple assertion: in a homogeneous medium, the field of quasi-uniform sources is quasi-uniform. The proof of this statement is straightforward.

2.2. GENERALIZED RADIANCE OF PLANE SOURCES

2.2.1. Definition of generalized radiance

The use of the Wigner function as a spectrum of fluctuations of a nonuniform wave field may be illustrated by an example of radiation of plane sources. Copious relevant results may be found in reviews cited in § 2.1. The generalized radiance of plane sources I^0 is represented in terms of the coherence function as:

$$I^0(\boldsymbol{n}, \boldsymbol{R}_\perp) = \left(\frac{k_0}{2\pi}\right)^2 n_z \int \Gamma^0(\boldsymbol{\rho}_\perp, \boldsymbol{R}_\perp) \exp(-\mathrm{i}k_0 \boldsymbol{n}_\perp \boldsymbol{\rho}_\perp) \, \mathrm{d}^2 \rho_\perp. \qquad (2.20)$$

For convenience, we consider here the case of monochromatic radiation (frequency dependencies are dropped), Γ^0 denotes the coherence function of the field $u^0(\boldsymbol{R}_\perp)$ of sources in the plane $z = 0$:

$$\Gamma^0(\boldsymbol{\rho}_\perp, \boldsymbol{R}_\perp) = \langle u^0(\boldsymbol{R}_\perp + \tfrac{1}{2}\boldsymbol{\rho}_\perp) u^{0*}(\boldsymbol{R}_\perp - \tfrac{1}{2}\boldsymbol{\rho}_\perp) \rangle, \qquad (2.21)$$

where $k_0 = \omega/c$ is the wave number, and $\boldsymbol{n} = (n_z, \boldsymbol{n}_\perp)$ is the unit vector in the direction of propagation.

Expression (2.20) is beyond the classical theory of radiation transfer. It combines the directivity of the radiation of plane sources (radiance I^0 as a function of $\boldsymbol{n}$) with the correlation function of the field in the plane $z = 0$. For radiance of monochromatic sources, this expression was first derived by Walther [1968]. Subsequently it was discussed by many authors from different points of view (see, e.g., the reviews of Baltes [1977] and Wolf [1978] and the literature cited therein). Walther [1973] revealed some drawbacks of eq. (2.20) and set forth an alternative definition of I^0. The possibility of various definitions for radiance is associated with the above lack of uniqueness in the local spectrum concept.

2.2.2. *Generalized radiance of plane sources and nonclassical radiometry*

In the final analysis, the source radiance (2.20) has a direct bearing on the choice of Wigner's function as a local spectrum of nonuniform fluctuations. Alternative definitions of local spectra may lead to different expressions for the radiance of the sources; however, in the quasi-uniform limit, all these definitions must give identical results and must reduce to a common limit in terms of the radiometric radiance.

Thus, the radiance of plane sources I^0 is expressed through the boundary value of the correlation function of the field in agreement with eq. (2.20). We consider some corollaries of this expression.

First, we note that the radiance (2.20) need not be positive, and in principle, can take on negative values. Therefore, in the general case, it can be treated as some *generalized radiance of the sources*. This conclusion does not seem strange if we observe that the field quasi-uniformity conditions do not require generally that the sources be quasi-uniform. Near appreciably irregular sources, the field is also appreciably nonuniform, thus invalidating the radiometric description in this region. At the same time, the quasi-uniformity conditions could be met far from the sources, where the radiometric description becomes valid. It is in this area that eq. (2.20) should be understood as the generalized source radiance. In the region of inapplicability of classical radiometry, eq. (2.20) does not possess all the properties of common radiometric radiance. Thus, the generalized radiance allows one to derive a correct description of the field in the quasi-uniformity region. Moreover, in conjunction with the radiative transfer equation, this formula covers the effects of diffraction – an unattainable goal in the framework of classical radiometry.

Thus we extended the applicability range of radiative transfer theory at the expense of the abandoned heuristic condition of nonnegative radiance. We will call this approach *nonclassical radiometry or diffraction radiative transfer theory*. Numerous applications of such a nonclassical approach may be found in reviews cited in § 2.1.

2.3. WOLF'S RED AND BLUE SHIFTS OF SPECTRAL LINES

The recent finding of Wolf [1986] that the spectral line of a radiation due to partially coherent sources can be deformed and shifted slightly to the red or blue end, even if the sources and the point of observation are not in relative motion, was unexpected evidence of the diffraction nature of radiative transfer from sources to a point of observation.

The essence of this effect may be illustrated as follows. At each point in space, the frequency spectrum of a source of an electromagnetic field is a set of intensities of frequency components. At the point of observation, the field and its frequency-angular spectrum (radiance) depends not only on the frequency distribution of the field of the sources, but also on their spatial correlation; that is, on two-point characteristics of the field of sources. In the final analysis, the spectral radiance of the sources observed at a distance from them is found to be dependent not only on the form of the radiation spectrum, but also on the spatial correlation of the sources. Alternatively, a change of the form of the correlation function of the sources will produce some distortion of the resultant spectral radiance as compared with the initial frequency spectrum. Thus, the Wolf effect is essentially correlation-induced changes of the radiation spectrum.

The Wolf effect, reflecting the nonlocality of the radiation process, first seemed rather unusual from the standpoint of classic radiometry, which disregards diffraction and interference phenomena altogether. Therefore, it attracted the attention of the scientific community. This effect stimulated a substantial literature. We note some representative examples without attempting to provide an exhaustive coverage of this literature.

We would like to note that the Wolf effect cannot bring about new radiation lines beyond those produced by the primary sources. This effect can only shift slightly (to the red or blue end, depending on the geometrical distribution of the sources and their spatial correlation) the center of a radiation line.

One more interesting modification of the Wolf effect is a red shift of radiation lines scattered at rapidly moving clusters of scatterers. This shift is proportional to the frequency of the primary irradiation as in the Doppler effect. This Doppler-like shift, substantial at relativistic velocities of scatterers, has some resemblance to cosmologic red shifts (Wolf [1986, 1987]).

§ 3. Phenomenological and Statistical-Wave Derivations of the Radiation Transfer Equation

3.1. RADIATION TRANSFER EQUATIONS IN SCATTERING MEDIA

3.1.1. Phenomenological derivation

In the phenomenological theory, the behavior of the radiance in a scattering medium is described by the well-known radiative transfer (or transport) equation (see, e.g., Chandrasekhar [1960]). In the simplest case of a scalar monochromatic field, it has the form

$$\frac{dI}{ds} + \alpha I = \int \sigma\left(n \leftarrow n'\right) I\left(n', r\right) d\Omega_{n'} \equiv \widehat{\sigma} I. \quad (3.1)$$

In contrast to the equation $dI/ds = 0$, describing the propagation in free space, eq. (3.1) contains two new addends with an obvious physical sense. The left-hand side of this equation describes the attenuation of the radiance I along the ray of length s due to scattering and absorption combined in the extinction coefficient α. The right-hand side describes the contribution due to scattering, where the parameter σ is the scattering cross section per unit volume. For weakly scattering media with continuous fluctuations, σ is usually evaluated in the Born approximation, and, in the case of sparse scatterers, it is calculated in the approximation of independent particles with a correction for the weak correlation of scatterer positions.

In order to derive the radiative transfer eq. (3.1) in the phenomenological theory, one may resort to simple arguments of energy balance in a physically infinitesimal volume of the scattering medium.

3.1.2. Heuristic applicability conditions for RTE

Photometry dates as far back as the 15th century and is associated with many famous scientists such as Leonardo da Vinci, Galileo (17th century), and especially Bouger [1729] and Lambert [1760]. The first rigorous mathematical formulation of radiative transport theory in turbid media owes its existence mainly to Khvolson [1890] and Schuster [1905], who derived the key equation for this theory – the equation of radiative transport. The history of this development and later advances of the phenomenological theory have been covered in the paper by Rozenberg [1977].

In the early days of photometry, the applicability of its concepts was not doubted, although from the very beginning it was obvious that these concepts were only approximate. We intend to formulate the basic heuristic conditions of applicability of classical radiometry remaining within the framework of modern phenomenological theory of radiative energy transfer.

The applicability of these approximations is associated primarily with the methods of measurement of light radiation whose wavelengths λ are exceedingly small compared with the characteristic dimensions L^* of recording devices. The condition $\lambda \ll L^*$ corresponds formally to the geometrical optics limit $\lambda \to 0$.

In addition, classic photometry has operated only with light sources of natural thermal origin which may be deemed chaotic to a good degree of accuracy owing to the small correlation radius. Now, the basic assumptions of classic photometry follow.

(i) *The ray approach.* It is assumed that the applicability conditions of the geometrical optics approximation are satisfied for every quasi-plane wave and so the *wave* field is treated as the *ray* field.

(ii) *Spatial incoherence of radiation.* It is assumed that rays arriving at a given point from different directions are totally incoherent, which, in a certain sense, corresponds to the statistical independence of the sources of this radiation.

(iii) *Averaged description.* Measured parameters are assumed to be not local or current values, but rather some time and space averaged, squared field characteristics, and radiative transfer theory operates with these characteristics.

(iv) *The ensemble average approach.* Radiation possesses the properties of stationarity and ergodicity and so the averaged characteristics coincide with the statistical mean.

Condition (i) is the principal one and allows one to speak of radiant fluxes propagating along the rays. In agreement with this condition, classical transport theory pays no attention to diffraction effects in wave propagation analyses. Condition (ii) allows one to sum energy quantities rather than fields, thus excluding the possibility of manifestation of interference effects. Conditions (iii) and (iv) are normally not formulated explicitly in photometric manuals, but they are assumed tacitly. These conditions put into correspondence radiometric quantities and parameters measured in typical experiments.

If we confine this consideration to the description of free, non-scattered radiation, then there will be no difficulties with the substantiation of radiometric concepts. In this case, the sufficiency of conditions (i)–(iv) is obvious, and radiometry acts as a theory linking the energy relations of the method of geometrical optics with the statistical assumption on incoherence of wave bundles.

Looking a little bit ahead, we remark that the sufficiency of conditions (i)–(iv) does not imply their necessity. It will be demonstrated later that the refusal of the requirement of non-negative radiance in some situations enables one to extend the framework of classical radiometry and partially take into account diffraction effects.

It will be much more difficult to evaluate the applicability limits of the radiometric description of the behavior of radiation in turbid media where scattering is significant. The scattering becomes significant when the medium contains inhomogeneities which are no longer smooth in the wavelength scale. However, near such inhomogeneities, the applicability condition of the method of geometrical optics breaks down and the applicability condition (i) of radiometric concepts is not satisfied. Notwithstanding this fact, the radiometric concepts are used widely in turbid media as well. This usage is justified since in turbid media, the applicability of geometrical optics is not required everywhere but only *in the mean* with respect to the lengths over which averaging is performed. Fine features, such as the behavior of the field near sharp inhomogeneities are excluded from radiometric considerations. Thus, in the case of turbid area, applicability of the radiometric description is not an easy matter and can be handled in the framework of the statistical-wave formalism, a more rigorous treatment than the phenomenological approach.

3.2. DYSON AND BETHE–SALPETER EQUATIONS

3.2.1. Discrete and continuous models of scattering media

The statistical approach to the description of radiation in scattering media hinges on the Dyson equation for the mean field and the Bethe–Salpeter equation for the second moment of the field (coherence function). Methods of analysis of these equations have been discussed in detail in the literature, including textbooks (see, e.g., Rytov, Kravtsov and Tatarskii [1989b]). The general form of these equations may be described as follows. We write the initial wave equation in the form of an integral equation:

$$u = u^0 + G^0 V u, \qquad (3.2)$$

where u^0 is the incident wave produced by sources of radiation, G^0 is the operator of propagation in free space, and V is the random perturbation operator. Then the Dyson equation for the mean field takes the form:

$$\langle u \rangle = u^0 + G^0 V^{\text{eff}} \langle u \rangle, \qquad (3.3)$$

and the Bethe–Salpeter equation for the coherence function $\Gamma_{1,2} = \langle u_1 u_2^* \rangle$ takes the form

$$\langle u_1 u_2^* \rangle = \langle u_1 \rangle \langle u_2^* \rangle + \langle G_1 \rangle \langle G_2^* \rangle K_{1,2} \langle u_1 u_2^* \rangle. \qquad (3.4)$$

The mass operator V^{eff} and the intensity operator $K_{1,2}$ entering in this equation can be represented only as infinite expansions in formal series of perturbation theory and, for engineering applications, only a few leading terms of these expansions are used.

The initial equation (3.2) may correspond to drastically different models of random media, which can be divided into discrete and continuous categories. The former represent particulate media consisting of individual scattering particles, whereas the latter correspond to media with continuous fluctuations such as a turbulent atmosphere. Discrete and continuous media are usually treated by different mathematical models because it is natural to solve the former case for a single scatterer. Nonetheless, in both situations, the Dyson and Bethe–Salpeter equations have similar forms and differ only in the explicit form of expansions of operators.

3.2.2. The Dyson equation and effective parameters of random media

In order to evaluate the mass operator V^{eff}, alternatively called the effective inhomogeneity operator, one must solve the problem of effective parameters of random inhomogeneous media. This problem is of great significance to many applications, and tackling it would lead us beyond the scope of this review. We confine ourselves to listing only a few approaches to the calculation of the effective parameters along with the relevant literature citations:
– coherent potential approximation (Landauer [1978]),
– effective medium approximation (Roth [1974]),
– quasi-crystalline approximation (Lax [1952]),
– iterated dilute approximation (Sen, Scala and Cohen [1981]),
– average field approximation (Polder and van Santen [1946]),
– a variety of cumulant expansions (Finkelberg [1964], Hori [1977], Felderhof, Ford and Cohen [1983], Ramshaw [1984]),
– Twersky's multiple scattering theory (Twersky [1962a,b], Tsolakis, Besieris and Kohler [1985]),
– Bergman's analytical representation for the effective permeability of two-phase composites (Bergman [1982]),
– a theory of strong fluctuations in the electromagnetic problem (Tatarskii and Gertsenshtein [1963], Ryzhov and Tamoikin [1970]).

We can also refer here to the various theories of feasible boundaries for effective medium parameters (Hashin and Shtrikman [1962], Bergman [1978], Milton [1981], Golden and Papanicolaou [1983], Kohler and Papanicolaou [1982]).

Since space limitations prevent us from commenting at length on the results of effective parameter evaluations, we mention only two main approximations used in calculations of the effective permittivity for media with discrete inclusions (cermet topology) and mixtures (aggregate topology). Both approximations extend beyond the frameworks of a simple perturbation theory and are valid in the quasistatic limit when the radiation wavelength is large compared to all internal scales of the medium.

The first approximation is constituted by the formulas of Clausius–Mossotti, Lorenz–Lorentz, or the Maxwell–Garnett formulas (for a history of this approximation, see Landauer [1978]). For a medium of particles with polarizability α populated in free space with a number density N, the Lorenz–Lorentz formula takes a familiar form:

$$\frac{\varepsilon - 1}{\varepsilon + 2} = \frac{4\pi}{3} N \alpha. \tag{3.5}$$

The second approach is the Bruggeman effective medium approximation, known also as the Maxwell–Odelevskii formula (Odelevskii [1951]):

$$\left\langle \frac{\tilde{\varepsilon}(r) - \varepsilon}{\tilde{\varepsilon}(r) + \varepsilon} \right\rangle = 0, \tag{3.6}$$

that describes the effective permittivity ε of the medium with fluctuating permittivity $\tilde{\varepsilon}(r)$. These two approximations are widely used in estimations of the permittivity of dispersive media and can be obtained from the same integral equation (Stroud [1975]).

While the Dyson equation defines the effective medium parameters and is not related directly to the energy characteristics of the field, the Bethe–Salpeter equation defines the transfer of the correlation of radiation and yields a transport equation for the generalized radiance in the asymptotic limit of the quasi-uniform field.

Many workers were active in deriving a radiative transport equation from first principles. Running the risk of missing some essential contributions, we list some reports of the 1960s and 1970s devoted to this problem. The earliest efforts are due to Bugnolo [1960] and Gnedin and Dolginov [1963]. These studies were followed by the contribution of Borovoi [1966] and a very important series of works by Barabanenkov [1967, 1969, 1975a,b], Barabanenkov and Finkelberg [1967], and Barabanenkov, Vinogradov, Kravtsov and Tatarskii [1972]. The last report already contained the main features of the modern theory. This achievement, however, had precursors in the papers by Walther [1968], Stott [1968], Watson [1969], Peacher and Watson [1970] and Galinas and Ott [1970]. The 1970s have seen also the publications of Ovchinnikov and Tatarskii [1972], Howe [1973], Ovchinnikov [1973], Apresyan [1973, 1974, 1975], Carter and

Wolf [1975], Acquista and Anderson [1977], Baltes [1977], Wolf and Carter [1977] and Wolf [1978]. Some results achieved in the substantiation of transfer theory have already been reflected in textbooks by Ishimaru [1978], Rytov, Kravtsov and Tatarskii [1989b] and Dolghinov, Gnedin and Silant'ev [1995].

In order to obtain a transfer equation for the radiance in a scattering medium from the Bethe–Salpeter equation, one may follow the scheme of Apresyan [1973], which differs from other derivations of the transfer equation in minor details.

In this approach the Bethe–Salpeter equation is written for the coherence function $\Gamma(R,\rho) = \langle u(x_1) u^*(x_2) \rangle$ in the variables $R=(x_1+x_2)/2$ and $\rho=x_1-x_2$. An asymptotic procedure is built in terms of the small parameter

$$\mu \approx \frac{|\partial_R \Gamma|}{|\partial_\rho \Gamma|}, \qquad (3.7)$$

where the smallness of μ implies the assumption of quasi-uniformity of Γ; namely, $|\partial_R \Gamma| \ll |\partial_\rho \Gamma|$. The Bethe–Salpeter equation is taken in one-group approximation (Barabanenkov and Finkelberg [1967]) which assumes, in a certain sense, that the scattering is weak and allows one to speak of the effective scattering inhomogeneities comparable in size to the correlation radius of the medium. It turns out that the Wigner function $W(R,K)$, obtained as the Fourier transform of $\Gamma(R,\rho)$ taken with respect to the difference argument ρ, satisfies the radiative transfer equation.

In the zero approximation in μ from the Bethe–Salpeter equation, it follows that the Wigner function is localized on an energy surface $\operatorname{Re} D(\boldsymbol{k}) = 0$. This means that the wave vector $\boldsymbol{k}$ satisfies the dispersion equation $\operatorname{Re} D(\boldsymbol{k}) = 0$ modified by the presence of scattering. Physically this implies a sufficiently weak scattering for the medium to conserve the types of propagating waves. It is worth noting that an increase of scattering expands the dispersion surface, which in the final analysis distorts the structure of modes in a strongly scattering medium.

This derivation of the transfer equation is similar to the Debye derivation scheme for the geometrical optics approximation (Kravtsov and Orlov [1990]). The difference is that now the small parameter $\mu \approx |\partial_R \Gamma|/|\partial_\rho \Gamma|$ associated with a weak statistical nonuniformity of the field, plays the role of the small geometrical optics parameter $\mu_1 \approx \lambda/L \ll 1$ ($\lambda \to 0$). The assumption of the quasi-uniformity of the field ($\mu \ll 1$) is less burdensome than the geometrical optics applicability condition $\mu_1 \ll 1$, since the quasi-uniformity of Γ follows from the condition $\mu_1 \to 0$. Therefore the transfer equation can be obtained from the Bethe–Salpeter equation. Barabanenkov and Finkelberg [1967] considered a less formal derivation of the transfer equation by iteration of the Bethe–Salpeter equation in the geometrical optics limit of the vanishingly small wavelength. They demonstrated that the solution of the transfer equation is equivalent to an approximate summation of ladder diagrams.

In the final analysis, multiple scattering theory leads to the transfer equation (3.1) and establishes simultaneously the statistical wave sense of the radiance $I_{\omega n}$. The latter is the angular spectrum of the wave field; that is, the Fourier transform of the correlation function of the field (Barabanenkov, Vinogradov, Kravtsov and Tatarskii [1972]).

Thus, we have revealed the niche of the classical transfer equation in the general theory of random wave fields: this equation is a corollary of a certain approximation of multiple scattering equations. To derive this equation, one may do without heuristic considerations about energy conservation which are used in the phenomenological derivation of the RTE.

§ 4. New Aspects of the RTE

4.1. TRANSFER EQUATIONS AND THE COHERENT FIELD

The classical transfer theory assumes that the coherent component of the field $\langle u \rangle$ is equal to zero, which implies that one deals with natural (i.e., incoherent), sources of radiation. Treatment of modern radar and laser applications often requires that the contributions due to coherent and incoherent parts of radiation be separated. This separation was first realized by Vinogradov, Kravtsov and Tatarskii [1973].

In transfer theory, one may take the coherent field $\langle u \rangle$ into account by assuming that both parts, coherent $\langle u_1 \rangle \langle u_2^* \rangle$ and incoherent $\Psi_{12} = \langle \tilde{u}_1 \tilde{u}_2^* \rangle$, of the full coherence function,

$$\Gamma_{12} = \Psi_{12} + \langle u_1 \rangle \langle u_2^* \rangle, \tag{4.1}$$

satisfy the quasi-uniformity condition (here $\tilde{u} = u - \langle u \rangle$ is the fluctuating part of the wave field). Accordingly, the full ray intensity (radiance $I_{\omega n}$) may be divided into the coherent and incoherent parts:

$$I_{\omega n} = I_{\omega n}^{\text{coh}} + I_{\omega n}^{\text{incoh}}. \tag{4.2}$$

The coherent component $I_{\omega n}^{\text{coh}}$ is proved to be localized around the direction n_R, that is, it introduces the delta function $\delta(n \cdot n_R)$ as follows:

$$I_{\omega n}^{\text{coh}} = I^0 \delta(n \cdot n_R). \tag{4.3}$$

This means that the coherent radiation arrives at each point R along a single ray coming from the source. In the absence of sources, the coherent radiance $I_{\omega n}^{\text{coh}}$ obeys the homogeneous transfer equation:

$$\left(\frac{d}{ds} + \alpha^{\text{ext}}\right) I_{\omega n}^{\text{coh}} = 0, \tag{4.4}$$

and varies along the ray as $\exp\left(-\int \alpha^{\text{ext}} ds\right)$, which reflects the mean field attenuation due to scattering and absorption. Since the full ray intensity $I_{\omega n}$ satisfies the transfer equation (3.1) and the coherent part $I_{\omega n}^{\text{coh}}$ obeys eq. (4.4), one may conclude that the incoherent part is governed by the equation:

$$\left(\frac{d}{ds} + \alpha^{\text{ext}}\right) I_{\omega n}^{\text{incoh}} = \widehat{\sigma}\left(I_{\omega n}^{\text{coh}} + I_{\omega n}^{\text{incoh}}\right). \tag{4.5}$$

Equations (4.4) and (4.5) are matched with one another in terms of energy. On the one hand, the coherent component $I_{\omega n}^{\text{coh}}$ is attenuated by absorption and scattering; that is, by conversion into the incoherent component [eq. (4.4)]. On the other hand, the incoherent component $I_{\omega n}^{\text{incoh}}$ is fed by the coherent field [eq. (4.5)]. Even a single scattering event transforms the coherent component into an incoherent component, which thereby acquires the sense of the ray intensity of singly scattered radiation. These processes have been discussed by Vinogradov, Kravtsov and Tatarskii [1973], and are indirectly reflected in the book by Ishimaru [1978], who took the attenuation of the primary field into consideration.

If one uses the formalism of perturbation theory to calculate the ray intensity of single scattering from eq. (4.5) and defines the correlation function of the scattering field from the base relation (2.7), then the result will be identical to the diffraction theory of single scattering. Such calculations, performed by Vinogradov, Kravtsov and Tatarskii [1973] and reproduced in the book of Rytov, Kravtsov and Tatarskii [1989b], prove the diffraction content of the radiative transfer equation.

4.2. DIFFRACTION RADIOMETRY. THE PARABOLIC EQUATION METHOD

The small-angle scattering formalism allows one to simplify the transfer equation by replacing the integration over a unit sphere with the integration with respect to angular variables θ_x and θ_y between infinite limits. This replacement has a small effect on the value of this integral, since the integrand (the product of the scattering cross section by the ray intensity) is confined within narrow angles only.

The resultant equation turns out to be identical to the parabolic equation for the coherence function $\Gamma_\perp = \Gamma_\perp(\rho_\perp, R)$ transverse to the direction of propagation (z axis); viz.,

$$\left[2ik\frac{d}{dz} + \Delta_1 - \Delta_2 + 2ikH_1(\rho_1 - \rho_2)\right]\Gamma_\perp = 0. \tag{4.6}$$

Here, Δ_1 and Δ_2 are the transverse Laplacians (with respect to $\rho_{\perp 1}$ and $\rho_{\perp 2}$) and H_1 is defined by

$$H_1(\rho_\perp) = \alpha^{abs} + \int \sigma(\boldsymbol{n}_\perp)[1 - \cos(k\boldsymbol{n}_\perp \rho_\perp)]d^2 n_\perp. \tag{4.7}$$

The scattering cross section $\sigma(\boldsymbol{n}_\perp)$ is proportional to the spatial spectrum of medium fluctuations:

$$\sigma(\boldsymbol{n}_\perp) = \frac{\pi k^4}{2}\phi_\varepsilon(k\boldsymbol{n}_\perp) = \frac{\pi k^4}{2}\int \langle \tilde{\varepsilon}(\rho)\tilde{\varepsilon}(0)\rangle \exp(-ik\boldsymbol{n}_\perp \rho)\frac{d^3\rho}{(2\pi)^3}.$$

In the small angle approximation, the coherence function $\Gamma_\perp$ is related to the radiance I by

$$\Gamma_\perp(\rho_\perp) = \int I(\boldsymbol{n}_\perp)\exp(ik\boldsymbol{n}_\perp \rho_\perp)d^2 n_\perp, \tag{4.8}$$

whereas the inverse transformation has the form

$$I(\boldsymbol{n}_\perp) = \left(\frac{k}{2\pi}\right)^2 \int \Gamma_\perp(\rho_\perp)\exp(-ik\boldsymbol{n}_\perp \rho_\perp)d^2\rho_\perp. \tag{4.9}$$

The identity of eq. (4.6) and the RTE was first established by Dolin [1964a, 1968], who revealed the diffraction sense of the transfer equation for narrow wave beams and paved the way to a statistical wave derivation of the RTE in the general sense. It should be noted that somewhat earlier Dolin [1964b] found an exact solution to the transfer equation in the small angle approximation. Using eqs. (4.8) and (4.9) and the solution to the RTE, one may obtain the solution of the parabolic eq. (4.6). This is a rare example of how the solution to a problem that is essentially one of diffraction can be found from the solution of the RTE.

4.3. CORRELATION SCATTERING CROSS SECTIONS

In discussing the physical meaning of the transition from the Bethe–Salpeter equation to the equation of radiative transfer, it is useful to somewhat extend the

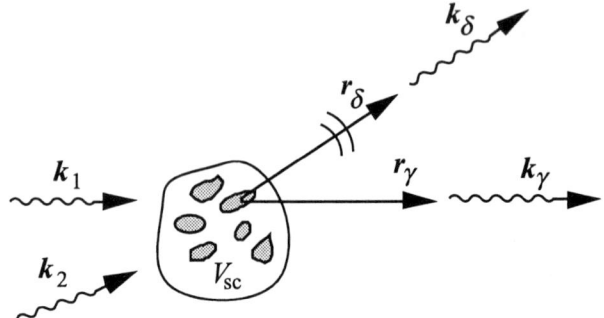

Fig. 4.1. Incidence of two plane waves on a bounded scattering volume.

conventional problem statement of scattering theory and to introduce the concept of the correlation scattering cross section.

Suppose that two plane waves with wave vectors k_1 and k_2, $k_{1,2} = k_0 n_{1,2}$, where $k_0 = \omega/c$ is the free-space wave number,

$$u^0(r) = A_1 e^{ik_1 r} + A_2 e^{ik_2 r}, \quad (4.10)$$

are incident upon a bounded scattering volume V_s (fig. 4.1). In this section, we restrict ourselves to the simplest case of a monochromatic field that satisfies the scalar wave equation [the factor $\exp(-i\omega t)$ is omitted].

In view of the linearity of the problem, at a point r_δ far from V_{sc}, the scattered field u_{sc} may be expressed in the form of two diverging spherical waves:

$$u_{sc}(r_\delta) = f(k_\delta \leftarrow k_1) A_1 r_\delta^{-1} \exp(ikr_\delta) + f(k_\delta \leftarrow k_2) A_2 r_\delta^{-1} \exp(ikr_\delta), \quad (4.11)$$

where $f(k_\delta \leftarrow k_j)$ are the random scattering amplitudes, and $k_\delta = k_0 r_\delta / r_\delta$. It is natural to treat this expression as a single spherical wave with the amplitude $f(k_\delta \leftarrow k_1) A_1 + f(k_\delta \leftarrow k_2) A_2$. Later, however, it will be more convenient to deal with two scattered waves corresponding to two incident waves.

Let us consider two points, r_δ and r_γ far from V_{sc}. From eq. (4.11) we have for the correlation function of the scattered field at these points:

$$\langle \tilde{u}_{sc}(r_\delta) u_{sc}^*(r_\gamma) \rangle = \frac{\exp[ik_0(r_\delta - r_\gamma)]}{r_\delta r_\gamma} \sum_{\alpha,\beta=1,2} \sigma(k_\delta, k_\gamma \leftarrow k_\alpha, k_\beta) A_\alpha A_\beta^*, \quad (4.12)$$

where $\tilde{u} = u - \langle u \rangle$ is the random part of u. It is natural to call the quantities

$$\sigma\langle k_\delta, k_\gamma \leftarrow k_\alpha, k_\beta \rangle = \langle \tilde{f}(k_\delta \leftarrow k_\alpha) \tilde{f}^*(k_\gamma \leftarrow k_\beta) \rangle \quad (4.13)$$

appearing in eq. (4.12) *the correlation scattering cross sections.*

The physical meaning of these parameters is evident: they describe the correlation of scattered plane waves with wave vectors k_δ and k_γ in the far zone (associated with the incidence of two plane waves with wave vectors k_α and k_β), or alternatively, the correlation of scattering processes $k_\alpha \to k_\delta$ and $k_\beta \to k_\gamma$. It follows that for determining the correlation scattering cross section (4.13) in the case of a stationary medium, the simultaneous incidence of two plane waves is unnecessary [the case of eq. (4.10)]. Instead, we may successively treat two scattering processes which are statistically correlated with one another owing to the scattering at the same fluctuations of the medium.

If the observation points are assumed to coincide, $(\delta = \gamma)$, and there is only one incident wave, $(A_2 = 0)$, the quantity $\delta(k, k \leftarrow k', k')$ is readily seen to be the usual cross section of incoherent scattering of a plane wave with wave vector k' into a wave with wave vector k:

$$\sigma\left(k, k \leftarrow k', k'\right) \equiv \sigma\left(k \leftarrow k'\right) = \sigma\left(\omega, n \leftarrow \omega', n'\right). \tag{4.14}$$

The concept of the correlation scattering cross section allows one to switch, in descriptions of scattered radiation, from its intensity to correlations. The convenience of this concept is especially evident in a diagram interpretation of the transfer equation (see, e.g., Barabanenkov and Finkelberg [1967], Apresyan and Kravtsov [1983]).

4.4. LIMITING RESOLVING POWER IN RADIOMETRIC MEASUREMENTS

The wave nature of radiation imposes fundamental constraints on the accuracy of radiometric measurements. These constraints may be viewed as a consequence of a wave uncertainty relation. The radiometric radiance $I(n, R)$ depends on the position vector R and on the direction of propagation n. Therefore it is expedient to evaluate the relation between angular and coordinate resolution. We solve this problem with a very simple measuring device – a lens of diameter D – by measuring the angular distribution of radiation in its focal plane (fig. 4.2).

Larger values of D improve the angular resolution of the lens $\Delta\theta = \lambda/D$, but impair the localization of radiance, for one has to refer the measured radiance $I(n, R)$ to a spot of diameter D rather than to a point R. In other words, the coordinate uncertainty ΔR is approximately equal to D. Consequently the angular and positional uncertainties are related by

$$\Delta\theta \cdot \Delta R \approx \frac{\lambda}{D} D = \lambda. \tag{4.15}$$

Thus, an increase in the angular resolution entails a loss in the spatial resolution and vice versa.

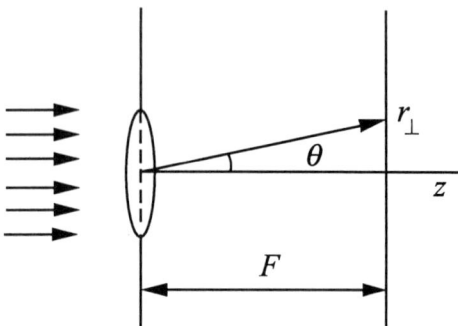

Fig. 4.2. Recording radiant intensity in the focal plane of a lens.

If the field measured in the plane of observation is statistically uniform, then one can achieve an arbitrary high angular resolution by increasing D. In a quasi-uniform field, however, it would be unwise to increase D above the quasi-uniformity scale L_R (Apresyan and Kravtsov [1983, 1984]), because for $D \geqslant L_R$, the distribution of intensity in the focal plane can no longer characterize the angular distribution of radiation. Consequently, the angular resolution limit is:

$$\Delta\theta_{\min} = \frac{\lambda}{L_R}. \tag{4.16}$$

Of course, this resolution requires that $D \leqslant L_R$.

When the angular distribution of radiation in the focal plane is known, the coherence function $\Gamma(\boldsymbol{R}, \boldsymbol{\rho})$ of the field can be determined with the aid of the Fourier transformation. The accuracy of determination of the coordinate of the center of gravity $\boldsymbol{R}$ (i.e., the accuracy of localization) is approximately equal to the lens diameter D. An increase in the difference variable ρ decreases the accuracy of reconstruction of the coherence function. The limiting value is estimated to be $\rho_{\max} \approx \lambda/\Delta\theta = D$; that is, the lens diameter.

Similar constraints also apply to the reconstruction of the temporal coherence function from a measured spectral density.

§ 5. New Application Fields of the RTE

5.1. ENHANCED BACKSCATTERING (WEAK LOCALIZATION)

The phenomenon of enhanced backscattering (weak localization) arouses interest for two reasons. First, this phenomenon limits the applicability range of the

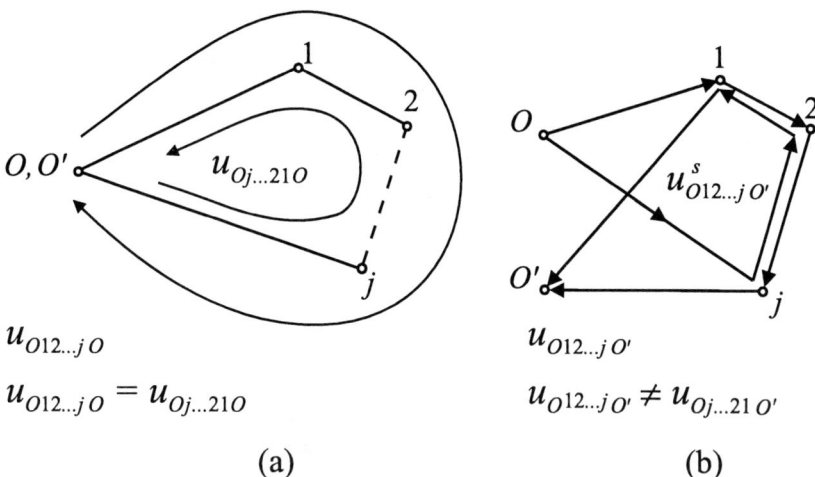

Fig. 5.1. (a) Watson's coherent backscattering channels occur when the receiver and the source are at one point. (b) When the receiver and the source are spaced apart, the fields passed over direct $(12\ldots j)$ and inverse $(j\ldots 21)$ sequences are no longer mutually coherent.

classical RTE, which becomes invalid in some rather small neighborhood of a coherent source. Second, using a straightforward modification one can describe the backscattering effect in terms of radiation transfer theory.

Backscattering enhancement owes its existence to the fact that a receiver placed near the source of radiation brings about specific coherent backscattering channels which are not covered by radiative transfer theory. The origin of these channels may be explained as follows.

Consider static discrete scatterers $S_1, S_2, \ldots, S_j$ which scatter, one by one, a monochromatic wave, radiated by a source at point O, so that the wave returns eventually to the receiver placed at the same point O, as shown in fig. 5.1a. In a continuous scattering medium, points $s_1, s_2, \ldots, s_j$ may be viewed as centers of elementary scattering volumes $dV_1, dV_2, \ldots, dV_j$.

Now let the scattered wave $u_{O12\ldots jO}$ correspond to the scattering sequence $Os_1s_2\ldots s_jO$. By the reciprocity theorem, a wave passing the same scatterers in the inverse order $s_j, s_{j-1}, \ldots, s_2, s_1$ produces the same field:

$$u_{O12\ldots jO} = u_{Oj\ldots 21O}. \tag{5.1}$$

We will call a sequence of scattering events at $s_1 \rightarrow s_2, \ldots, \rightarrow s_j$ a *scattering channel*. With this convention, eq. (5.1) tells us that the direct and inverse channels contribute equally to the scattered field, or that these scattering channels are mutually coherent.

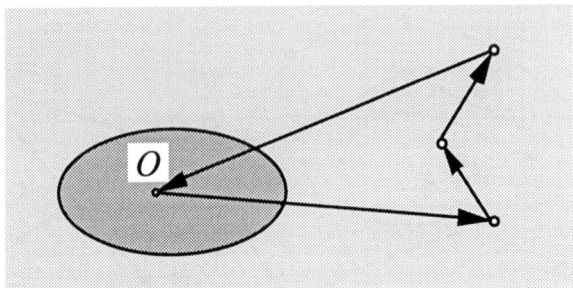

Fig. 5.2. Zone of coherence of backscattering channels is formed aroud the source O. Within this zone, transfer theory is not applicable directly.

Coherent backscattering channels will be referred to as Watson's channels, in deference to Watson [1969] who was the first to demonstrate their existence in radiation transfer theory. It is important to emphasize that mutual coherence of the fields $u_{O12...jO}$ and $u_{Oj...21O}$ holds for any arrangement of scatterers, however complex.

Equation (5.1) holds in view of the reciprocity theorem; i.e., in view of the system symmetry relative to time reversal. All factors destroying such symmetry (motion of scatterers, gyrotropy of the medium, etc.) invalidate relationship (5.1) and, consequently, the mutual coherence of Watson's channels.

The coherence of fields scattered forward and backward is characteristic only of situations where the positions of the receiver and the source coincide. If receiver O' and source O are separated in space, as in fig. 5.1b, then beginning from a certain separation OO', the fields $u_{O12...jO'}$ and $u_{Oj...21O'}$ lose mutual coherence:

$$u_{O12...jO'} \neq u_{Oj...21O'}. \tag{5.2}$$

Thus, the source is surrounded by a certain zone where coherent backscattering channels can occur, as shown in fig. 5.2.

In the case of a point source and a uniform distribution of scatterers, the coherence zone covers only distances in the order of the wave length. However, if an extended source of size l_s radiates a narrow beam of angular dimension $\Delta\theta \sim \lambda/l_s \ll 1$, then the coherence zone extends: its size across the beam is comparable to the dimension of the source, $l_\perp^{coh} \sim l_s$ and its size along the beam extends to a distance $l_\parallel^{coh} \sim l_s/\Delta\theta \sim l_s^2/\lambda$.

Of course, one cannot directly apply transfer theory to describe coherent effects. However, after a small modification, transfer theory effectively takes these effects into consideration. When scattering occurs strictly backward and the location of the point of observation coincides with that of the source, the intensities of multiply scattered fields double, whereas the singly scattered field does not change because, for this field, no Watson coherent partner-channels exist. Therefore, for the ray intensity of the backscattered field we may write:

$$I_{\text{bsc}} = I_{\text{single}} + 2I_{\text{multiple}} = I_{\text{single}} + 2\left(I_{\text{RTE}} - I_{\text{single}}\right) = 2I_{\text{RTE}} - I_{\text{single}}, \quad (5.3)$$

where I_{RTE} is the ray intensity calculated from the radiative transfer equation, I_{single} is the intensity of the singly scattered field (it can be obtained after a single iteration of the RTE with the unperturbed ray intensity I_0 on the right hand side), and $I_{\text{multiple}} = I_{\text{RTE}} - I_{\text{single}}$ is the contribution of multiple scattering in I_{RTE}.

Equation (5.3) admits a nice formulation in terms of Feynman diagrams. Barabanenkov [1973] demonstrated that, near the source, the term $2I_{\text{multiple}}$ in the first line of eq. (5.3) should be replaced with the sum of all ladder diagrams and a similar sum of all cyclic (or maximal crossed) diagrams; i.e.,

$$I_{\text{bsc}} = I_{\text{single}} + \sum I_{\text{ladder}} + \sum I_{\text{cyclic}}. \quad (5.4)$$

In scattering strictly backward, the contribution of cyclic diagrams exactly equals the contribution of ladder diagrams and together both types of diagram yield $2I_{\text{multiple}}$.

In contrast to this setting, far away from the source, the contribution of cyclic diagrams reduces sharply (due to dephasing of direct and reverse Watson channels), and in place of eq. (5.4) we obtain an ordinary ray intensity I_{RTE} satisfying the RTE:

$$I_{\text{RTE}} = I_{\text{single}} + \sum I_{\text{ladder}} = I_{\text{single}} + I_{\text{multiple}}. \quad (5.5)$$

In his pioneering work, Watson [1969] discussed the effect of multiple scattering on the effective backscattering cross section of electrons in a randomly inhomogeneous plasma. This phenomenon of the mutual coherence of Watson's pairs is frequently referred to as the *weak localization* of scattered waves, in contrast to the strong, or Anderson, localization which we shall touch upon briefly in § 5.2. The term "weak localization" originates from solid state physics, where the mutual coherence of Watson's channels was found to be significant for electrons interacting with impurities in metals. Inclusion of weak localization in the analysis results in a decrease in the conductivity of the metal at low temperatures as compared with that predicted by classic kinetic theory (Altshuler, Aronov, Khmelnitskii and Larkin [1982]).

De Wolf [1971] used this idea to estimate the backscattering enhancement for light and microwaves scattered by smooth inhomogeneities of a turbulent atmosphere, when the dimensions of inhomogeneities are large when compared with the wavelength. However, the backscattering intensity was found to be very small (exponentially decaying) and practically inaccessible for experimental measurements.

An allied enhancement effect of backscattering from bodies embedded in a turbulent medium is much stronger, because the field scattered backward by a body are many times stronger than the field scattered backward by the main inhomogeneities of the turbulent medium. This effect was analyzed by Belenkii and Mironov [1972] and Vinogradov, Kravtsov and Tatarskii [1973]. It was the latter authors who introduced the term backscattering enhancement.

Almost simultaneously, Barabanenkov [1973, 1975a,b] considered the backscattering from a half-space filled with random inhomogeneities and showed the presence of a peak of scattering in the direction opposite to the incidence of the primary wave. Many results of the early Barabanenkov papers on coherent backscattering enhancement were reiterated later by other researchers.

Enhanced backscattering from bodies in a turbulent medium was measured experimentally by Kashkarov and Gurvich [1977]. As to the peak of backscattering from suspensions and emulsions predicted by Barabanenkov, its observation was retarded by almost ten years. It was observed only in 1984-1985, simultaneously, by several research groups including Kuga and Ishimaru [1984], Kuga, Tsang and Ishimaru [1985], Wolf and Maret [1985], and Van Albada and Lagendijk [1985]. These authors used the term *weak localization* by analogy to the related effect in solid state physics.

At present, the literature abounds in theoretical and experimental studies dealing with investigations of coherent optical reflection. The results cover the reflection from a semi-infinite medium solved in the diffusion approximation (Akkerman, Wolf and Maynard [1986], Edrei and Kaveh [1987]) and a more rigorous radiative transfer theory (Gorodnichev, Dudarev and Rogozkin [1989], Ozrin [1992]), an allowance for moving scatterers (Golubentsev [1984a]), the effect of a magnetic field in a gyrotropic medium (Golubentsev [1984b]), the effect of reflection from interfaces (Freund and Berkovits [1990]), polarization effects (Wolf and Maret [1985]), to list but a few. The reader may find a detailed list of pertinent references in a review by Barabanenkov, Kravtsov, Ozrin and Saichev [1991]. Some other mechanisms exist in addition to those mentioned above; a classification of the entire multitude of these mechanisms may be found in Apresyan [1993].

Among experimental works on backscatter enhancement, an interesting paper

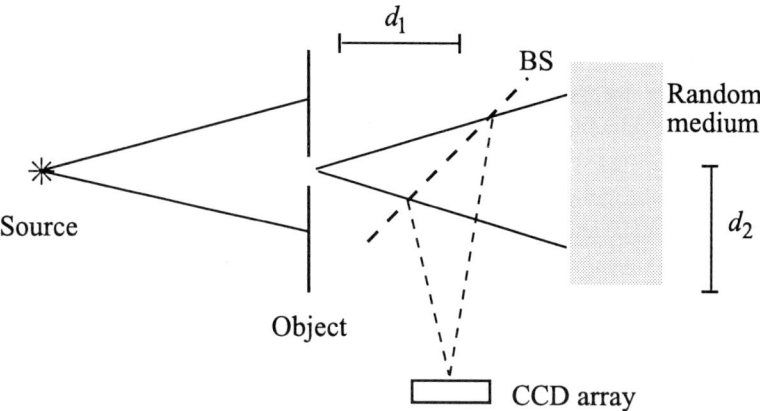

Fig. 5.3. Experimental configuration for the observation of an image due to backscattering without any focusing optics. The image is best focused when $d_1 = d_2$ (after Rochon and Bissonnette 1990).

by Rochon and Bissonnette [1990] deserves mention. The backscattering enhancement effect was used to image a transparency without a lens (fig. 5.3). In this experiment, a weak image of a transparency was recovered from a relatively uniform background of back-scattered light (a 512×256 array of charge-coupled devices was used as a detector).

5.2. NONLINEAR RTE AND STRONG LOCALIZATION

Up to this point, we have considered linear radiative transfer theory whose principal validity is restricted by the condition $\lambda \ll L_{\text{ext}}$, or, in the absence of losses, $\lambda \ll L_{\text{sc}}$. We have already pointed out that if this condition fails (i.e., in the region of high disorder, where the Ioffe–Regel condition $\lambda \sim L_{\text{sc}}$ holds), a qualitatively new phenomenon of strong localization arises, whose description is beyond the scope of the theory under consideration. This phenomenon was predicted by Anderson [1958] in his classic work on the behavior of a quantum particle in a random potential.

Anderson demonstrated that a new localized regime arises for wave field in a scattering medium beginning with a certain level of disorder. In this regime, wave packets do not spread out; i.e., the eigenfunctions of the scattering system which describe free propagation of the wave function are standing waves vanishing exponentially at infinity, in contrast to traveling quasi-plane waves considered in transfer theory. This is the reason why the description of Anderson's localization is far beyond the classic transfer theory.

Anderson's results are of importance for quantum theory of solids, particularly after Mott [1974] used Anderson's ideas to describe the metal-to-dielectric transition that occurs in systems with electric conductivity at low temperatures. Further development of the theory revealed that Anderson's transition to a strong localization regime is a continuous phase transition that can be described in terms of the general theory of phase transitions such as diverging length, critical exponent, and order parameters (see, e.g., Ma [1976]). The importance of Anderson's results was recognized by a 1977 Nobel Prize in physics.

Later it was found that Anderson's transition is not a peculiar quantum effect and is related directly to the use of the Schrödinger equation. This transition owes its existence to wave interference phenomena and can appear in different kinds of waves. Moreover, for one-dimensional and, probably, two-dimensional systems, the strong localization turns out to be a universal property that occurs at an arbitrary low level of disorder.

Linear transfer theory is not applicable to such systems when scattering occurs without frequency variations, thus forcing researchers to use other, more complicated methods. For linear transfer theory, it is likely that in the three-dimensional case, the localization occurs only under the condition of strong disorder $\lambda \sim L_{sc}$. For systems with weak fluctuations, the eigenfunctions are found to be delocalized, which relates them to familiar plane waves. A curve showing the transition to localized states in the space of disorder-describing parameters plus frequency is called the *mobility edge* (Mott [1974]).

For electrons in a solid, the condition of strong localization $\lambda \sim L_{sc}$ is realized with relative ease either by a small scattering length, or by a large wavelength λ. However, it is not a simple matter to construct the region of strong disorder for classic waves of different nature. Notwithstanding repeated attempts, no Anderson transition has yet been observed for classic (acoustic or electromagnetic) waves in a three-dimensional system. Simple estimates of the scattering length in the independent-particle approximation ($L_{sc} \approx 1/v_0 \sigma_1$; σ_1 is the cross section of an isolated scatterer, and v_0 is the concentration of scatterers) show that the condition $\lambda \sim L_{sc}$ can be satisfied only under resonant conditions where the scattering cross section is large enough.

Some investigators have cast doubt on the principal feasibility of the Anderson transition for classic waves because of pronounced losses in the medium which are absent for the wave function of electrons in a solid. Nevertheless, we consider it useful to elucidate certain features of strong localization to provide an insight into the limited possibilities of linear transfer theory.

The development of the theory of strong localization turned out to be a challenging problem. It has been approached by a variety of methods including numerical analysis, different forms of self-consistent theories, computational schemes based on perturbation theory, theories which describe localization by calculating Lyapunov's exponents, methods of percolation theory, methods of the field theory such as the σ nonlinear model, and supersymmetric theories. An extensive literature is devoted to attempts to describe strong localization. We refer the reader to the reviews of Kirkpatrick and Dorfman [1985] and Lee and Ramakrishnan [1985], to multiauthor volumes edited by Nagaoka and Fukuyama [1982] and by Kramers, Bergman and Bruynseraede [1985] which deal with strong localization of electrons, and to a multiauthor volume edited by Sheng [1990] devoted to the search for localized states of classic waves.

Despite much attention drawn to the problem, rigorous results are scarce. They are related mainly to one-dimensional problems which may be described in terms of 2×2 transition matrices (Erdős and Herndon [1982]). The best progress in the general theory of strong localization is usually associated with the work of Abrahams, Anderson, Licciardello and Ramakrishnan [1979], who suggested a scaling theory.

The results of the self-consistent theory of Vollhardt and Wölfle [1980, 1982] are also accepted widely. Starting from the earlier self-consistent theories (Goetze [1978]) and using a perturbation theory in the diagram representation together with some auxiliary heuristic assumptions, these authors derived a self-consistent, nonlinear equation for a generalized diffusion coefficient that offers a reasonable pattern for Anderson's localization in one- and two-dimensional systems and allows the boundaries of the localization region to be estimated in the case of three-dimensional problems. The results of that theory are used widely to estimate the conditions for an incipient Anderson's transition in both quantum and classic systems (see, e.g., Kirkpatrick [1985], Zhang, Chu, Xue and Sheng [1990]).

Apresyan [1989, 1990] has demonstrated that a nonlinear equation, drawing heavily on the diagram expansion, can be constructed for the intensity operator in the Bethe–Salpeter equation. Treated in the quasi-uniform limit, such a nonlinear equation leads to a self consistent equation for the diffusion coefficient derived by Vollhardt and Wölfle [1980]. As in other known strong localization theories, this nonlinear approach uses a certain extrapolation of perturbation theory that still requires mathematical corroboration. Nonetheless, the results of an analysis allow us to conclude that an efficient method to describe strong localization consists in constructing a nonlinear self consistent transfer theory based on an allowance for far phase correlations. Realization of this approach requires further research efforts.

5.3. RTE WITH FLUCTUATING PARAMETERS. THE EFFECT OF TRANSLUCENCE

The concept of turbid media with random parameters may seem strange at first, because turbid media are themselves objects with random characteristics. Nevertheless, the concept of *random turbid media* becomes justified if one takes into account the *scale hierarchy* in natural media.

The scattering cross-section σ and extinction coefficient α are introduced into the transfer equation as local parameters; i.e., averaged over a certain volume V. Dimensions of the averaging volume $l_{av} \sim V^{1/3}$ must evidently be larger than the correlation length l_{micro} of microinhomogeneities:

$$l_{av} \gg l_{micro}. \tag{5.6}$$

Consider a turbid medium, whose local parameters α and σ exhibit slow variations with macroscales l_{macro} much larger than the averaging scale l_{av}:

$$l_{macro} \gg l_{av} \gg l_{micro}. \tag{5.7}$$

If these variations of σ and α are of an irregular nature and follow a statistical pattern, then it is appropriate to consider σ and α as random functions of coordinates in the range of microscales.

This switching from micro- to macro-scales in transfer theory has its analogy in phenomenological optics. A passage from a microscopic to a macroscopic (phenomenological) description implies consideration of quantities averaged over a certain region satisfying the condition (5.6). In this way macroscopic characteristics of transparent media, first of all, their dielectric permittivity, evolve. Nevertheless, averaged (i.e., macroscopic) characteristics may, in turn, become the subject of statistical analysis when large-scale processes satisfying the condition (5.7) are considered. The statistical theory of propagation of light in turbulent media is build in this way. In view of these facts, one may consider the effects caused by fluctuations of the transfer equation parameters (extinction coefficient and scattering cross section) to derive an equation for the moments this time of the random radiance of radiation.

Taking this into account, consider the random values of extinction coefficient α and scattering cross-section σ by setting:

$$\alpha(r) = \bar{\alpha} + \tilde{\alpha}, \qquad \sigma(r, n \leftarrow n') = \bar{\sigma} + \tilde{\sigma}. \tag{5.8}$$

We write the transfer equation with random parameters in the symbol form:

$$\frac{dI}{ds} = AI, \tag{5.9}$$

where $dI/ds = \mathbf{n}\nabla I$ and A is the operator, introduced for brevity, that operates in accordance with the rule

$$AI = -\alpha(\mathbf{r})I + \int \sigma(\mathbf{r}, \mathbf{n} \leftarrow \mathbf{n}') I(\mathbf{n}') \, d\Omega'. \tag{5.10}$$

The radiance I in eq. (5.9) now becomes a random quantity whose fluctuations are caused by the fluctuations of the operator A. Considering eq. (5.9) as an initial *stochastic* equation, we could follow the procedure of Rytov, Kravtsov and Tatarskii [1989b] for deriving the Dyson equation and immediately write the equation for the mean $\bar{I}$. We give here only simplified calculations leading to equation for $\bar{I}$.

The operator A is characterized by its mean value $\bar{A}$ and the fluctuation component $\tilde{A}$:

$$A = \bar{A} + \tilde{A}, \tag{5.11}$$

which are expressed through $\bar{\alpha}$, $\tilde{\alpha}$, $\bar{\sigma}$, and $\tilde{\sigma}$ in an obvious manner. The radiance I can also be represented as the sum of its mean and fluctuating parts:

$$I = \bar{I} + \tilde{I}. \tag{5.12}$$

By substituting eqs. (5.11) and (5.12) into the transfer eq. (5.9), we single out the mean component,

$$\frac{d\bar{I}}{ds} = \bar{A}\bar{I} + \langle \tilde{A}\tilde{I} \rangle, \tag{5.13}$$

and the fluctuating component with zero mean,

$$\frac{d\tilde{I}}{ds} = \tilde{A}\bar{I} + \bar{A}\tilde{I} + \tilde{A}\tilde{I} - \langle \tilde{A}\tilde{I} \rangle. \tag{5.14}$$

The system of eqs. (5.13) and (5.14) cannot be decoupled into independent equations for $\bar{I}$ and $\tilde{I}$. Therefore the evaluation of the mean radiance $\bar{I}$, the correlation function $\langle \tilde{I}_1 \tilde{I}_2 \rangle$, and higher moments $\langle \tilde{I}_1 \ldots \tilde{I}_m \rangle$ is possible only by using some approximate methods. We now consider a method that reduces the transfer equation with random parameters to an effective transfer equation for the mean radiance $\bar{I}$; i.e., to a transfer equation with effective parameters by an example of the bleaching effect (effect of translucence) in turbid media with random parameters. This effect was apparently first discussed by Dolin [1984]

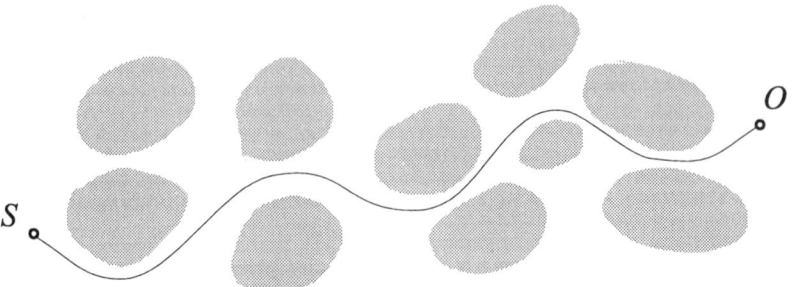

Fig. 5.4. Illustration of the least attenuation principle: the radiation arrives at the observation point over a variety of propagation channels, but the channel corresponding to the least attenuation prevails.

and Borovoi [1984]. Here we will follow a recent paper by Kleeorin, Kravtsov, Mereminskii and Mirovskii [1989].

The bleaching effect manifests itself in that the scattering medium with random variations of its parameters becomes more transparent on average than the same medium without fluctuations. To put this another way, the medium exhibits its lowest transparency when its scattering inhomogeneities are distributed uniformly.

This peculiarity of radiative transport in a random turbid medium may be related to the physically obvious principle of least attenuation. According to this principle, in the presence of many competitive paths (or propagation channels) in the medium, radiation comes to the observer over the path with the lowest attenuation, because this path contributes most effectively to the observed intensity.

Figure 5.4 illustrates the least attenuation principle. The dark areas represent sites of high attenuation. Among all possible paths from the source S to the observer O, the path A yields the maximum intensity contribution because it is mainly through light regions, whereas other virtual paths cause higher attenuation. The enhancement of intensity at the observation point due to the contributions from paths with lower attenuation is greater than the loss of intensity due to paths with higher attenuation.

To make this point clear, we consider the following simple model. Let the absorption coefficient α be a fluctuating quantity that can take on two values: $\langle\alpha\rangle+\Delta$ and $\langle\alpha\rangle-\Delta$ with equal probability $P_+ = P_- = \frac{1}{2}$. Then the average value of the respective transmittance (i.e., the attenuation factor) $e^{-\alpha}$ is:

$$\langle e^{-\alpha}\rangle = \tfrac{1}{2}e^{-\langle\alpha\rangle}\left(e^{\Delta}+e^{-\Delta}\right),$$

which is always greater than its unperturbed value $e^{-\langle\alpha\rangle}$.

We support the above qualitative considerations by a simple calculation that can provide insight into the physical mechanism of the bleaching effect, but cannot give its complete universal description. We now neglect the scattering and intrinsic radiation of the medium and consider a special case with fluctuations of the absorption factor α_{abs} only. The transfer equation (5.9) then takes the form

$$\frac{dI}{ds} = -\alpha_{\text{abs}}I, \qquad (5.15)$$

because $A = -\alpha_{\text{abs}}$ under the assumptions above.

Assuming that the fluctuations $\tilde{\alpha}$ are small when compared with $\bar{\alpha}$ and the fluctuations of radiance $\tilde{I}$ are small in comparison with $\bar{I}$, we neglect the difference $\tilde{\alpha}\tilde{I} - \langle\tilde{\alpha}\tilde{I}\rangle$ in eq. (5.14) as a term of the second order of smallness compared with the main term $\bar{\alpha}\tilde{I}$. Neglecting $\tilde{\alpha}\tilde{I} - \langle\tilde{\alpha}\tilde{I}\rangle$ is equivalent to the Bourret approximation for the Dyson equation for $\bar{I}$. True, here all the quantities are energy rather than field variables. Note that the derivation of the Dyson equation for $\bar{I}$ from eqs. (5.13) and (5.14) was considered by Vardanyan [1988] and Manning [1989].

Neglecting the difference $\tilde{\alpha}\tilde{I} - \langle\tilde{\alpha}\tilde{I}\rangle$ carries eq. (5.14) to the form (with allowance for $A = -\alpha$):

$$\frac{d\tilde{I}}{ds} = -\tilde{\alpha}\bar{I} - \bar{\alpha}\tilde{I},$$

that admits an explicit solution

$$\tilde{I} = -\int_{\infty}^{s} \tilde{\alpha}\bar{I}\exp\left[\bar{\alpha}(s'-s)\right]ds - \int_{s}^{\infty} \tilde{\alpha}\bar{I}\exp\left[\bar{\alpha}(s-s')\right]ds', \qquad (5.16)$$

which satisfies the reasonable requirement that $\tilde{I} \to 0$ as $\tilde{\alpha} \to 0$. Substituting eq. (5.16) into eq. (5.13), we obtain a closed integro-differential equation for the mean radiance $\bar{I}$: the Dyson equation in the Bourret approximation:

$$\frac{d\bar{I}}{ds} = -\bar{\alpha}\bar{I} + \int_{-\infty}^{\infty} \langle\tilde{\alpha}(r)\tilde{\alpha}(r')\rangle \exp\left(-\bar{\alpha}|s-s'|\right)\bar{I}(r')\,ds'. \qquad (5.17)$$

Here, s corresponds to the point r, and s' corresponds to the point r'.

Define the correlation length of the absorption factor by the integral formula

$$l_c = \int_{-\infty}^{\infty} \frac{\langle \tilde{\alpha}(r)\tilde{\alpha}(r')\rangle}{\langle \tilde{\alpha}^2 \rangle} d\xi', \qquad \xi' = |r - r'|,$$

and assume that the correlation length l_c of its inhomogeneities is small when compared with the absorption length, $l_{\text{abs}} = 1/\bar{\alpha}$, coinciding with the characteristic variation length of the mean radiance $\bar{I}$. Thus, we assume that:

$$\bar{\alpha} l_c \ll 1. \tag{5.18}$$

Then the exponential factor in eq. (5.17) may be replaced with unity, and all the integral terms may be replaced with the local value:

$$\int \langle \tilde{\alpha}(r)\tilde{\alpha}(r')\rangle \exp(-\bar{\alpha}|s - s'|) \bar{I}(r') \, ds' \approx \langle \tilde{\alpha}^2 \rangle l_c \bar{I}(r).$$

As a result, eq. (5.7) takes the form

$$\frac{d\bar{I}}{ds} = -\alpha_{\text{eff}} \bar{I}, \tag{5.19}$$

where the quantity

$$\alpha_{\text{eff}} = \bar{\alpha} - \langle \tilde{\alpha}^2 \rangle l_c \tag{5.20}$$

represents the effective absorption factor in the medium with fluctuating parameters. This quantity is always smaller than the mean value $\bar{\alpha}$: $\alpha_{\text{eff}} < \bar{\alpha}$; i.e., the turbid medium with fluctuating parameters appears *more transparent* than the nonfluctuating medium with the mean absorption factor $\bar{\alpha}$.

The simple analysis given above shows that the bleaching of turbid media with fluctuating parameters is associated with the correlation, or more correctly, with the anticorrelation between the fluctuations of the radiance $\tilde{I}$ and the absorption $\tilde{\alpha}$: the higher the absorption α, the lower the radiance $\tilde{I}$. It is due to this anticorrelation that radiation tends to select a path corresponding to the smallest absorption.

If fluctuating parameters include not only the absorption factor α_{abs} but also the cross-section per unit volume σ, then mean-radiance calculations become more complicated. The equation for the mean radiance $\bar{I}$ does not reduce, in the general case, to a radiative transfer equation. However, it becomes a transfer equation with effective parameters if the condition (5.18) holds; i.e., if the optical thickness of an isolated inhomogeneity is small.

Suppose that the fluctuations in absorption and scattering are brought about by the same cause; namely, by the fluctuations in the concentration of absorbing and scattering particles. In this case, we may write:

$$\tilde{\alpha} = B\tilde{n}, \quad \tilde{\sigma} = D\tilde{n}, \quad \bar{\alpha} = B\bar{n}, \quad \bar{\sigma} = D\bar{n}, \tag{5.21}$$

where $\bar{n}$ is the average concentration of particles, $\tilde{n}$ is the fluctuating part of the concentration, and B and D are proportionality coefficients. Calculations identical to those of the previous section, with the substitution of $-\tilde{A} = \tilde{\alpha} - \tilde{\sigma} = (B-D)\tilde{n}$ for $\tilde{\alpha}$ and $-\bar{A} = (B-D)\bar{n}$ for $\bar{\alpha}$, yield the transfer equation (5.19), where the operator

$$A_{\text{eff}} = -(B-D)\bar{n} - (B-D)^2 \langle \tilde{n}^2 \rangle l_c \tag{5.22}$$

replaces $-\alpha_{\text{eff}}$.

Taking eqs. (5.21) into account we may represent the operator relationship (5.22) in the equivalent form $A_{\text{eff}} = -\alpha_{\text{eff}} + \sigma_{\text{eff}}$, where $\alpha_{\text{eff}} = \bar{\alpha}(1 - \bar{\alpha}\gamma l_c)$, and the operator σ_{eff} has the kernel

$$\sigma_{\text{eff}}(\mathbf{n}\leftarrow\mathbf{n}') = (1 - 2\bar{\alpha}\varepsilon l_c)\bar{\sigma}(\mathbf{n}\leftarrow\mathbf{n}') + \gamma l_c \int \sigma(\mathbf{n}\leftarrow\mathbf{n}'')\sigma(\mathbf{n}''\leftarrow\mathbf{n}')\,\mathrm{d}\Omega''. \tag{5.23}$$

Here, $\gamma = \langle \tilde{n}^2 \rangle / \bar{n}^2$ denotes relative concentration fluctuations.

We see that the fluctuations of medium parameters result in a familiar decrease of the extinction factor and in some change of the scattering cross-section (5.23). First, in the expression for the average cross-section $\bar{\sigma}$, the factor $1 - 2\gamma l_c \bar{\alpha}$ appears that is even smaller than the factor $1 - \gamma l_c \bar{\alpha}$ in the expression for α_{eff}. Second, in the relation (5.23) for the kernel, the integral term that corresponds to "double scattering" of radiance by inhomogeneities of the turbid medium, appears. In the general case, the scattering pattern of that term is broader than that of the first term. The first and second terms taken together produce weaker scattering as compared with the case of a homogeneous medium. Thus, the above trend to decrease the absorption finds its analog in a reduction of the effective scattering.

The considered effect may manifest itself first in astrophysics. Inhomogeneities in plasma density and in the density of interstellar gas clouds may be the reason for an enhancement of the brightness temperature of space objects. Vardanyan [1988] and Manning [1989] estimated the magnitude of this effect. Of course, in this case, it is required that eq. (5.9) include an additional term describing the strength of sources.

Inhomogeneous snow cover is another physical system of practical interest. The brightness temperature of the earth's surface where covered with snow, is decided mainly by the transparency of the snow cover that shields the relatively warm earth from the air. Inhomogeneities increase the transparency of the snow cover and thereby raise its brightness temperature. This problem was considered theoretically by Kleeorin, Kravtsov, Mereminskii and Mirovskii [1989], whereas experimental evidence in support of the bleaching effect may be found in the paper of Wen, Tsang, Winebrenner and Ishimaru [1990].

We also mention the optical effect connected with inhomogeneities in a continuous cloud cover: inhomogeneities, even small ones, may considerably increase the radiance of the earth's surface.

Finally, we notice one implication for systems of protection from neutron fluxes: the homogeneity of screening layers may be a more important factor than the total mass of the protective cover. This may be deduced from the heuristic principle of least attenuation: any inhomogeneities of the protecting screen, not necessarily random, act as "loop-holes" for neutrons leaving the reactor.

5.4. WARMING AND COOLING EFFECTS IN SCATTERING MEDIA

Remote sensing is now a widely accepted technique for the study of the atmosphere, ocean, and land using receivers of thermal radiation aboard airplane and satellites. Brightness temperature measured with such receivers depends on the parameters of these media and above all on their refraction and absorption indexes. It has been found recently that random volume inhomogeneities and irregularities of the interface can contribute considerably to the brightness temperature. Therefore, studies of the thermal radiation of natural media are an important source of information on regular and random parameters of these media.

A quantitative analysis of the effect of random inhomogeneities on thermal radiation can be carried out in principle using transfer theory. For scattering media, this was done, for example, by England [1974] and Tsang and Kong [1975]. For random boundaries, calculations on transfer theory are technically difficult, especially if the dimensions of surface irregularities are small in comparison with the wavelength.

Under these circumstances, it is more convenient to resort to the electromagnetic theory of thermal fluctuations, especially using the diffraction generalization of Kirchhoff's law. According to this generalization, the power of thermal radiation from a body or a medium at a given frequency and at a given point (or in a given direction) is proportional to the power of an auxiliary wave absorbed by the body or the medium. The auxiliary wave is assumed to be transmitted from the desired point (or in desired direction) and has the desired polarization and frequency.

In this formulation, the Kirchhoff statement – that the radiating and absorbing powers of a body are proportional to each other, as was initially derived in the limit of geometrical optics – is also valid for bodies with dimensions below or above the wavelength. This extension of the Kirchhoff law is given, in particular, in the book of Rytov, Kravtsov and Tatarskii [1989a].

A general statement can be derived from the diffraction formulation of the Kirchhoff law: this statement concerns the conditions for "warming" and "cooling" of radiating bodies and media with random inhomogeneities. Let T_0 be the physical temperature of a material medium, and $T_b(n)$ be its brightness temperature in the direction n. In accordance with the extended Kirchhoff law, the brightness temperature $T_b(n)$ differs from T_0 by the factor $\chi_{\text{eff}}(n_0)$ that characterizes the effective absorbing power; i.e., the proportion of absorbed power of a plane wave incident at the interface in the direction $n_0 = -n$:

$$T_b(n) = T_0 \chi_{\text{eff}}(n_0). \tag{5.24}$$

This formula is often written in the form

$$T_b(n) = T_0 [1 - R_{\text{eff}}(n_0)], \tag{5.25}$$

where $R_{\text{eff}}(n_0) = 1 - \chi_{\text{eff}}(n_0)$ is the effective proportion of the reflected and scattered power when the medium is irradiated from the direction n_0.

The scattering of the primary field below the interface increases the effective reflection coefficient, thus reducing the brightness temperature. This reduction of T_b corresponds to the effect of "cooling" of the scattering medium as compared with the transparent medium (Gurvich, Kalinin and Matveev [1973]).

Straightforward but cumbersome calculations helped explain the frequency profile of the brightness temperature of glaciers for various hypothesis about ice parameters (Gurvich, Kalinin and Matveev [1973]).

The above effect of cooling of a scattering medium with small-scale inhomogeneities does not contradict the warming effect that is caused by large scale fluctuations of the parameters α and σ in the radiative transfer equation (see § 5.3). These effects are simply related to different causes. In order to describe these effects simultaneously, one should consider an optically thin layer of the medium ($\alpha H \leqslant 1$) and take into account additional terms caused by the propagation of thermal radiation from a warm layer underlying the ice. Large-scale fluctuations of medium parameters result in a higher transparency of the layer with respect to the radiation coming from below.

5.5. THERMAL RADIO EMISSION OF ROUGH SURFACES

Studies of the brightness temperature of the ocean surface are an important source of data about oceanic processes. Calculations of the brightness temperature may be carried out by the formula (5.25) using some approximation for R_{eff} (Rytov, Kravtsov and Tatarskii [1989b]). Most applications use the perturbation method (which describes small-scale irregularities), the Kirchhoff method (which describes large-scale irregularities), and the combined approach that deals with both large- and small-scale irregularities.

The pioneering calculations of thermal radiation (Stogryn [1967], Wu and Fung [1972]) were predominantly based on the perturbation method. Genchev [1984] and Raizer and Filonovich [1989] calculated thermal radiation with the Kirchhoff method for a sinusoidal surface. Finally, Wentz [1975] analyzed thermal radiation for a surface described by a two-scale model.

Calculations showed that surface roughness may both increase and decrease the brightness temperature. The sign of the increment in the brightness temperature relative to the brightness temperature $T_{b0} = T_0 (1 - R_0)$ of a plane boundary depends on the nature of roughness, observation angle, and polarization. The temperature contrast, ΔT_b, between the land and sea may be as large as 5–10°C.

The foam on the ocean surface decreases R_{eff} and, hence, acts toward "warming" of the ocean surface. Bubbles in the foam act as matching elements between the air and water, producing the effect of an oil film (Raizer and Sharkov [1981]). In contrast, splashes over the ocean surface cause a "cooling" of the ocean (Raizer [1992], Dombrovskii and Raizer [1992]) because splashes give rise to an additional scattering and decrease the brightness temperature in accordance with eq. (5.25).

Quasi-periodic disturbances on the surface of a fluid whose characteristics scales are of the order of the wavelength may result in a decrease of the relative temperature in some definite, critical directions θ_{cr} dictated by the condition

$$K = k_0 (1 \pm \sin \theta_{\text{cr}}). \tag{5.26}$$

Here, $K = 2\pi/\Lambda$ is the wave number for a sinusoidal disturbance with period Λ and k_0 is the wave number of the electromagnetic wave.

When the condition (5.26) is satisfied, one of the diffraction spectra due to the diffraction at a sinusoidal surface propagates along the surface and generates (if polarized vertically) a surface wave. Such waves are absorbed effectively in the fluid and, consequently, they increase the brightness temperature in agreement with eq. (5.24). As a result, the brightness temperature increases in the critical directions θ_{cr} defined by eq. (5.26) (Kravtsov, Mirovskaya, Popov, Troitskii and Etkin [1978]).

Critical phenomena are also described by Kong, Lin and Chuang [1984] and Etkin, Irisov and Trokhimovskii [1992]. Allowance for these phenomena markedly increases the accuracy of the agitated ocean surface parameters, specifically the sea surface temperature and the wind speed and direction over the surface.

§ 6. New Effects in Statistical Radiative Transfer

6.1. SCALAR MEMORY EFFECT

In classic radiative transfer theory it is usually assumed that, in multiple scattering, radiation "forgets" its initial characteristics such as the direction of propagation or the angular spectrum and the initial frequency if the scattering is accompanied by changes in frequency. However, this assumption is only partially correct. The information on the initial characteristics of radiation disappears not completely, rather it is "encoded" in the process of multiple scattering.

Consider a standard problem of multiple scattering of a plane monochromatic wave by a statistically homogeneous layer of a scattering medium. For the sake of definiteness, we will consider the transmitted wave, though similar conclusions may be drawn for the reflected radiation. To simplify the problem, we will assume the inhomogeneities in the layer to be small-scale. Then, if the layer thickness L greatly exceeds the mean free path of radiation L_{sc}, the transmitted radiation will be nearly isotropic; i.e., the radiation pattern behind the layer will be nearly isotropic regardless of the direction of propagation of the incident wave at the incidence on the layer. It is in this sense that one may speak of "forgetting" of the initial wave characteristics. In reality, however, the propagating wave "memorizes" partially the incident wave. Let us consider some of such "memories", and, above all, the *scalar memory effect*.

The scalar memory effect was first predicted in the theoretical paper of Feng, Kane, Lee and Stone [1988] and was borne out experimentally by Freund, Rosenbluh and Feng [1988] (see also Berkovits, Kaveh and Feng [1989]). The essence of this effect may be formulated as follows.

Consider a plane wave with a wave vector k_a incident on a scattering layer. This wave is transformed in transmitted and reflected waves, which have generally wide angular spectra (fig. 6.1). Keeping in mind that the radiation is usually observed in the statistical Fresnel zone relative to scatterers, we consider

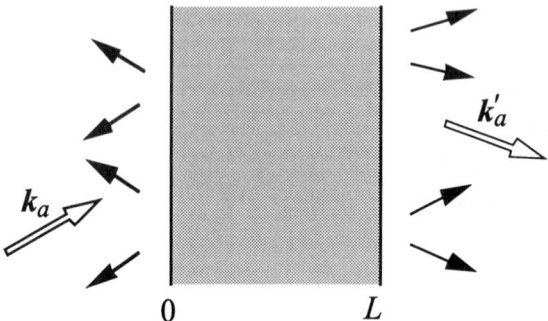

Fig. 6.1. Scattering of a plane wave by a statistically homogeneous layer (wave vectors for scattering process $k_a \to k'_a$ under consideration are indicated by double shafted arrows).

the transmitted wave with wave vector k'_a. We will call such a scattering process, $k_a \to k'_a$, process "a".

Let us look now at another scattering process $k_b \to k'_b$, process "b", that differs from process "a" in that the directions of the wave vectors k_a and k'_a are replaced with k_b and k'_b, respectively. If the scattering layer thickness L is large when compared with the scattering length L_{sc} and the layer is statistically uniform in plane $z =$ const, then the correlation of the two processes under consideration, "a" and "b", prove to be not very small only if:

$$q_{a\perp} = q_{b\perp}, \tag{6.1}$$

and

$$|k_{a\perp} - k_{b\perp}| \leqslant \frac{1}{L}. \tag{6.2}$$

Here $k_\perp$ denotes the component of any vector k parallel to the layer, and $q_a = k'_a - k_a$ and $q_b = k'_b - k_b$ are the scattering vectors for processes "a" and "b", respectively. The qualitative sense of the memory effect is as follows.

In the case of a medium with continuous fluctuation of permittivity ε, the first condition (6.1) may be rearranged in the Bragg form. Indeed, applying the Fourier transformation over coordinates $r_\perp$ transverse with respect to z, we can represent the field in the layer as a superposition of all possible interacting waves of opposite directions $u^{(+)}(k'_\perp, z)$ and $u^{(-)}(k''_\perp, z)$, where $k'_\perp$ and $k''_\perp$ are wave vectors transverse with respect to z (see, for example, Malakhov and Saichev [1979]). The strongest resonant interaction occurs for waves which satisfy the condition of synchronism: $k'_\perp - k''_\perp = q_\perp$, where $q_\perp$ is the wave vector of fluctuations in the medium $\varepsilon(q_\perp, z)$.

Let us assume that the resonant fluctuations of the medium are, on average, most important for the scattering process. This means that the fluctuations $\varepsilon(q_{a_\perp}, z)$ with $q_{a_\perp} = k'_{a_\perp} - k_{a_\perp}$ are essential for the process $k_a \to k'_a$ and, similarly, $\varepsilon(q_{b_\perp}, z)$ with $q_{b_\perp} = k'_{b_\perp} - k_{b_\perp}$ for the process $k_b \to k'_b$. Therefore, the correlation of amplitudes of the two processes is proportional – with allowance for the statistical uniformity of fluctuations in the layer – to the quantity $\langle \varepsilon(q_{a_\perp}, z') \varepsilon^*(q_{b_\perp}, z'') \rangle \propto \delta(q_{a_\perp} - q_{b_\perp})$, whose singularity explains the sense of the condition (6.1).

To comprehend the second condition (6.2), we move slightly the incident wave vector k_a to find out from what amplitude this motion becomes noticeable for the complex speckle structure of the propagating wave. Consider an arbitrary point R at the far side of the layer at $z = L$. To this point there corresponds an "influence region" A at the near side of the layer ($z = 0$) such that a variation of the incident wave in A changes appreciably the magnitude of $u(r)$. This region is similar to the Fresnel zone for a wave in a free space. If L_A denotes the characteristic size of A, then changing from the wave vector k_a to its new value k_b is of little effect until the phase difference on the edges of this region remains small in comparison with unity; i.e., until $|k_{a_\perp} - k_{b_\perp}| L_A < 1$. In the problem under consideration, with its thick layer and small-scale fluctuations of scatterers, the wave evolution in the layer has a diffuse behavior, which results in L_A of the order of the layer thickness $L : L_A \sim L$. Thus, we have arrived at the condition (6.2).

This reasoning is valid not only for the transmitted wave but also for the reflected wave with the difference that, for a thick layer $L \gg L_{\text{ext}}$, the "influence region" size L_A is comparable to the extinction length L_{ext} and is independent of the layer thickness L. This means that, for the reflected wave, condition (6.1) keeps its form and condition (6.2) transforms into:

$$|k_{a_\perp} - k_{b_\perp}| \leqslant \frac{1}{L_{\text{ext}}}. \tag{6.3}$$

We emphasize that the memory effect bears substantially on the assumption of statistical uniformity of the layer in the plane $z = \text{const}$. Therefore any factor destroying this uniformity (e. g., a finite size of the layer in the plane $z = \text{const}$.) breaks down the memory effect (Eliyahu, Berkovits and Kaveh [1991]).

If the incident wave propagates near the normal to the layer, then the memory effect admits a simpler interpretation. Indeed, in this case, $|k_{a_\perp} - k_{b_\perp}| \approx k_0 \theta_{ab}$, where θ_{ab} is the angle between k_a and k_b so that the condition (6.2) denotes $\theta_{ab} = \theta'_{ab}$ where θ'_{ab} is the angle between k'_a and k'_b (see fig. 6.1). In other words, deflection of the incident wave on a small angle $\theta_{ab} < \lambda/L$ results, on average, in deflection of the speckle structure of the propagating wave as a whole on the same angle. For the reflected wave, the angle θ_{ab} is limited by the condition $\theta_{ab} < \lambda/L_{ext}$ that follows from eq. (6.3). Note that in the single-scattering approximation, such a deflection occurs not only on average, but in each particular realization.

6.2. TIME REVERSED MEMORY EFFECT

The scalar memory effect considered in the foregoing section defines the region of correlation of scattering processes "a" and "b" in the simplest case when the difference between the wave vectors of these processes is small ($|k_{a\perp} - k_{b\perp}| \leqslant 1/L$ for the transmitted wave, and $|k_{a\perp} - k_{b\perp}| \leqslant 1/L_{ext}$ for the reflected wave). There is an additional, more complex mechanism of such a correlation – the *time-reversed memory effect* that is connected with symmetry under time reversal; i.e., with the reciprocity theorem.

If scatterers satisfy the conditions of this theorem (excluding from consideration gyrotropy, nonstationarity of medium, and other effects violating reciprocity), then the transmission factor through the layer $t(k' \leftarrow k)$ satisfies the reciprocity condition $t(k \leftarrow k') = t(-k' \leftarrow -k)$ in the scalar approximation. A similar condition holds for the reflection factor. Hence, equal transmission factors correspond to the scattering processes $k_a \rightarrow k'_a$ and $-k'_a \rightarrow -k_a$; i.e., these processes are strongly correlated (Berkovits and Kaveh [1990a]).

In the case of transmission, this new correlation can be observed only if scattering process "b" is due to the wave incident on the layer from the opposite side (fig. 6.2a). In the case of reflection, there is no need to change the direction of propagation of the incident wave because the wave vectors of direct and time-reversed processes lie on the one side of the scattering layer. In the last case, the time-reversed memory effect gives rise to an additional correlation peak that occurs when process "b" coincides with time-reversed process "a".

Polarization complicates the time-reversed memory effect. In this case, the transmission and reflection factors become tensor quantities, and the scalar reverse of time discussed above should be completed by the transposition of related tensors. As a result, the pattern of correlations becomes substantially dependent upon the polarizations of incident and scattered waves. These vector memory effects were described in Berkovits and Kaveh [1990b].

6.3. LONG CORRELATION EFFECTS FOR INTENSITIES

The most interesting results in this range are associated with the departure in the statistics of the scattered field from Gaussian. Even if scattering has a

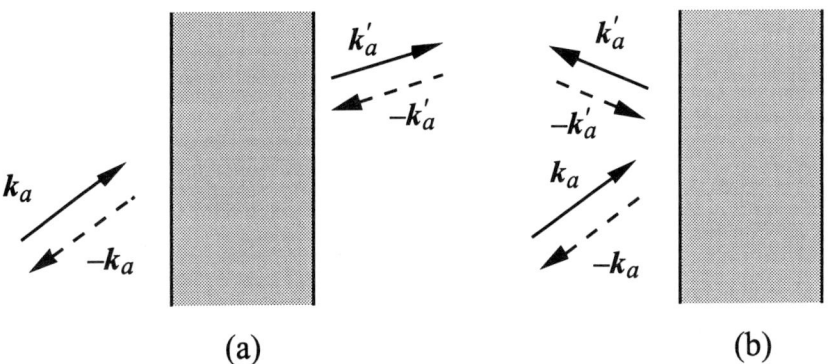

Fig. 6.2. Strongly correlated direct and time-reversed waves for a scattering process symmetric with respect to time reversal: (a) transmission; (b) reflection.

diffuse character and the average intensity can be described in the diffusion approximation of transfer theory, the correlations of intensities may include appreciable non-Gaussian corrections. These corrections result in a slower decay of correlations when compared with the results of the simple "quasi-Gaussian" approximation. We outline some results in this field, avoiding complicated calculations.

In the range of weak disorder $\lambda \ll L_{sc}$, the standard perturbation theory is applicable in the form of diagram expansions. The diagram technique gives the fourth moments in the form of four-deck diagrams similar to the two-deck diagrams used to calculate second moments. Such diagrams can be partially summed by extracting internal two-deck blocks which can be estimated using the conventional "correlation" transfer theory. With this approach, Feng, Kane, Lee and Stone [1988] showed that the normalized correlation function $C_I = \langle \tilde{I}(k_1) \tilde{I}(k_2) \rangle$ of intensity fluctuations $\tilde{I}(k)$ of the transmitted wave in a waveguide is representable as a sum of three terms:

$$C_I \approx C^{(1)} + C^{(2)} + C^{(3)}. \tag{6.4}$$

These terms are given in diminishing order with respect to the small parameter $(\lambda/L_\perp)^2 \ll 1$; i.e., $C^{(j)} \propto (\lambda/L_\perp)^{2(j-1)}$ where $L_\perp$ is the transverse size of the waveguide. Wang and Feng [1989] extended this result to reflected waves.

The first term, $C^{(1)}$, corresponds to the "quasi-Gaussian" approximation, where the fourth moment is represented in the form of the sum of pair products of second moments, as in the case with a circular random Gaussian quantity. This term describes the memory effect considered in the foregoing section. In an infinite medium, the term $C^{(1)}$ decays exponentially with the separation of the observation points. The characteristic scale of this decay is of the order of the extinction length L_{ext} (Shapiro [1986]).

The second and third terms in eq. (6.4) correspond to non-Gaussian correlations. The term $C^{(2)}$ is characterized by a long-range power-law behavior and describes slowly decaying correlations of intensities. These correlations were first found by Stephen and Cwilich (1987), who interpreted them as a result of an interference interaction of diffusion modes. Later it was shown that the term $C^{(2)}$ can be described using the Langeven approach (Zuzin and Spivak [1987]), describing the long-range correlations if the close-range correlations are considered as extraneous fluctuation currents (Pnini and Shapiro [1989]).

Finally, the smallest term, $C^{(3)}$, corresponds to the so-called universal or permanent correlations. In k-space, this term turns out to be independent of the difference $k_a - k_b$ between the wave vectors of the two scattering processes "a" and "b". The diagrams associated with this term have a fairly complicated structure (Feng, Kane, Lee and Stone [1988], Wang and Feng [1989]) and may be explained as a result of a more complicated interaction of diffusion modes than in the case of $C^{(2)}$. In the physics of solids, such correlations are related to "universal fluctuations of conductivity" which have been observed at low temperatures in "mesoscopic" samples of size $L \ll L_{sc}$ (Altshuler, Kravtsov and Lerner [1986]).

6.4. POLARIZATION EFFECTS

In addition to scalar correlation effects, the literature also describes specific polarization effects associated with polarization of radiation (Freund [1990a, 1991], Freund, Kaveh, Berkovits and Rosenbluh [1990]). Some of these effects can be obtained simply by extending the results of scalar theory to polarization characteristics. For instance, the scalar memory effect deals with the correlation of intensities of two scattering processes with different directions of incident and scattered waves. Extending these results to polarization characteristics, it is reasonable to explore how the correlations of intensities would change if we alter the polarization of the incident and scattered waves.

We present only the result of Freund, Kaveh, Berkovits and Rosenbluh [1990]. Consider two scattering processes "a" and "b" with identical directions of incidence and scattering, but with orientations of the polarization vectors differing by an angle $\Delta\theta$. Freund, Kaveh, Berkovits and Rosenbluh [1990] showed that the normalized correlation function of intensities is represented as $C(\Delta\theta) = \cos^2(\Delta\theta)$ in the diffusion regime; i.e, under the conditions of scattering with strong depolarization. It follows that the scattered waves are uncorrelated if the incident waves had orthogonal polarizations, whereas the correlation of scattered waves is nonzero for nonorthogonal polarizations of the incident waves.

6.5. CORRELATION EFFECTS OF IMAGING IN DIFFUSE SCATTERING MEDIA

The problem of image transfer through highly turbid media, where the multiple scattering is predominant, is an interesting and practically important problem that involves correlation effects. This problem is of special importance in astronomy, where the effect of a turbulent atmosphere limits considerably the possibilities of ground based telescopes. Special techniques of stellar speckle interferometry and adaptive optics were developed to compensate for this effect (Dainty [1984], Beichman and Ridgway [1991]). Unfortunately, these techniques have been developed for large-scale media, which mainly produce phase distortions, and the model of phase distortions appears to be unable to describe the diffuse scattering in the small-scale media considered in this section.

It is common lore that nothing can be seen through turbid media. In physical terms this means that multiple scattering prevents us from transferring clear optical images. It blurs the image and makes indistinguishable separate points of an object.

However, this does not mean that multiple scattering annihilates all the information on the primary wave. It would be more correct to say that this information is "encoded" in the process of multiple scattering into a complex speckle structure of the scattered wave. The fine details of this structure contain the imprint of both the incident wave and the parameters of the medium. This fact suggests an idea regarding how to "decode" the speckle structure to extract the useful information on the properties of the primary wave. This idea can be realized, in particular, by analyzing the higher statistical moments and multiple-point correlations of the scattered field.

The study of this problem is in its infancy. Nevertheless, there are some results which are worth mentioning. We give a simple example that is of itself of little practical significance, but is useful to prove the possibility to transfer optical images through turbid media by input–output matched filtering of the transmitted signal (V.V. Karavaev, private communication).

Suppose that we need to image a transparency through a thick layer of a turbid medium. This problem, difficult at first glance, is solved easily if the transmission is point by point rather than for the whole object at once. The simplest way to effect this sequential transmission is to use two identical screens with small holes, one of which is put on the transparency and the other in front of the photographic film right behind the scattering layer. The wave incident on the transparency passes through the hole in the screen. Because of multiple scattering in the layer, the transmitted wave becomes diffusive and almost uniformly illuminates the screen, obscuring the film. This arrangement images a point to the film, the image intensity being proportional to the transmittance of the slide at the respective point. Thus identically moving the holes over the film and slide and repeating the registration process yields a correct image of the transparency on the film. This image can be enlarged or reduced by either increasing or decreasing the amplitude of the motion of the second hole.

This example seems to be trivial. However, it includes the basic features of optical information transfer through turbid media. To be more specific, (i) it is impossible to directly transfer an optical image without distortions and (ii) it is necessary to establish an additional coupling between the input and output. This coupling is set up by a reference signal.

Speaking of the proposed methods of information transfer through turbid media, we should note all the possibilities of memory effects considered above. Using these effects, one may estimate, for example, a variation in the direction of propagation or in the polarization state of the incident wave (Freund [1990a]). However, the possibilities of memory effects are rather limited. A development of the theory in this direction may be found in the work of Freund [1990b], who considered some characteristic properties of optical data transfer through scattering media by using the correlation analysis with reference signals. Berkovits and Kaveh [1990c] reported the effect of scattering at an isolated scatterer on the mean transmission factor. In his later work, Berkovits [1991] found that fluctuations of transmission factor are highly sensitive to motion of isolated scatterers.

Most of the results in this field are yet tentative in nature, so that the possibilities of progress in studies of the correlation phenomena are far from being exhausted. Note that, because of limited potentialities of measuring instruments, the object of investigations in classic transfer theory was for long time the energy characteristics alone; i.e., one-point second moments of the field. Such a description is incomplete in the statistical sense and does not include higher and and many-point correlations of the field. However, these correlations may carry definite additional information on the properties of the medium.

It is not a simple problem to measure the correlation characteristics of a field (especially in the optical range), but these difficulties are step-by-step overcome with the development of experimental instruments. Clearly it is the study of higher moments and statistical distributions of scattered radiation from which a significant advancement in both theory and experiment should be expected.

§ 7. Conclusion

The last three decades have seen how the phenomenological theory of radiative transfer has been endowed with a statistical wave content. The early theory was a rather isolated branch of physics, but it has gradually become a part of the general physical stream. Today we are in a position to state that transfer theory is a spectral form of the correlation theory of quasi-uniform random fields. Therefore, nowadays one would not wonder, as one might have two or three decades ago, that transfer theory is able to describe a number of phenomena generated by diffraction. One should wonder rather why the RTE statistical wave background has evolved over such a long time.

Despite considerable advance in the understanding of the diffraction origin of the RTE, a number of relevant topics have not been elucidated adequately. First, notwithstanding a known limited applicability range of the RTE (weak scattering events and a quasi-uniformity of the field), no universal applicability criterion valid for the entire multitude of particular cases has been formulated.

Second, one cannot be certain about the applicability of the RTE to densely packed systems of scatterers when the near (induction) zone of each scatterer embraces many adjacent particles.

Third, one cannot be entirely certain about the deviation of the field statistics from the Gaussian law and about the relation of Gaussian statistics when multiple scattering is taken into account.

Finally, further studies of radiation transfer theory in nonlinear media, in optical waveguides with random inhomogeneities, and in similar systems would be useful.

Recognizing a steady advance in the statistical wave perception of the RTE, we do not doubt that the remaining problems will be gradually resolved.

References

The titles of Russian papers cited have been translated for convenience. Notice that some Soviet (Russian) journals have been translated into English on a cover-to-cover basis; e.g., Akust. Zh. [Sov. Phys.-Acoust., now Acoust. Phys.], Dokl. Akad. Nauk SSSR [Sov. Phys.-Dokl.], Izv. VUZ Radiofiz. [Radiophys. Quantum Electron.], Radiotekh. Electron. [Radio Eng. Electron. Phys. until 1986; Sov. J. Commun. Electron. Technol. after 1986], Kvant. Elektron. [Sov. J. Quant. Electron.], Zh. Eksp. Teor. Fiz. [Sov. Phys.-JETP], Kratk. Soobsh. Fiz. [Sov. Phys. Lebedev Inst. Reports], Zh. Tekhnich. Fiz. [Sov. Phys.-JTP], Izv. Akad. Nauk SSSR, Fiz. Atm. Okeana [Izv. Acad. Sci. USSR, Atm. Ocean Phys.], Astronom. Zh. [Astron. J.], Opt. Spektrosk.[Opt. Spectrosc.].

Abrahams, E., P.W. Anderson, D.C. Licciardello and T.V. Ramakrishnan, 1979, Scaling theory of localization: absence of quantum diffusion in two dimensions, Phys. Rev. Lett. **42**(10), 673–676.

Acquista, C., and J.L. Anderson, 1977, A derivation of the radiative transfer equation for partially polarized light from quantum electrodynamics, Ann. Phys. **106**(2), 435–443.

Akkerman, E., P.E. Wolf and R. Maynard, 1986, Coherent backscattering of light by disordered media: Analysis of the peak line shape, Phys. Rev. Lett. **56**(14), 1471–1474.

Altshuler, B.L., A.G. Aronov, D.E. Khmelnitskii and A.I. Larkin, 1982, Coherent effects in disordered conductors, in: Quantum Theory of Solids, ed. E.M. Lifshits (Mir, Moscow) pp. 130–237.

Altshuler, B.L., V.E. Kravtsov and I.V. Lerner, 1986, Statistics of mesoscopic fluctuations and single-parameter scaling instability, Zh. Eksp. Teor. Fiz. **91**(6), 2276–2302.

Anderson, P.W., 1958, Absence of diffusion in certain random lattices, Phys. Rev. **109**(5), 1492–1505.

Apresyan, L.A., 1973, Radiative transfer equation with allowance for longitudinal waves, Izv. Vyssh. Uchebn. Zaved. Radiofiz. **16**(3), 461–472.

Apresyan, L.A., 1974, Methods of statistical perturbation theory, Izv. Vyssh. Uchebn. Zaved. Radiofiz. **17**(2), 165–184.

Apresyan, L.A., 1975, Application of the radiative transfer equation to a free electromagnetic field, Izv. Vyssh. Uchebn. Zaved. Radiofiz. **18**(12), 1870–1873.

Apresyan, L.A., 1989, Nonlinear radiative transport equation and Anderson localization in scattering media. Available from VINITI, Moscow, no. 3504–B89. Summary in: Izv. Vyssh. Uchebn. Zaved. Radiofiz. **32**(6), 721.

Apresyan, L.A., 1990, Nonlinear Bethe–Salpeter equation: diagrammatic approach, Izv. Vyssh. Uchebn. Zaved. Radiofiz. **33**(7), 882–884.

Apresyan, L.A., 1993, Classification of the mechanisms of backward scattering, Izv. Vyssh. Uchebn. Zaved. Radiofiz. **34**(10–12), 1125–1134.

Apresyan, L.A., and Yu.A. Kravtsov, 1983, Radiation Transfer (Nauka, Moscow).

Apresyan, L.A., and Yu.A. Kravtsov, 1984, Photometry and coherence: wave aspects of the theory of radiation transport, Sov. Phys. Usp. **27**(4), 301–313. Reprinted 1993, in: Selected Papers on Coherence and Radiometry, ed. A.T. Friberg, SPIE Milestone Series, Vol. 69 (SPIE Optical Engineering Press, Bellingham).

Apresyan, L.A., and Yu.A. Kravtsov, 1996, Radiation Transfer: Statistical and Wave Aspects (Gordon and Breach, London, NY).

Balescu, R., 1963, Statistical Mechanics of Charged Particles (Wiley Interscience, New York).

Baltes, H.P., 1977, Coherence and radiation laws, Appl. Phys. **12**(3), 221.

Baltes, H.P., ed., 1978, Inverse Source Problems in Optics (Springer, Berlin).

Barabanenkov, Yu.N., 1967, Radiative transfer equation in the model of isotropic point scatterers, Dokl. Akad. Nauk SSSR **174**(1), 53–55.

Barabanenkov, Yu.N., 1969, Spectral theory of the transport equation, Zh. Eksp. Teor. Fiz. **56**(4), 1262–1272.

Barabanenkov, Yu.N., 1973, Wave corrections to the transport equation for the backscattering direction, Izv. Vyssh. Uchebn. Zaved. Radiofiz. **16**(1), 88–96.

Barabanenkov, Yu.N., 1975a, Multiple scattering of waves by ensembles of particles and the theory of radiation transport, Sov. Phys. Usp. **18**(5), 673–689.

Barabanenkov, Yu.N., 1975b, Resolvent method of error estimation of the Bethe–Salpeter equation in the Fraunhofer approximation, Izv. Vyssh. Uchebn. Zaved. Radiofiz. **18**(5), 716–723.

Barabanenkov, Yu.N., and V.M. Finkelberg, 1967, Radiative transfer equation for correlated scatterers, Zh. Eksp. Teor. Fiz. **53**(3), 978–985.

Barabanenkov, Yu.N., and V.M. Finkelberg, 1968, Optical theorem in the theory of multiple scattering, Izv. Vyssh. Uchebn. Zaved. Radiofiz. **11**(5), 719–725.

Barabanenkov, Yu.N., Yu.A. Kravtsov, V.D. Ozrin and A.I. Saichev, 1991, Enhanced backscattering in optics, in: Progress in Optics, Vol. 29, ed. E. Wolf (North-Holland, Amsterdam) pp. 65–197.

Barabanenkov, Yu.N., A.G. Vinogradov, Yu.A. Kravtsov and V.I. Tatarskii, 1972, Application of the theory of multiple scattering of waves to the derivation of the radiation transfer equation for a statistically inhomogeneous medium, Radiophys. Quantum Electron. **15**(12), 1420–1425.

Bartelt, H.O., K.N. Brenner and A.W. Lohmann, 1980, The Wigner distribution function and its optical production, Opt. Commun. **32**(1), 32–38.

Beichman, C.A., and S. Ridgway, 1991, Adaptive optics and interferometry, Phys. Today **44**(4), 48–51.

Bekefi, G., 1966, Radiation Processes in Plasmas (Wiley, New York).

Belenkii, M.S., and V.L. Mironov, 1972, Diffraction of optical radiation at a mirror disc in a turbulent atmosphere, Kvant. Elektron. **5**(11), 38–45.

Bergman, D.J., 1978, The dielectric constant of a composite material. A problem in classical physics, Phys. Rep. **43C**, 377.

Bergman, D.J., 1982, Resonances in the bulk properties of composite media, Theory and applications. Lecture Notes in Physics, Vol. 154 (Springer, New York) pp. 10–37.

Berkovits, R., 1991, Sensitivity of multiple-scattering speckle pattern to the motion of a single scatterer, Phys. Rev. B **43**(10), 8638–8640.

Berkovits, R., and M. Kaveh, 1990a, Time reversed memory effects, Phys. Rev. B **41**(4), 2635–2638.

Berkovits, R., and M. Kaveh, 1990b, The vector memory effect for waves, Europhys. Lett. **13**(2), 97–101.

Berkovits, R., and M. Kaveh, 1990c, Theory of speckle-pattern tomography in multiple-scattering media, Phys. Rev. Lett. **65**(25), 3120–3123.

Berkovits, R., M. Kaveh and S. Feng, 1989, Memory effect of waves in disordered systems: A real-space approach, Phys. Rev. B **40**(1), 737–739.

Born, M., and E. Wolf, 1980, Principles of Optics, 6th Ed. (Pergamon Press, New York).

Borovoi, A.G., 1966, Iteration technique in multiple scattering. The transport equation, Izv. Vyssh. Uchebn. Zaved. Fizika **6**, 50–54.

Borovoi, A.G., 1984, Radiative transfer in inhomogeneous media, Dokl. Akad. Nauk SSSR **276**(6), 1374–76.

Bouger, P., 1729, Essai d'optique sur la gradation de la lumière, Paris.

Bugnolo, D., 1960, Transport equation for the spectral density of a multiple-scattered electromagnetic field, Appl. Phys. **31**(7), 1176–1182.

Carter, W.H., and E. Wolf, 1975, Coherence properties of lambertian and nonlambertian sources, J. Opt. Soc. Am. **65**(9), 1067–1071.

Carter, W.H., and E. Wolf, 1977, Coherence and radiometry with quasi-homogeneous planar sources, J. Opt. Soc. Am. **67**(4), 785–796.

Chandrasekhar, S., 1960, Radiative Transfer (Dover, New York).

Cohen, L., 1989, Time-Frequency Distributions. A Review, Proc. IEEE **77**(7), 941–981.

Dainty, J.C., ed., 1984, Laser Speckle and Related Phenomena, 2nd Ed. (Springer, Berlin).

De Wolf, D.A., 1971, Electromagnetic reflection from an extended turbulent medium: cumulative forward-scatter single back-scatter approximation, IEEE Trans. Antennas Propag. **AP-19**(2), 254–262.

Dolghinov, A.Z., Yu.N. Gnedin and N.A. Silant'ev, 1995, Propagation and Polarization of Radiation in Cosmic Media (Gordon and Breach, Reading).

Dolin, L.S., 1964a, Ray description of weakly inhomogeneous wave fields, Izv. VUZ Radiofiz. **7**(3), 559–562.

Dolin, L.S., 1964b, Scattering of light in a layer of turbid medium, Izv. VUZ Radiofiz. **7**(2), 380–382.

Dolin, L.S., 1968, Equations for correlation functions of a wave beam in a random inhomogeneous medium, Izv. VUZ Radiofiz. 11(6), 840–849.

Dolin, L.S., 1984, On the randomization of a radiation field in a medium with fluctuating extinction coefficient, Dokl. Akad. Nauk SSSR 227(1), 77–80.

Dombrovskii, L.A., and V.Yu. Raizer, 1992, Microwave model of two-phase media in near-water ocean layer, Izv. Russ. Acad. Sci. 28(2), 863–872.

Eberly, J.H., and K. Wodkiewicz, 1977, The time dependent physical spectrum of light, J. Opt. Soc. Am. 67(9), 1252–1261.

Edrei, I., and M. Kaveh, 1987, Weak localization of photons and backscattering from finite systems, Phys. Rev. B 35(12), 6461–6463.

Eliyahu, D., R. Berkovits and M. Kaveh, 1991, Long-range angular correlations of waves in a tube geometry, Phys. Rev. B 43(16), 13501–13505.

England, A.W., 1974, Thermal microwave emission from a halfspace containing scatterers, Rad. Sci. 9(4), 447–454.

Erdős, P., and R.C. Herndon, 1982, Theories of electrons in one-dimensional disordered systems, Adv. Phys. 31(2), 65–163.

Etkin, V.S., V.G. Irisov and Yu.G. Trokhimovskii, 1992, Resonant radiothermal emission of water surface with non-small periodic roughness, Int. Geosci. and Remote Sensing Symposium IGARSS '92, Houston, TX, May 26–29, 1992, Vol. 2, pp. 1457–1459.

Felderhof, B.U., G.W. Ford and E.G.D. Cohen, 1983, The Clausius–Mossotti formula and its nonlocal generalization for a dilute suspension of spherical inclusions, J. Stat. Phys. 33(2), 241–260.

Feng, S., C. Kane, P.A. Lee and A.D. Stone, 1988, Correlations and fluctuations of coherent wave transmission through disordered media, Phys. Rev. Lett. 61(7), 834–837.

Finkelberg, V.M., 1964, Dielectric permittivity of mixtures, Zh. Tekh. Fiz. 34(3), 503–518.

Finkelberg, V.M., 1967, Propagation of waves in a random inhomogeneous medium. The method of correlation groups, Zh. Eksp. Teor. Fiz. 53(17), 401–415.

Freund, I., 1990a, Stokes-vector reconstruction, Opt. Lett. 15(24), 1425–1427.

Freund, I., 1990b, Correlation imaging through multiply scattering media, Phys. Lett. A 147(8, 9), 502–506.

Freund, I., 1991, Optical Intensity fluctuations in multiply scattering media, Opt. Commun. 81(3, 4), 251–258.

Freund, I., and R. Berkovits, 1990, Surface reflection and transport through random media: coherent backscattering, optical memory effect, frequency and dynamical correlations, Phys. Rev. B 41(1), 496–502.

Freund, I., M. Kaveh, R. Berkovits and M. Rosenbluh, 1990, Universal polarization correlations and microstatistics of optical waves in random media, Phys. Rev. B 42(4), 2613–2616.

Freund, I., M. Rosenbluh and S. Feng, 1988, Memory effects in propagation of optical waves through disordered media, Phys. Rev. Lett. 61(20), 2328–2331.

Friberg, A.T., ed., 1993, Selected papers on Coherence and Radiometry, SPIE Milestone Series, Vol. 69 (SPIE Optical Engineering Press, Bellingham).

Galinas, R.J., and R.L. Ott, 1970, Quantum theoretical deduction of radiative transfer equations in the spectral line regime, Ann. Phys. 59(2), 323–374.

Genchev, J.D., 1984, Electromagnetic wave scattering on a surface with small and smooth inhomogeneities, Izv. VUZ Radiofiz. 27(1), 48–55.

Gnedin, Yu.N., and A.Z. Dolginov, 1963, Theory of multiple scattering, Zh. Eksp. Teor. Fiz. 45(4), 1136–1149.

Goetze, W., 1978, An elementary approach towards the Anderson transition, Solid State Commun. 27(12), 1393–1395.

Golden, K., and G.C. Papanicolaou, 1983, Bounds for effective parameters of heterogeneous media by analytic continuation, Commun. Math. Phys. **90**, 473.
Golubentsev, A.A., 1984a, Suppression of interference effects in multiple scattering, Zh. Eksp. Teor. Fiz. **86**(1), 47–59.
Golubentsev, A.A., 1984b, Contribution of interference into the albedo of a strongly gyrotropic medium with random inhomogeneities, Izv. VUZ Radiofiz. **27**(6), 734–745.
Gorodnichev, E.E., S.L. Dudarev and D.B. Rogozkin, 1989, Coherent enhancement of backscattering under weak localization of waves in disordered two- and three-dimensional systems, Zh. Eksp. Teor. Phys. **96**(3), 847–864.
Gurvich, A.S., V.I. Kalinin and D.H. Matveev, 1973, Influence of the internal structure of glaciers on their thermal radio emission, Izv. Acad. Sci. USSR, Atm. Ocean Phys. **9**(12), 713.
Hashin, Z., and S. Shtrikman, 1962, A variational approach to the theory of the effective magnetic permeability of multiphase materials, J. Appl. Phys. **33**, 3125.
Hori, H., 1977, Theory of effective dielectric, thermal and magnetic properties of random heterogeneous materials. VIII. Comparison of different approaches, J. Math. Phys. **18**(3), 487–501.
Howe, M.S., 1973, On the kinetic theory of wave propagation in random media, Philos. Trans. R. Soc. London **274**(1242), 523–549.
Ishimaru, A., 1978, Wave Propagation and Scattering in Random Media (Academic, New York) 2 volumes.
Kashkarov, S.S., and A.S. Gurvich, 1977, On the enhanced scattering in a turbulent medium, Izv. VUZ Radiofiz. **20**(5), 794–800.
Khvolson, O.D., 1890, Grundzüge einer mathematischen Theorie der inneren Diffusion des Lichtes, Bull. St. Petersburg Acad. Sci. **33**(2), 221–256.
Kirkpatrick, T.R., 1985, Localization of acoustic waves, Phys. Rev. B **31**(9), 5746–5755.
Kirkpatrick, T.R., and J.R. Dorfman, 1985, Aspects of localization, Fundamental Problems in Statistical Mechanics, Vol. VI, ed. E.G.D. Cohen (North-Holland, Amsterdam) pp. 365–398.
Kleeorin, N.I., Yu.A. Kravtsov, A.E. Mereminskii and V.G. Mirovskii, 1989, Effect of translucence and radiative transfer in a media with large-scale fluctuation of scatterers concentration, Izv. VUZ Radiofiz. **32**(9), 1072–1080.
Kohler, W., and G.C. Papanicolaou, 1982, Bounds for effective conductivity of random media, in: Macroscopic Properties of Disordered Media, eds R. Burridge, S. Childress and G.Papanicolaou (Springer, Berlin) p. 111.
Kong, J.A., S.L. Lin and S.L. Chuang, 1984, Microwave thermal emission from periodic surface, IEEE Trans. Geosci. Electron. **GE-22**(4), 377–381.
Kramers, B., C. Bergman and Y. Bruynseraede, eds, 1985, Localization, Interaction and Transport Phenomena (Springer, Berlin).
Kravtsov, Yu.A., E.A. Mirovskaya, A.E. Popov, I.A. Troitskii and V.S. Etkin, 1978, Critical phenomena in thermal radiation of periodically uneven water surface, Izv. Akad. Nauk SSSR, Fiz. Atm. Ocean **14**(7), 733–739.
Kravtsov, Yu.A., and Yu.I. Orlov, 1990, Geometrical Optics of Inhomogeneous Media (Springer, Berlin).
Kuga, Y., and A. Ishimaru, 1984, Retroreflectance from a dense distribution of spherical particles, J. Opt. Soc. Am. A **1**(8), 831–839.
Kuga, Y., L. Tsang and A. Ishimaru, 1985, Depolarization effects of the enhanced retroreflectance from a dense distribution of spherical particles, J. Opt. Soc. Am. A **2**(6), 616–618.
Lambert, J.H., 1760, Photometria, sive de mensura et gradibus luminis, colorum et umbral, Augsburg.
Lampard, D.G., 1954, Generalization of the Wiener–Khinchin theorem to nonstationary processes, J. Appl. Phys. **25**(6), 802–803.
Landauer, R., 1978, Electrical conductivity in inhomogeneous media, in: Electrical Transport and

Optical Properties of Inhomogeneous Media, Part 2, eds J.C. Garland and D.B. Tanner (American Institute of Physics, New York).

Lax, M., 1952, Multiple scattering of waves, II. The effective field in dense systems, Phys. Rev. **85**(3), 621–629.

Lee, A.P., and T.V. Ramakrishnan, 1985, Disordered electronic systems, Rev. Mod. Phys. **57**(2), 287–338.

Levin, M.L., and S.M. Rytov, 1967, Theory of Equilibrium Thermal Fluctuations in Electrodynamics (Nauka, Moscow). In Russian.

Ma, S.-K., 1976, Modern Theory of Critical Phenomena (Benjamin-Cummings, Reading, MA).

Malakhov, A.N., and A.I. Saichev, 1979, Representation of a wave reflected from a random inhomogeneous layer as a series satisfying the causality condition, Izv. VUZ Radiofiz. **22**(7), 1324–1330.

Manning, R.M., 1989, Radiative transfer for inhomogeneous media with random extinction and scattering coefficient, J. Math. Phys. **30**(8), 2432–2440.

Marc, W.D., 1970, Spectral analysis of the convolution and filtering of non-stationary stochastic processes, J. Sound. Vibr. **11**(1), 19–63.

Milton, G.W., 1981, Bounds on the complex permittivity of a two-component composite material, J. Appl. Phys. **52**(18), 5286–5293.

Mori, H., I. Oppenheim and J. Ross, 1962, Wigner function and transport theory studies, in: Studies in Statistical Mechanics, Vol. 1 (North-Holland, Amsterdam) pp. 217–302.

Mott, N.F., 1974, The Metal–Insulator Transition (Taylor & Francis, London).

Nagaoka, Y., and H. Fukuyama, eds, 1982, Anderson Localization (Springer, Berlin).

Odelevskii, V.I., 1951, Calculation of generalized conductivity of heterogeneous systems, 2. Statistical combinations of nonextended particles, Zh. Tekh. Fiz. **21**(5), 667–678.

Ovchinnikov, G.I., 1973, Equation for the field spectral density in medium with random inhomogeneities, Radiotekh. Elektron. **18**(10), 2044–2049.

Ovchinnikov, G.I., and V.I. Tatarskii, 1972, On the relation between coherence theory and radiative transfer equation, Izv. VUZ Radiofiz. **15**(9), 1419–1421.

Ozrin, V.D., 1992, Exact solution for coherent backscattering from a semi-infinite random medium of anisotropic scattering, Phys. Lett. A **162**, 341–345.

Page, C.H., 1952, Instantaneous power spectra, J. Appl. Phys. **23**(1), 103–106.

Peacher, J.I., and K.M. Watson, 1970, Doppler shift in frequency in the transport of electromagnetic wave through an underdense plasma, J. Math. Phys. **11**(4), 1496–1504.

Pnini, R., and B. Shapiro, 1989, Fluctuations in transmission of waves through disordered slabs, Phys. Rev. B **39**(10), 6986–6994.

Polder, D., and J.H. van Santen, 1946, The effective permeability of mixture of solids, Physica **12**, 257–271.

Raizer, V.Yu., 1992, Two-phase ocean-surface structures and microwave remote sensing, in: Int. Geoscience and Remote Sensing Symp. IGARSS '92, Houston, TX, May 26–29, 1992, pp. 1460–1462.

Raizer, V.Yu., and S.R. Filonovich, 1989, Radiative characteristics of sinusoidal surfaces in the Kirchhoff approximation, Izv. VUZ Radiofiz. **32**(7), 940–942.

Raizer, V.Yu., and E.A. Sharkov, 1981, Electrodynamic description of densely packed dispersed systems, Radiophys. Quantum. Electron. **24**(7), 553–560.

Ramshaw, J.D., 1984, Dielectric polarization in random media, J. Stat. Phys. **35**(1/2), 49–75.

Rochon, P., and D. Bissonnette, 1990, Lensless imaging due to backscattering, Nature **348**(6303), 708–710.

Roth, L.M., 1974, Effective medium approximation for liquid metals, Phys. Rev. B **9**(6), 2476–2484.

Rozenberg, G.V., 1970, Statistical and electrodynamical content of the photometric quantities and basic concepts of radiative transfer theory, Opt. Spektrosk. **28**(2), 392–398.
Rozenberg, G.V., 1977, The light ray (contribution to theory of light field), Sov. Phys. Usp. **20**, 55–79.
Rytov, S.M., 1953, Theory of Electric Fluctuations and Thermal Radiation (Izd. Akad. Nauk SSSR, Moscow). In Russian.
Rytov, S.M., Yu.A. Kravtsov and V.I. Tatarskii, 1989a, Principles of Statistical Radiophysics, Vol. 3: Random Fields (Springer, Berlin).
Rytov, S.M., Yu.A. Kravtsov and V.I. Tatarskii, 1989b, Principles of Statistical Radiophysics, Vol. 4: Wave Propagation Through Random Media (Springer, Berlin).
Ryzhov, Y.A., and V.V. Tamoikin, 1970, Radiation and propagation of electromagnetic waves in randomly inhomogeneous media, Radiophys. Quantum Electron. **13**(3), 273–300.
Schuster, A., 1905, Radiation through a foggy atmosphere, Astrophys. J. **21**(1), 1–22. Reprinted: 1966, in: Selected Papers on the Transfer of Radiation, ed. D.H. Menzel (Dover, New York).
Sen, P.N., C. Scala and M.H. Cohen, 1981, A self-similar model for sedimentary rocks with application to the dielectric constant of fused glass beads, Geophysics **46**(5), 781–795.
Shapiro, B., 1986, Large intensity fluctuations for wave propagation in random media, Phys. Rev. Lett. **57**(17), 2168–2171.
Sheng, P., ed., 1990, Scattering and Localization of Classical Waves in Random Media (World Science, Singapore).
Sobolev, V.V., 1956, Light Scattering in Planetary Atmospheres (Nauka, Moscow) [English transl.: 1975 (Pergamon Press, Oxford)].
Sobolev, V.V., 1975, A Course in Theoretical Astrophysics (Nauka, Moscow). In Russian.
Stephen, M.J., and G. Cwilich, 1987, Intensity correlation functions and fluctuations in light scattered from a random media, Phys. Rev. Lett. **59**(3), 285–287.
Stogryn, A., 1967, The apparent temperature of sea at microwave frequencies, IEEE Trans. Antennas Propag. **AP-15**(2), 278–286.
Stott, P.E., 1968, A transport equation for the multiple scattering of electromagnetic waves by a turbulent plasma, J. Phys. A **1**(6), 675–689.
Stroud, D., 1975, Generalized effective-medium approach to the conductivity of an inhomogeneous medium, Phys. Rev. B **12**(11), 3368–3373.
Tatarskii, V.I., 1983, Wigner representation in quantum mechanics, Usp. Fiz. Nauk **139**(4), 587–620.
Tatarskii, V.I., and M.E. Gertsenshtein, 1963, Propagation of waves in a medium with strong fluctuations of the refractive index, Sov. Phys. JETP **17**(3), 458–463.
Tsang, L., and J.A. Kong, 1975, The brightness temperature of a half space random medium with nonuniform temperature profile, Rad. Sci. **10**(12), 1025–1033.
Tsolakis, A.I., I.M. Besieris and W.E. Kohler, 1985, Two-frequency radiative transfer equation for scalar waves in a random distribution of discrete scatterers with pair correlations, Radio Sci. **20**(12), 1037–1052.
Twersky, V., 1962a, On scattering of waves by random distributions. I. Free space scatterer formalism, J. Math. Phys. **3**(4), 700–715.
Twersky, V., 1962b, On scattering of waves by random distributions. II. Two space scattering formalism, J. Math. Phys. **3**(4), 724–734.
Van Albada, M.P., and A. Lagendijk, 1985, Observation of weak localization of light in a random medium, Phys. Rev. Lett. **55**(24), 2692–2695.
Vardanyan, R.C., 1988, Radiation transfer in stochastic media, Astrophys. Spac. Sci. **141**(2), 375–383.
Vinogradov, A.G., Yu.A. Kravtsov and V.I. Tatarskii, 1973, Enhancement of backscatter from bodies immersed in a medium with random inhomogeneities, Izv. VUZ Radiofiz. **16**(7), 1064–1070.

Vollhardt, D., and P. Wölfle, 1980, Diagrammatic self-consistent treatment of the Anderson localization problem, Phys. Rev. B **22**(10), 4666–4679.

Vollhardt, D., and P. Wölfle, 1982, Scaling equations from a self-consistent theory of Anderson localization, Phys. Rev. Lett. **48**(10), 699–702.

Walther, A., 1968, Radiometry and coherence, J. Opt. Soc. Am. **58**(9), 1256–1259.

Walther, A., 1973, Radiometry and coherence, J. Opt. Soc. Am. **63**(12), 1622–1623.

Wang, L., and S. Feng, 1989, Correlations and fluctuations in reflection coefficients for coherent wave propagation in disordered scattering media, Phys. Rev. B **40**(12), 8284–8289.

Watson, K.M., 1969, Multiple scattering of electromagnetic waves in an underdense plasma, J. Math. Phys. **10**(4), 688–702.

Wen, B., L. Tsang, D.R. Winebrenner and A. Ishimaru, 1990, Dense medium radiative transfer theory: comparison with experiment and application to microwave remote sensing and polarimetry, IEEE Trans. Geosci. Remote Sensing **28**(1), 46–58.

Wentz, F.J., 1975, A two-scale scattering model for foam-free sea microwave brightness temperatures, J. Geophys. Res. **20**(24), 3441–3446.

Wigner, E.P., 1932, On quantum corrections for thermodynamic equilibrium, Phys. Rev. **40**(6), 749–759.

Wolf, E., 1978, Coherence and radiometry, J. Opt. Soc. Am. **68**(1), 6–16.

Wolf, E., 1986, Invariance of spectrum of light on propagation, Phys. Rev. Lett. **56**(12), 1370.

Wolf, E., 1987, Redshifts and blueshifts of spectral lines emitted by two correlated sources, Phys. Rev. Lett. **58**(24), 2646.

Wolf, E., and W.H. Carter, 1977, Coherence and radiometry with quasihomogeneous planar sources, J. Opt. Soc. Am. **67**(6), 785–796.

Wolf, P.E., and G. Maret, 1985, Weak localization and coherent backscattering of photons in disordered media, Phys. Rev. Lett. **55**(24), 2696–2699.

Wu, S.T., and A.K. Fung, 1972, A noncoherent model for microwave emissions and backscattering from the sea surface, J. Geophys. Res. **77**(30), 5917–5928.

Zhang, Z.-Q., Q.-J. Chu, W. Xue and P. Sheng, 1990, Anderson localization in anisotropic random media, Phys. Rev. B **42**(7), 4613–4630.

Zheleznyakov, V.V., 1970, Radioemission of Sun and Planets (Pergamon Press, Oxford).

Zuzin, A.Yu., and B.Z. Spivak, 1987, Langevin description of mesoscopic fluctuations in disordered media, Zh. Eksp. Teor. Fiz. **93**(3), 994–1006.

E. WOLF, PROGRESS IN OPTICS XXXVI
© 1996 ELSEVIER SCIENCE B.V.
ALL RIGHTS RESERVED

V

PHOTON WAVE FUNCTION

BY

Iwo BIALYNICKI-BIRULA

*Center for Theoretical Physics, Polish Academy of Sciences,
Al. Lotników 32, 02-668 Warsaw, Poland*

AND

*Rochester Theory Center for Optical Science and Engineering,
University of Rochester, Rochester, NY 14627, USA*

CONTENTS

		PAGE
§ 1.	INTRODUCTION.	248
§ 2.	WAVE EQUATION FOR PHOTONS	254
§ 3.	PHOTON WAVE FUNCTION IN COORDINATE REPRESENTATION	259
§ 4.	PHOTON WAVE FUNCTION IN MOMENTUM REPRESENTATION	263
§ 5.	PROBABILISTIC INTERPRETATION	267
§ 6.	EIGENVALUE PROBLEMS FOR THE PHOTON WAVE FUNCTION	273
§ 7.	RELATIVISTIC INVARIANCE OF PHOTON WAVE MECHANICS.	278
§ 8.	LOCALIZABILITY OF PHOTONS	279
§ 9.	PHASE-SPACE DESCRIPTION OF A PHOTON	281
§ 10.	HYDRODYNAMIC FORMULATION	283
§ 11.	PHOTON WAVE FUNCTION IN NON-CARTESIAN COORDINATE SYSTEMS AND IN CURVED SPACE	285
§ 12.	PHOTON WAVE FUNCTION AS A SPINOR FIELD	286
§ 13.	PHOTON WAVE FUNCTIONS AND MODE EXPANSION OF THE ELECTROMAGNETIC FIELD	288

§ 14. SUMMARY 290
ACKNOWLEDGMENTS 291
REFERENCES 292

§ 1. Introduction

The photon wave function is a controversial concept. The controversies stem from the fact that photon wave functions cannot have all the properties of the Schrödinger wave functions of nonrelativistic wave mechanics. Insistence on those properties which, owing to the peculiarities of photon dynamics, cannot be rendered, led some physicists to the extreme opinion that the photon wave function does not exist. I reject such a fundamentalist point of view in favor of a more pragmatic approach. In my view, the photon wave function exists as long as it can be defined precisely and made useful. Many authors whose papers are quoted in this review share the same opinion and had no reservations about using the name 'photon wave function' when referring to a complex vector-function of space coordinates r and time t that adequately describes the quantum state of a single photon.

The notion of the photon wave function is certainly not new, but strangely enough it has never been explored systematically and fully. Some textbooks on quantum mechanics start the introduction to quantum theory with a discussion of photon polarization measurements (cf., for example, Dirac [1958], Baym [1969], Lipkin [1973], Cohen-Tannoudji, Diu and Laloë [1977]), but in all these expositions a complete photon wave function never takes on a specific mathematical form. Even Dirac, who wrote: "The essential point is the association of each of the translational states of the photon with one of the wave functions of ordinary wave optics", never expresses this association in an explicit form. In this context he also uses the now famous phrase: "Each photon interferes only with itself", which implies the existence of photon wave functions whose superposition leads to interference phenomena.

In the textbook analysis of polarization, only simple prototype two-component wave functions are used to describe various polarization states of the photon, and with their help the preparation and the measurement of polarization is analyzed. However, it is not explained why a wave function should not be used to describe also the "translational states of the photon" mentioned by Dirac. After such a heuristic introduction to quantum theory, the authors go on to the study of massive particles, and if they ever return to quantum theory of photons, it is always within the formalism of second quantization with creation and annihilation operators. In some textbooks (cf., for example, Bohm [1954], Power [1964]), one may even find statements which negate completely the possibility of introducing a wave function for the photon.

A study of the photon wave function should be preceded by an explanation of what a photon is, and why a description of the photon in terms of a wave function must exist. According to modern quantum field theory, photons, together with all other particles (and also quasiparticles, phonons, excitons, plasmons, etc.), are the *quantum excitations* of a field. In the case of photons, these are the excitations of the electromagnetic field. The lowest field excitation of a *given type* corresponds to one photon and higher field excitations involve more than one photon. This concept of a photon (called the "modern photon" in a tutorial review by Kidd, Ardini and Anton [1989]) enables one to use the photon wave function to describe not only quantum states of an excitation of the free field, but also of the electromagnetic field interacting with a medium. Conceptually, the difference between free space and a medium is not essential since the physical vacuum is like a polarizable medium. It is filled with all the virtual pairs – zero point excitations of charged quantum fields. Therefore, even in free space, photons can be also viewed as the excitations of the vacuum made mostly of virtual electron–positron pairs (Bialynicki-Birula [1963], Bjorken [1963]).

Even though, in principle, all particles can be treated as field excitations, photons are much different from massive particles. They are also different from massless neutrinos since the photon number does not obey a conservation law. There are problems with the photon localization, and as a result the position operator for the photon is ill-defined, but the similarities between photons and other quantum particles are so ample that the introduction of the photon wave function seems to be fully justified and even necessary in order to achieve a complete unification of our description of all particles.

1.1. COORDINATE VS. MOMENTUM REPRESENTATION

In nonrelativistic quantum mechanics, the term *coordinate representation* is used to denote the representation in which the wave function $\psi(r)$ is defined as a projection of the state vector $|\psi\rangle$ on the eigenstates $|r\rangle$ of the components $\hat{x}, \hat{y}$, and $\hat{z}$ of the position operator $\hat{r}$,

$$\psi(r) = \langle r|\psi\rangle. \tag{1.1}$$

The wave function in coordinate representation, therefore, becomes automatically a function of the eigenvalues of the position operator $\hat{r}$. The position operators act on the wave function simply through a multiplication. In quantum mechanics of photons, this approach does not work due to difficulties with the definition of the photon position operator (cf. § 8). One may still, however, introduce functions of the coordinate vector r to describe quantum states of the photon. By adopting this less stringent point of view that does not tie the wave function in coordinate representation with eq. (1.1), one avoids the consequences of the nonexistence of the photon position operator $\hat{r}$. In principle, any function of r that adequately describes photon states may be called a photon wave function in coordinate representation, and it is a matter of taste and convenience regarding which one to use. It should be pointed out that in a relativistic quantum theory, even for particles with nonvanishing rest mass, the position operator and the localization associated with it do not live up to our nonrelativistic expectations. The differences in localization of photons and, say electrons, are more quantitative than qualitative since they amount to the "spilling of the wave function" beyond the localization region governed by a power law versus an exponential decay.

The photon wave function in *momentum representation* has not stirred any controversy since the photon momentum operator $\hat{p}$ is well defined. Its existence, as the generator of translations, follows directly from the general theory of representations of the Poincaré group developed by Wigner [1939]. It has always been taken for granted by all physicists working in relativistic quantum electrodynamics that the notion of the photon wave function in momentum representation is well founded. Such wave functions describing initial and final states of photons appear in all formulas for transition amplitudes in the S-matrix theory of scattering phenomena (cf., for example, Schweber [1961], Akhiezer and Berestetskii [1965], Bialynicki-Birula and Bialynicka-Birula [1975], Cohen-Tannoudji, Dupont-Roc and Grynberg [1989]). Thus, one may safely assert that the photon wave function in momentum representation is a well-defined and fully established object.

1.2. PHASE REPRESENTATION

The photon wave functions discussed in this review are distinct from the *one-mode* wave functions which have been introduced in the past (London [1927], Bialynicki-Birula and Bialynicka-Birula [1976], Pegg and Barnett [1988]) to describe *multi-photon states*. These functions depend on the phase φ of the field and were called the wave functions in the phase representation by Bialynicki-Birula and Bialynicka-Birula [1976]. The wave functions $\Psi(\varphi)$ characterize quantum states of a *selected mode* of the quantized electromagnetic field and, in general, they describe a superposition of states with different numbers of photons. All spatial characteristics of these states are contained in the mode function that defines the selected mode of the electromagnetic field. One-mode wave functions $\Psi(\varphi)$ describe properties of multi-photon states of the quantized electromagnetic field with all photons being in the same quantum-mechanical state. In contrast, the photon wave function in the coordinate representation can be identified with the mode function itself (cf. § 13). It describes a state of a *single* photon and not a state of the quantized field.

1.3. LANDAU–PEIERLS WAVE FUNCTION

The concept of the photon wave function in coordinate representation was introduced for the first time by Landau and Peierls [1930]. The same function has been independently rediscovered more recently by Cook [1982a,b], and Inagaki [1994]. The Landau–Peierls proposal has not been met with great enthusiasm, since their wave function is a highly nonlocal object.

The nonlocality of the Landau–Peierls wave function is introduced by operating on the local electromagnetic field with the integral operator $(-\Delta)^{-1/4}$:

$$((-\Delta)^{-1/4} f)(r) = \pi \int \frac{d^3 r'}{(2\pi |r - r'|)^{5/2}} f(r'). \tag{1.2}$$

This integral operator corresponds to a division by $\sqrt{|k|}$ of the Fourier transform, and it changes the dimension of the wave function from L^{-2}, characteristic of the electromagnetic field, to $L^{-3/2}$. Therefore the modulus squared of the Landau–Peierls wave function has the right dimensionality to be interpreted as a probability density to find a photon. In particular, these wave functions can be normalized to one with the standard definition of the norm since the integral of the modulus squared of the wave function is dimensionless. However, as has been already noted by Pauli [1933], despite its right dimensionality, the nonlocal wave function has serious drawbacks. First, it does not transform under Lorentz transformations as a tensor field or any other geometric object. Second, a nonlocal wave function taken at a point in one coordinate system depends on the values of this wave function in *all of space* in another coordinate system. Third, the probability density defined with the use of a nonlocal wave function does not correspond to the probability of interaction of localized charges with the electromagnetic field. The vanishing of the wave function at a definite point, in Pauli's words (Pauli [1933]), has "no direct physical significance" because the electromagnetic field does act on charges at the points where the probability of finding a photon vanishes. The Landau–Peierls wave functions can not be used as primary objects in the presence of a medium since one is unable to impose proper boundary conditions on such nonlocal objects. These functions can be introduced, if one wishes to do so, as secondary objects related to the local wave function by a nonlocal transformation (§ 5). Scalar products and expectation values look simpler when they are expressed in terms of the Landau–Peierls wave functions, but that is, perhaps, their only advantage.

1.4. RIEMANN–SILBERSTEIN WAVE FUNCTION

The mathematical object that fully deserves the name of the photon wave function can be traced back to a complex version of Maxwell's equations that was already known at the turn of the century. The earliest reference is the second volume of the lecture notes on the differential equations of mathematical physics by Riemann, which were edited and published by Weber [1901]. Various applications of this form of Maxwell's equations were given by Silberstein [1907a,b, 1914] and Bateman [1915] in the framework of classical physics.

The complex form of Maxwell's equations is obtained by multiplying the first pair of these equations,

$$\partial_t \mathbf{D}(\mathbf{r},t) = \nabla \times \mathbf{H}(\mathbf{r},t), \quad \nabla \cdot \mathbf{D}(\mathbf{r},t) = 0, \tag{1.3}$$

by the imaginary unit and then by subtracting from it the second pair

$$\partial_t \mathbf{B}(\mathbf{r},t) = -\nabla \times \mathbf{E}(\mathbf{r},t), \quad \nabla \cdot \mathbf{B}(\mathbf{r},t) = 0. \tag{1.4}$$

In the SI units that are used here, the vectors $\mathbf{D}$ and $\mathbf{B}$ have different dimensions and prior to subtraction one must equalize the dimensions of both terms. The resulting equations in empty space are:

$$i\partial_t \mathbf{F}(\mathbf{r},t) = c\nabla \times \mathbf{F}(\mathbf{r},t), \tag{1.5}$$

$$\nabla \cdot \mathbf{F}(\mathbf{r},t) = 0, \tag{1.6}$$

where

$$\mathbf{F}(\mathbf{r},t) = \left(\frac{\mathbf{D}(\mathbf{r},t)}{\sqrt{2\epsilon_0}} + i\frac{\mathbf{B}(\mathbf{r},t)}{\sqrt{2\mu_0}} \right), \tag{1.7}$$

and $c = 1/\sqrt{\epsilon_0\mu_0}$. Around the year 1930, Majorana (unpublished notes quoted by Mignani, Recami and Baldo [1974]) arrived at the same complex vector exploring the analogy between the Dirac equation and the Maxwell equations. Kramers [1938] made extensive use of this vector in his treatment of quantum radiation theory. This vector is also a natural object to use in the quaternionic formulation of Maxwell's theory (Silberstein [1914]). With the advent of spinor calculus that superseded the quaternionic calculus, the transformation properties of the Riemann–Silberstein vector have become even more transparent. When Maxwell's equations were cast into the spinor form (Laporte and Uhlenbeck [1931], Oppenheimer [1931]), this vector turned into a symmetric second-rank spinor. The use of the Riemann–Silberstein vector as the wave function of the photon has been advocated by Oppenheimer [1931], Molière [1949], Good [1957], Bialynicki-Birula [1994], Sipe [1995], and Bialynicki-Birula [1996a].

It was already noticed by Silberstein [1907a] that the important dynamical quantities associated with the electromagnetic field – the energy density and the Poynting vector – can be represented as *bilinear expressions* built from the complex vector F. Using modern terminology, one would say that the formulas for the energy E, momentum P, angular momentum M, and the moment of energy N of the electromagnetic field look like quantum-mechanical expectation values; i.e.,

$$E = \int d^3 r \, F^* \cdot F, \tag{1.8}$$

$$P = \frac{1}{2ic} \int d^3 r \, F^* \times F, \tag{1.9}$$

$$M = \frac{1}{2ic} \int d^3 r \, r \times (F^* \times F), \tag{1.10}$$

$$N = \int d^3 r \, r (F^* \cdot F), \tag{1.11}$$

evaluated in a state described by the wave function F. All these quantities are invariant under the multiplication of F by a phase factor $\exp(i\alpha)$. Such a multiplication results in the so called duality rotation (Misner and Wheeler [1957]) of the field vectors

$$\frac{D'}{\sqrt{\epsilon_0}} = \frac{\cos \alpha \, D}{\sqrt{\epsilon_0}} - \frac{\sin \alpha \, B}{\sqrt{\mu_0}}, \tag{1.12}$$

$$\frac{B'}{\sqrt{\mu_0}} = \frac{\cos \alpha \, B}{\sqrt{\mu_0}} + \frac{\sin \alpha \, D}{\sqrt{\epsilon_0}}. \tag{1.13}$$

The coupling of the electromagnetic field with charges fixes the phase α but for a free photon one has the same complete freedom in choosing the overall phase of the photon wave function as in standard wave mechanics of massive particles.

The Riemann–Silberstein vector also has many other properties which one would associate with a one-photon wave function, except for a somewhat modified probabilistic interpretation. Insistence on exactly the same form of the expressions for transition probabilities as in nonrelativistic wave mechanics leads back to the Landau–Peierls wave function with its highly nonlocal transformation properties.

§ 2. Wave Equation for Photons

The wave equation for the photon is taken to be the complexified form [eq. (1.5)] of Maxwell's equations. In order to justify this choice, one may show (cf. § 4) that the Fourier decomposition of the solutions of this wave equation leads to the same photon wave functions in momentum representation that can be introduced without any reference to a wave equation directly from the general theory of representations of the Poincaré group (Bargmann and Wigner [1948], Lomont and Moses [1962]). There is also a heuristic argument indicating that eq. (1.5) is the right choice. Namely, the wave equation (1.5) can be written in the same form as the Weyl equation for the neutrino wave function. As a matter of fact, all wave equations for massless particles with arbitrary spin can be cast into the same form (cf. § 12).

2.1. WAVE EQUATION FOR PHOTONS IN FREE SPACE

In order to see a correspondence between Maxwell's equations and quantum-mechanical wave equations, one may follow Oppenheimer [1931] and Molière [1949] and rewrite eq. (1.5) with the use of the spin-1 matrices s_x, s_y, s_z well known from quantum mechanics (see, for example, Schiff [1968]). The matrices which will be used here are in a different representation from the one usually used in quantum mechanics, since they act on the Cartesian vector components of the wave function and not on the components labeled by the eigenvalues of s_z. That is the reason why the matrix s_z is not diagonal:

$$s_x = \begin{bmatrix} 0 & 0 & 0 \\ 0 & 0 & -i \\ 0 & i & 0 \end{bmatrix}, \quad s_y = \begin{bmatrix} 0 & 0 & i \\ 0 & 0 & 0 \\ -i & 0 & 0 \end{bmatrix}, \quad s_z = \begin{bmatrix} 0 & -i & 0 \\ i & 0 & 0 \\ 0 & 0 & 0 \end{bmatrix}. \tag{2.1}$$

Equation (1.5) can be written in terms of these spin matrices if the following conversion rule from vector notation to matrix notation is applied:

$$\mathbf{a} \times \mathbf{b} = -i(\mathbf{a} \cdot \mathbf{s})\mathbf{b}. \tag{2.2}$$

The resulting equation,

$$i\hbar\partial_t F(r,t) = c\left(s\cdot\frac{\hbar}{i}\nabla\right)F(r,t), \tag{2.3}$$

is of a Schrödinger type but with a different Hamiltonian. The divergence condition (1.6) can also be expressed in terms of spin matrices, either as

$$(s\cdot\nabla)s_j F = \nabla_j F, \tag{2.4}$$

or equivalently (Pryce [1948]) as

$$\left(s\cdot\nabla\right)^2 F = \Delta F. \tag{2.5}$$

The form (2.3) of the Maxwell equations compares directly with the Weyl equation for neutrinos (Weyl [1929]):

$$i\hbar\partial_t \phi(r,t) = c\left(\sigma\cdot\frac{\hbar}{i}\nabla\right)\phi(r,t). \tag{2.6}$$

Equation (2.6) differs from eq. (2.3) only in having the Pauli matrices, appropriate for spin-$\frac{1}{2}$ particles, instead of the spin-1 matrices which are appropriate for photons. Of course, one may cancel the factors of $\hbar$ appearing on both sides of eqs. (2.3) and (2.6), but their presence makes the connection with quantum mechanics more transparent.

Some authors (Oppenheimer [1931], Ohmura [1956], Moses [1959]) introduced a different, although equivalent, form of the photon wave equation for a *four-component* wave function. The inclusion of the fourth component enables one to incorporate the divergence condition in a natural way. This approach is related directly to the spinorial representation of the photon wave function, and will be discussed in § 12.

In quantum mechanics the stationary solutions of the wave equation play a distinguished role. They are the building blocks from which all solutions can be constructed. Stationary solutions of the wave equation are obtained by separating the time variable and solving the resulting eigenvalue problem. The eigenvalue equation resulting from the photon wave equation (2.3) is:

$$c\left(s\cdot\frac{\hbar}{i}\nabla\right)F(r) = \hbar\omega F(r). \tag{2.7}$$

Assuming that the photon energy $\hbar\omega$ is positive, one reads from eq. (2.7) that the projection of the spin on the direction of momentum (helicity) is positive.

It can be easily checked that one can reverse this sign by changing i into −i in the definition (1.7) of the Riemann–Silberstein vector. Thus, the choice of sign in the definition of this vector is equivalent to choosing positive or negative helicity, corresponding to left-handed or right-handed circular polarization. In order to account for both helicity states of the photons, one must consider both vectors; one may call them F_+ and F_-. This doubling of the vectors F had already been considered by Silberstein [1914] in the context of the classical Maxwell equations.

Of course, if one is interested only in translating the Maxwell equations into a complex form, one can restrict oneself to either F_+ or F_-. In both cases, one obtains a one-to-one correspondence between the real field vectors D and B and their complex combination $F_\pm$. However, to have a bona fide photon wave function one must be able to superpose different helicity states without changing the sign of the energy (frequency). This can be done only when both helicities are described by different components of *the same wave function* as in the theory of spin-$\frac{1}{2}$ particles. One can see even more clearly the need to use both complex combinations when one deals with the propagation of photons in a medium.

2.2. WAVE EQUATION FOR PHOTONS IN A MEDIUM

In free space, the two vectors $F_\pm$ satisfy two separate wave equations:

$$i\partial_t F_\pm(r,t) = \pm c\nabla \times F_\pm(r,t). \tag{2.8}$$

In a homogeneous medium, using the values of ϵ and μ for the medium in the definition of the vectors $F_\pm$,

$$F_\pm(r,t) = \left(\frac{D(r,t)}{\sqrt{2\epsilon}} \pm i\frac{B(r,t)}{\sqrt{2\mu}}\right), \tag{2.9}$$

one also obtains two separate wave equations. The new vectors $F_\pm$ are linear combinations of the free-space vectors $F_\pm^0$,

$$F_+ = \frac{1}{2}\left[\left(\sqrt{\frac{\epsilon_0}{\epsilon}} + \sqrt{\frac{\mu_0}{\mu}}\right) F_+^0 + \left(\sqrt{\frac{\epsilon_0}{\epsilon}} - \sqrt{\frac{\mu_0}{\mu}}\right) F_-^0\right], \tag{2.10}$$

$$F_- = \frac{1}{2}\left[\left(\sqrt{\frac{\epsilon_0}{\epsilon}} - \sqrt{\frac{\mu_0}{\mu}}\right) F_+^0 + \left(\sqrt{\frac{\epsilon_0}{\epsilon}} + \sqrt{\frac{\mu_0}{\mu}}\right) F_-^0\right]. \tag{2.11}$$

Thus, the positive and negative helicity states in a medium are certain linear superpositions of such states in free space. The necessity to form linear superpositions of both helicity states shows up even more forcefully in an inhomogeneous medium, because then it is not possible to split the wave equations into two independent sets. For a linear, time-independent, isotropic medium, characterized by space-dependent permittivity and permeability, one obtains the following coupled set of wave equations:

$$i\partial_t F_+(r,t) = v(r)(\nabla \times F_+(r,t)) \tag{2.12}$$
$$- \frac{1}{2v(r)} F_+(r,t) \times \nabla v(r) - \frac{1}{2h(r)} F_-(r,t) \times \nabla h(r),$$

$$i\partial_t F_-(r,t) = -v(r)(\nabla \times F_-(r,t)) \tag{2.13}$$
$$- \frac{1}{2v(r)} F_-(r,t) \times \nabla v(r) - \frac{1}{2h(r)} F_+(r,t) \times \nabla h(r),$$

where $F_\pm(r,t)$ are built with the values of $\epsilon(r)$ and $\mu(r)$ in the medium, $v(r) = 1/\sqrt{\epsilon(r)\mu(r)}$ is the value of the speed of light in the medium, and $h(r) = \sqrt{\mu(r)/\epsilon(r)}$ is the "resistance of the medium" (the sole justification for the use of this name is the right dimensionality of Ohm). The divergence condition (1.6) in an inhomogeneous medium takes on the form

$$\nabla \cdot F_+(r,t) = \frac{1}{2v(r)} F_+(r,t) \cdot \nabla v(r) + \frac{1}{2h(r)} F_-(r,t) \cdot \nabla h(r), \tag{2.14}$$

$$\nabla \cdot F_-(r,t) = \frac{1}{2v(r)} F_-(r,t) \cdot \nabla v(r) + \frac{1}{2h(r)} F_+(r,t) \cdot \nabla h(r). \tag{2.15}$$

The quantum-mechanical forms of eqs. (2.12) and (2.13) are:

$$i\hbar\partial_t F_+(r,t) = \sqrt{v(r)}\left(s \cdot \frac{\hbar}{i}\nabla\right)\sqrt{v(r)} F_+(r,t)$$
$$- i\hbar \frac{v(r)}{2h(r)} (s \cdot \nabla h(r)) F_-(r,t), \tag{2.16}$$

$$i\hbar\partial_t F_-(r,t) = \sqrt{v(r)}\left(s \cdot \frac{\hbar}{i}\nabla\right)\sqrt{v(r)} F_-(r,t)$$
$$+ i\hbar \frac{v(r)}{2h(r)} (s \cdot \nabla h(r)) F_+(r,t). \tag{2.17}$$

In view of the coupling in the evolution equations (2.16) and (2.17) between the vectors F_+ and F_-, one must combine them to form one wave function, $\mathcal{F}$, with six components:

$$\mathcal{F} = \begin{bmatrix} F_+ \\ F_- \end{bmatrix}. \tag{2.18}$$

The wave equation for this function can be written in a compact form,

$$i\hbar\partial_t \mathcal{F}(r,t) = \sqrt{v(r)}\rho_3\left(s\cdot\frac{\hbar}{i}\nabla\right)\sqrt{v(r)}\,\mathcal{F}(r,t)$$
$$+ \hbar\frac{v(r)}{2h(r)}\rho_2(s\cdot\nabla h)\mathcal{F}(r,t), \qquad (2.19)$$

where the spin matrices s_i operate separately on upper and lower components,

$$s_i\mathcal{F} = \begin{bmatrix} s_i F_+ \\ s_i F_- \end{bmatrix}, \qquad (2.20)$$

and the three Pauli matrices ρ_i act on $\mathcal{F}$ as follows:

$$\rho_1\mathcal{F} = \begin{bmatrix} F_- \\ F_+ \end{bmatrix}, \quad \rho_2\mathcal{F} = \begin{bmatrix} -iF_- \\ iF_+ \end{bmatrix}, \quad \rho_3\mathcal{F} = \begin{bmatrix} F_+ \\ -F_- \end{bmatrix}. \qquad (2.21)$$

The divergence conditions (2.14) and (2.15) can also be written in this compact form:

$$\nabla\cdot\mathcal{F}(r,t) = \frac{1}{2v(r)}\mathcal{F}(r,t)\cdot\nabla v(r) + \rho_1\frac{1}{2h(r)}\mathcal{F}(r,t)\cdot\nabla h(r). \qquad (2.22)$$

One would not be able to write a linear wave equation for an inhomogeneous medium in terms of just one three-dimensional complex vector, without doubling the number of components. Note that the speed of light v may vary without causing necessarily the mixing of helicities. This happens, for example, in the gravitational field (cf. § 11). It is only the space-dependent resistance $h(r)$ that causes mixing.

It is worth stressing that the study of the propagation of photons in an inhomogeneous medium separates clearly local wave functions from nonlocal ones. In free space there are many wave functions satisfying the same set of equations. For example, differentiations of the wave functions or integral operations of the type (1.2) do not change the form of these equations. The essential difference between various wave functions shows up forcefully in the study of the wave equation in an inhomogeneous medium, or in curved space (§ 11). All previous studies, except Bialynicki-Birula [1994], were restricted to propagation in *free space*, and this very important point was missed completely. The photon wave equations in an inhomogeneous medium are not very simple but that is due, perhaps, to a phenomenological character of macroscopic electrodynamics. The propagation of a photon in a medium is a succession of absorptions and subsequent emissions of the photon by the charges which form the medium. The number of photons of a given helicity is, in general, not conserved in these processes, and that accounts for all the complications. The photon wave equations in an inhomogeneous medium describe in actual fact a propagation of some collective excitations of the whole system and not just of free photons.

2.3. ANALOGY WITH THE DIRAC EQUATION

The analogy with the relativistic electron theory, mentioned in § 1, becomes the closest when the photon wave equation is compared with the Dirac equation written in the chiral representation of the Dirac matrices. In this representation the bispinor is made of two relativistic spinors

$$\psi(r,t) = \begin{bmatrix} \phi_A(r,t) \\ \chi^{A'}(r,t) \end{bmatrix}, \tag{2.23}$$

and the Dirac equation may be viewed as two Weyl equations coupled by the mass term:

$$i\hbar\partial_t\phi(r,t) = c\left(\boldsymbol{\sigma}\cdot\frac{\hbar}{i}\nabla\right)\phi(r,t) + mc^2\chi(r,t), \tag{2.24}$$

$$i\hbar\partial_t\chi(r,t) = -c\left(\boldsymbol{\sigma}\cdot\frac{\hbar}{i}\nabla\right)\chi(r,t) + mc^2\phi(r,t). \tag{2.25}$$

These equations are analogous to eqs. (2.16) and (2.17) for the photon wave function. For photons, the role of the mass term is played by the inhomogeneity of the medium.

§ 3. Photon Wave Function in Coordinate Representation

Despite a formal similarity between the wave equations for the photon [eqs. (2.16) and (2.17)] and for the electron [eqs. (2.24) and (2.25)], there is an important difference. Photons, unlike electrons, do not have antiparticles and this fact influences the form of solutions of the wave equation and their interpretation.

3.1. PHOTONS HAVE NO ANTIPARTICLES

Elementary, plane-wave solutions of relativistic wave equations in free space are of two types: they have positive or negative frequency,

$$\exp(-i\omega t + \boldsymbol{k}\cdot\boldsymbol{r}), \quad \text{or} \quad \exp(i\omega t - \boldsymbol{k}\cdot\boldsymbol{r}). \tag{3.1}$$

According to relativistic quantum mechanics, solutions with positive frequency correspond to particles, and solutions with negative frequency correspond to antiparticles. Thus, positive and negative frequency parts of the same solution of the wave equation describe two different physical entities: particle and antiparticle.

Photons do not have antiparticles, or to put it differently, antiphotons are identical with photons. Hence, the information carried by the negative frequency solutions must be the same as the information already contained in the positive frequency solution. Therefore, one may disregard completely the negative frequency part as redundant. An alternative method is to keep also the negative frequency part, but to impose an additional condition on the solutions of the wave equation. This condition states that the operation of particle–antiparticle conjugation,

$$\mathcal{F}^c(\boldsymbol{r},t) = \rho_1 \mathcal{F}^*(\boldsymbol{r},t), \tag{3.2}$$

leaves the function $\mathcal{F}$ invariant:

$$\mathcal{F}^c(\boldsymbol{r},t) = \mathcal{F}(\boldsymbol{r},t). \tag{3.3}$$

This condition is compatible with the evolution equation (2.19), and it eliminates the unwanted degrees of freedom (cf. Bialynicki-Birula [1994]). Equation (3.3) is satisfied automatically if the wave function is constructed according to the definition (2.18). This follows from the fact that F_+ and F_- are complex conjugate to each other. The information carried by the six-component function $\mathcal{F}$ satisfying the condition (3.3) is contained in its positive energy part, and is the same as that carried by the initial Riemann–Silberstein vector $\boldsymbol{F}$. It follows from eq. (3.3) that the negative frequency part $\mathcal{F}^{(-)}$ can always be obtained by complex conjugation and by an interchange of the upper and lower components of the positive frequency part $\mathcal{F}^{(+)}$,

$$\mathcal{F}^{(-)}(\boldsymbol{r},t) = \rho_1 \mathcal{F}^{(+)*}(\boldsymbol{r},t). \tag{3.4}$$

In this review, I shall use the symbol Ψ to denote the properly normalized, *positive energy* (positive frequency) part of the function $\mathcal{F}$:

$$\Psi(r,t) = \mathcal{F}^{(+)}(r,t). \tag{3.5}$$

This is the true *photon wave function*. Proper normalization of the photon wave function is essential for its probabilistic interpretation and is discussed in § 5. Note that the function Ψ carries the same amount of information as the original Riemann–Silberstein vector since Ψ can be constructed from F by splitting this vector into positive and negative frequency parts and then using the first part as the upper components of Ψ and the complex conjugate of the second part as the lower components:

$$\Psi(r,t) = \begin{bmatrix} F^{(+)}(r,t) \\ F^{(-)*}(r,t) \end{bmatrix}. \tag{3.6}$$

The positive frequency part of the solutions of wave equations is also a well-known concept in classical electromagnetic theory, where it is called the analytic signal (Born and Wolf [1980], Mandel and Wolf [1995]).

3.2. TRANSFORMATION PROPERTIES OF THE PHOTON WAVE FUNCTION IN COORDINATE REPRESENTATION

In free space, the components of the electromagnetic field form a tensor, and that allows one to establish the transformation properties of $\mathcal{F}$. Transformation properties of the photon wave function Ψ are the same as those of $\mathcal{F}$. Under rotations, the upper and the lower half of Ψ transform as three-dimensional vector fields. Under Lorentz transformations, the upper and the lower part also transform independently and the corresponding rules can be inferred directly from classical electrodynamics (cf., for example, Jackson [1975]). Under the Lorentz transformation characterized by the velocity v, the vectors $F_\pm$ change as follows:

$$F'_\pm = \gamma\left(F \mp i\frac{v \times F}{c}\right) - \frac{\gamma^2}{\gamma+1}\frac{v(v \cdot F)}{c^2}, \tag{3.7}$$

where γ is the standard relativistic factor $\gamma = \sqrt{1 - v^2/c^2}$. One may check that this transformation preserves the square of these vectors,

$$(F')^2 = (F)^2. \tag{3.8}$$

This is understood easily if one observes that $F_\pm^2$ is a combination of the well-known scalar invariant $\mathcal{S} = (\epsilon_0 E^2 - B^2/\mu_0)/2$ and the pseudoscalar invariant $\mathcal{P} = \sqrt{\epsilon_0/\mu_0}\, E \cdot B$ of the electromagnetic field,

$$F_\pm^2 = \mathcal{S} \pm i\mathcal{P}. \tag{3.9}$$

Thus, rotations and Lorentz transformations act on vectors $F_\pm$ as elements of the orthogonal group in three dimensions (Kramers [1938] p. 429) leading to the following transformation properties of the wave function:

$$\Psi'(r',t') = \begin{bmatrix} C & 0 \\ 0 & C^* \end{bmatrix} \Psi(r,t), \tag{3.10}$$

where C is a three-dimensional, complex orthogonal matrix and C^* is its complex conjugate. The unification of rotations and Lorentz boosts into one complex orthogonal transformation is even more transparent for infinitesimal transformations,

$$\Psi' = \Psi + \left((i\delta\gamma + \rho_3\delta\upsilon)\cdot s\right)\Psi, \tag{3.11}$$

where $\delta\gamma$ is the vector of an infinitesimal rotation. Under the space reflection $r \to -r$, the upper and lower part of Ψ do not transform independently but are interchanged because D and B transform as a vector and a pseudovector, respectively:

$$\Psi'(-r,t) = \rho_1 \Psi(r,t). \tag{3.12}$$

All these transformation properties can also be simply stated in terms of second-rank spinor fields (cf. § 12).

3.3. PHOTON HAMILTONIAN

The operator appearing on the right hand side of the evolution equation (2.19) for the photon wave function is the Hamiltonian operator $\widehat{H}$ for the photon:

$$\widehat{H} = \sqrt{\upsilon(r)}\rho_3\left(s\cdot\frac{\hbar}{i}\nabla\right)\sqrt{\upsilon(r)} + \hbar\frac{\upsilon(r)}{2h(r)}\rho_2\left(s\cdot\nabla h(r)\right). \tag{3.13}$$

In free space, this expression reduces to:

$$\widehat{H}_0 = c\rho_3\left(s\cdot\frac{\hbar}{i}\nabla\right). \tag{3.14}$$

The formulas (3.13) and (3.14) define a Hermitian operator with continuous spectrum extending from $-\infty$ to ∞. Hermiticity is defined here with respect to the standard (mathematical) scalar product,

$$(\Psi_1 | \Psi_2) = \int d^3r \, \Psi_1^\dagger(r) \, \Psi_2(r). \tag{3.15}$$

Wave functions of physical photons are built from positive-energy solutions of the eigenvalue problem for the Hamiltonian,

$$\widehat{H}\Psi(r) = E\Psi(r). \tag{3.16}$$

There is a simple relation between positive-energy solutions and negative-energy solutions. One may obtain all solutions for the negative energies just by an interchange of upper and lower components, since $\rho_1 \widehat{H} \rho_1 = -\widehat{H}$. Such a simple symmetry of solutions is a result of photons being identical with antiphotons and it is not found, in general, for particles whose antiparticles are physically distinct. For example, the solutions of the wave functions describing electrons in the Coulomb potential of the proton are quite different from the wave function of positrons moving in the same potential. In the first case the potential is attractive (bound states), while in the second case it is repulsive (only scattering states).

Explicit solutions of the energy eigenvalue problem for photons are obtained easily in free space, but in the presence of a medium this can be done only in special cases. In this respect, wave mechanics of photons are not much different from wave mechanics of massive particles, where explicit solutions can also be found only for special potentials.

§ 4. Photon Wave Function in Momentum Representation

The most thorough textbook treatment of quantum mechanics of photons has been given by Akhiezer and Berestetskii [1965], who devoted a long chapter to this problem. Their discussion is limited to momentum representation except for a brief subsection under a characteristic title: "Impossibility of introducing a photon wave function in the coordinate representation". This impossibility will be addressed in § 8.

Wave mechanics of photons in momentum representation can be derived directly from relativistic quantum kinematics and group representation theory, but here the analysis will be based on the Fourier representation of the photon wave function in coordinate representation.

In this review, I shall use traditionally the wave vector, k, instead of the photon momentum vector, $p = \hbar k$, as the argument of the wave function in momentum space. The explicit introduction of Planck's constant is necessary only when the proper normalization of the wave function is needed.

4.1. PHOTON WAVE FUNCTION AS A FOURIER INTEGRAL

The standard procedure for solving the wave equations (1.5) or (2.3) is based on Fourier transformation. Since every solution of eqs. (1.5) and (1.6) is a solution of the d'Alembert equation,

$$\left(\frac{1}{c^2}\frac{\partial^2}{\partial t^2} - \Delta\right) F_\pm(r,t) = 0, \tag{4.1}$$

the vectors $F_\pm$ can be represented as superpositions of plane waves

$$F_+(r,t) = \sqrt{\hbar c} \int \frac{d^3k}{(2\pi)^3} \left[f_+(k) e^{-i\omega t + i k \cdot r} + f_-^*(k) e^{i\omega t - i k \cdot r} \right], \tag{4.2}$$

$$F_-(r,t) = \sqrt{\hbar c} \int \frac{d^3k}{(2\pi)^3} \left[f_-(k) e^{-i\omega t + i k \cdot r} + f_+^*(k) e^{i\omega t - i k \cdot r} \right], \tag{4.3}$$

where $\omega = c|k|$ and the factor $\sqrt{\hbar c}$ has been introduced for future convenience. The remaining integral has the dimension of 1/length^2, so that the Fourier coefficients $f_\pm$ have the dimension of 1/length. It has already been taken into account in eqs. (4.2) and (4.3) that the vectors F_+ and F_- are complex conjugate of each other. In order to fulfill Maxwell's equations, the two complex vectors $f_+(k)$ and $f_-(k)$ must satisfy the set of linear, algebraic equations which result from eqs. (1.5) and (1.6); respectively,

$$i k \times f_\pm(k) = \pm |k| F_\pm(k), \tag{4.4}$$

$$k \cdot f_\pm(k) = 0. \tag{4.5}$$

Actually, the second equation is superfluous since it follows from the first. Solutions of these equations are determined up to a complex factor. Denoting by $e(k)$ a normalized solution of the first equation taken with a plus sign

$$i k \times e(k) = |k| e(k), \tag{4.6}$$

$$e^*(k) \cdot e(k) = 1, \tag{4.7}$$

one can express the vectors $f_\pm(k)$ in the form:

$$f_+(k) = e(k) f(k, 1), \quad f_-(k) = e^*(k) f(k, -1). \tag{4.8}$$

The two complex functions $f(k, \lambda)$, where $\lambda = \pm 1$, describe the independent degrees of freedom of the free electromagnetic field. The vector $e(k)$ can be decomposed into two real vectors $l_i(k)$ which form, together with the unit vector $n(k) = k/|k|$, an orthonormal set:

$$e(k) = (l_1(k) + il_2(k))/\sqrt{2}, \quad l_i(k) \cdot l_j(k) = \delta_{ij}, \tag{4.9}$$

$$n(k) \cdot l_i(k) = 0, \quad l_1(k) \times l_2(k) = n(k). \tag{4.10}$$

The only freedom left in the definition of $e(k)$ is its phase. A multiplication by a phase factor amounts to a rotation of the vectors $l_i(k)$ around the vector $n(k)$. The same freedom characterizes the coefficient functions $f(k, \lambda)$. This phase may, in general, depend on k and it plays an important role in the study of the photon wave function in momentum representation. The final form of the Fourier representation for vectors $F_\pm$ is:

$$F_+(r,t) = \sqrt{\hbar c} \int \frac{d^3k}{(2\pi)^3} e(k) \left[f(k, 1) e^{-i\omega t + ik \cdot r} + f^*(k, -1) e^{i\omega t - ik \cdot r} \right], \tag{4.11}$$

$$F_-(r,t) = \sqrt{\hbar c} \int \frac{d^3k}{(2\pi)^3} e^*(k) \left[f(k, -1) e^{-i\omega t + ik \cdot r} + f^*(k, 1) e^{i\omega t - ik \cdot r} \right]. \tag{4.12}$$

4.2. INTERPRETATION OF FOURIER COEFFICIENTS

In free space, the energy, momentum, angular momentum, and moment of energy of the classical electromagnetic field are given by the expressions (1.8)–(1.11). With the help of eqs. (4.11) of (4.12) they can be expressed in terms of the coefficient functions $f(k, \lambda)$ (cf., for example, Bialynicki-Birula and Bialynicka-Birula [1975]):

$$E = \sum_\lambda \int \frac{d^3k}{(2\pi)^3 |k|} \hbar \omega f^*(k, \lambda) f(k, \lambda), \tag{4.13}$$

$$P = \sum_\lambda \int \frac{d^3k}{(2\pi)^3 |k|} \hbar k f^*(k, \lambda) f(k, \lambda), \tag{4.14}$$

$$M = \sum_\lambda \int \frac{d^3k}{(2\pi)^3 |k|} f^*(k, \lambda) \left(\hbar k \times \frac{1}{i} D_k + \lambda \hbar \frac{k}{|k|} \right) f(k, \lambda), \tag{4.15}$$

$$N = \sum_\lambda \int \frac{d^3k}{(2\pi)^3 |k|} f^*(k, \lambda) i\hbar \omega D_k f(k, \lambda), \tag{4.16}$$

where

$$D_k = \partial_k + i\lambda \alpha(k), \tag{4.17}$$

and

$$a(k) = l_1(k) \cdot \partial_k l_2(k) - l_2(k) \cdot \partial_k l_1(k). \tag{4.18}$$

The operation D_k is a natural covariant derivative on the light cone (Starusz-kiewicz [1973], Bialynicki-Birula and Bialynicka-Birula [1975, 1987]). Note that this operation depends, through the vector $a(k)$, on the phase convention for the polarization vector $e(k)$. It is interesting to note that the vector $a(k)$ has properties simular to the electromagnetic vector potential. It changes by a gradient under a change of the phase, but its curl is defined uniquely. Indeed, it follows from the definition (4.18) of $a(k)$ and from the orthonormality conditions (4.10) that the vector $a(k)$ obeys the equation:

$$\partial_i a_j - \partial_j a_i = -\epsilon_{ijk} n_k / k^2. \tag{4.19}$$

This equation determines the Berry phase (Bialynicki-Birula and Bialynicka-Birula [1987]) in the propagation of photons.

The formulas (4.13) and (4.14) indicate that $f(k, \pm 1)$ describe field amplitudes with energy $\hbar\omega$ and momentum $\hbar k$. The formula (4.15) shows that $f(k, 1)$ and $f(k, -1)$ describe field amplitudes with positive and negative helicity, since their contribution to the component of angular momentum in the direction of momentum is equal to $\pm\hbar$, respectively.

The functions $f(k, \lambda)$ actually have a dual interpretation. In classical theory they yield full information about the electromagnetic field. In wave mechanics of the photon, these functions are the components of the photon wave function in momentum representation. In order to distinguish between these two cases, I shall denote the two components of the photon wave function in momentum representation by a new symbol, $\phi(k, \lambda)$. Wave function must be normalized and the proper normalization of the photon wave function $\phi(k, \lambda)$ is discussed in § 5.

The expansion of the photon wave function into plane waves has the following form:

$$\Psi(r,t) = \sqrt{\hbar c} \int \frac{d^3 k}{(2\pi)^3} \begin{bmatrix} e(k, 1)\phi(k, 1) \\ e(k, -1)\phi(k, -1) \end{bmatrix} e^{-i\omega t + i k \cdot r}, \tag{4.20}$$

where

$$e(k, 1) = e(k), \quad e(k, -1) = e^*(k). \tag{4.21}$$

The integral (4.20) defines a certain (continuous) superposition of the wave functions $\phi(k, \lambda)$ with different values of the wave vector. If the photon wave function in momentum representation is accepted as a legitimate concept, then the superpositions of such functions must also be accepted. Those who are not sure about the meaning of a continuous superposition, in the form of an integral, may restrict the electromagnetic field to a box and replace the integral by a discrete sum.

4.3. TRANSFORMATION PROPERTIES OF THE PHOTON WAVE FUNCTION IN MOMENTUM REPRESENTATION

Transformation rules for the wave function in momentum representation may be derived from the transformation properties of the Riemann–Silberstein vector. They are the same in the classical theory of the electromagnetic field and in the quantum theory of photons. Under space and time translations, functions $\phi(\mathbf{k}, \lambda)$ are multiplied by the phase factors

$$\phi'(\mathbf{k}, \lambda) = \exp(-i\omega t_0 + i\mathbf{k}\cdot \mathbf{r}_0)\,\phi(\mathbf{k}, \lambda), \tag{4.22}$$

where $(\mathbf{r}_0, t_0)$ is the four-vector of translation. From expressions (4.13) and (4.14) one may deduce that under rotations and Lorentz transformations, the functions $\phi(\mathbf{k}, \lambda)$ are also multiplied by phase factors,

$$\phi'(\mathbf{k}', \lambda) = \exp(-i\lambda\Theta(\mathbf{k}, \Lambda))\,\phi(\mathbf{k}, \lambda), \tag{4.23}$$

where the phase function $\Theta(\mathbf{k}, \Lambda)$ depends on the Poincaré transformation Λ. This transformation property can be derived easily from the transformation law of the energy-momentum four-vector for the electromagnetic field. The right hand side in the formulas (4.13) and (4.14) has three factors: the integration volume $d^3k/(2\pi)^3|\mathbf{k}|$, the four-vector $(\mathbf{k}, \omega)$, and the moduli squared of $\phi(\mathbf{k}, \lambda)$. The integration volume is an invariant (cf., for example, Weinberg [1995] p. 67), and therefore $|\phi(\mathbf{k}, \lambda)|^2$ must also be invariant. An explicit form of the phase $\Theta(\mathbf{k}, \Lambda)$ corresponding to a given Poincaré transformation Λ can be given (cf., for example, Amrein [1969]), but it is not very illuminating.

§ 5. Probabilistic Interpretation

Probabilistic interpretation of wave mechanics requires, first of all, a definition of the scalar product $\langle \Psi_1 | \Psi_2 \rangle$ that is to be used in the calculation of transition probabilities. The modulus squared of the scalar product of two normalized wave functions $|\langle\Psi_1|\Psi_2\rangle|^2$ determines the probability of finding a photon in the state Ψ_1 when the photon is in the state Ψ_2. The probability, of course, must be a pure number and – as a true observable – must be invariant under all Poincaré transformations. The most obvious definition of the scalar product (3.15) cannot be used, because it is neither Poincaré invariant nor dimensionally correct. There is essentially only one candidate for the correct scalar product. Its heuristic derivation is the easiest in momentum representation.

5.1. SCALAR PRODUCT

According to quantum mechanics, each photon with momentum $\hbar\boldsymbol{k}$ carries energy $\hbar\omega$. Thus, the total number of photons N present in the electromagnetic field is obtained by dividing the integrand in the formula (4.13) by $\hbar\omega$,

$$N = \sum_\lambda \int \frac{\mathrm{d}^3 k}{(2\pi)^3 |\boldsymbol{k}|} |f(\boldsymbol{k},\lambda)|^2. \tag{5.1}$$

A photon wave function describes just one photon. The normalized wave function must, therefore, satisfy the condition $N = 1$. Normalized photon wave functions in momentum representation satisfy the normalization condition:

$$\sum_\lambda \int \frac{\mathrm{d}^3 k}{(2\pi)^3 |\boldsymbol{k}|} |\phi(\boldsymbol{k},\lambda)|^2 = 1. \tag{5.2}$$

The form of the scalar product can be deduced from the expression for the norm and it reads:

$$\langle\Psi_1|\Psi_2\rangle = \sum_\lambda \int \frac{\mathrm{d}^3 k}{(2\pi)^3 |\boldsymbol{k}|} \phi_1^*(\boldsymbol{k},\lambda)\phi_2(\boldsymbol{k},\lambda). \tag{5.3}$$

This scalar product can be also expressed in terms of the photon wave functions in coordinate representation by inverting the Fourier transformation in eq. (4.20):

$$\phi(\boldsymbol{k},\lambda) = \frac{1}{\sqrt{\hbar c}} e^*(\boldsymbol{k},\lambda) \cdot \int \mathrm{d}^3 r \exp(-\mathrm{i}\boldsymbol{k}\cdot\boldsymbol{r}) \Psi(\boldsymbol{r},t), \tag{5.4}$$

where the scalar product with $e^*(\boldsymbol{k},\lambda)$ is evaluated separately for upper and lower components, and for each λ only one of them does not vanish. Upon substituting this expression into eq. (5.3), interchanging the order of integrations, and using the following properties of the vectors $e(k, \lambda)$,

$$e^*(k, \lambda) \cdot e(k, \lambda') = \delta_{\lambda, \lambda'}, \tag{5.5}$$

$$\sum_\lambda e_i^*(k, \lambda) e_j(k, \lambda) = \delta_{ij} - n_i(k) n_j(k), \tag{5.6}$$

one obtains the following expression for the scalar product in coordinate representation

$$\langle \Psi_1 | \Psi_2 \rangle = \frac{1}{2\pi^2 \hbar c} \int d^3 r \int d^3 r' \, \Psi_1^\dagger(r) \frac{1}{|r - r'|^2} \Psi_2(r'). \tag{5.7}$$

The norm associated with this scalar product is:

$$N = \|\Psi\|^2 = \frac{1}{2\pi^2 \hbar c} \int d^3 r \int d^3 r' \, \Psi^\dagger(r) \frac{1}{|r - r'|^2} \Psi(r'). \tag{5.8}$$

The scalar product (5.7) and the associated norm (5.8) for photon wave functions have been arrived at by numerous authors starting from various premises. Gross [1964] has proven that this scalar product and this norm are invariant not only under Poincaré transformations but also under conformal transformations. Zeldovich [1965] derived the formula (5.8) for the number of photons in terms of the electromagnetic field vectors. Recently, the norm (5.8) has been found very useful in the formulation of wavelet electrodynamics (Kaiser [1992]). The same expression (5.3) for the scalar product can also be derived by considering quantum-mechanical expectation values. That approach has been used by Good [1957] and is presented below.

5.2. EXPECTATION VALUES OF PHYSICAL QUANTITIES

In wave mechanics of photons, the normalized photon wave function $\phi(k, \lambda)$ replaces the classical field amplitudes $f(k, \lambda)$. The classical expressions for the energy, momentum, angular momentum, and moment of energy become the formulas for the quantum-mechanical expectation values:

$$\langle E \rangle = \langle \Psi | \hat{H} | \Psi \rangle, \quad \langle P \rangle = \langle \Psi | \hat{P} | \Psi \rangle, \tag{5.9}$$

$$\langle M \rangle = \langle \Psi | \hat{J} | \Psi \rangle, \quad \langle N \rangle = \langle \Psi | \hat{K} | \Psi \rangle. \tag{5.10}$$

These equations compared with the formulas (4.13)–(4.16) enable one to identify the operators $\hat{H}$, $\hat{P}$, $\hat{J}$, and $\hat{K}$ in momentum representation as:

$$\hat{H} = \hbar \omega, \tag{5.11}$$

$$\widehat{\boldsymbol{P}} = \hbar \boldsymbol{k}, \tag{5.12}$$

$$\widehat{\boldsymbol{J}} = \boldsymbol{k} \times \frac{\hbar}{i}\boldsymbol{D}_k + \lambda\hbar\frac{\boldsymbol{k}}{|\boldsymbol{k}|}, \tag{5.13}$$

$$\widehat{\boldsymbol{K}} = \hbar\omega\frac{\hbar}{i}\boldsymbol{D}_k. \tag{5.14}$$

The operators $\widehat{H}$, $\widehat{\boldsymbol{P}}$, $\widehat{\boldsymbol{J}}$, and $\widehat{\boldsymbol{K}}$ are Hermitian with respect to the scalar product given by eq. (5.3).

The formulas (5.11)–(5.14) are fully consistent with the interpretation of $\phi(\boldsymbol{k}, \lambda)$ as the probability amplitude in momentum representation. The probability density to find the photon with the momentum $\hbar \boldsymbol{k}$ and the helicity λ is:

$$\text{Probability density} = \frac{|\phi(\boldsymbol{k}, \lambda)|^2}{(2\pi)^3|\boldsymbol{k}|}. \tag{5.15}$$

In order to express the quantum-mechanical expectation values (5.11)–(5.14) in coordinate representation, one must identify the proper form of the scalar product for the photon wave function Ψ. This identification was made by Good [1957] who compared the classical formula for the energy of the electromagnetic field with the quantum-mechanical expression involving the Hamiltonian

$$\langle E \rangle = \langle \Psi | \widehat{H} | \Psi \rangle, \tag{5.16}$$

and came to the conclusion that the scalar product for the photon wave function has to be modified as follows:

$$\langle \Psi_1 | \Psi_2 \rangle = \int d^3 r \, \Psi_1^\dagger \frac{1}{\widehat{H}} \Psi_2. \tag{5.17}$$

It is assumed here that the wave functions are built from positive energy states only, and that assumption guarantees the positive definiteness of the norm,

$$\|\Psi\|^2 = \int d^3 r \, \Psi^\dagger \frac{1}{\widehat{H}} \Psi, \tag{5.18}$$

associated with that scalar product. This form of the scalar product leads to the following expectation values of the energy, momentum, angular momentum, and moment of energy operators:

$$\langle E \rangle = \int d^3 r \, \Psi^\dagger \widehat{H}^{-1} \left(\boldsymbol{s} \cdot \frac{\hbar}{i} \nabla \right) \Psi, \tag{5.19}$$

$$\langle \boldsymbol{P} \rangle = \int d^3 r \, \Psi^\dagger \widehat{H}^{-1} \frac{\hbar}{i} \nabla \Psi, \tag{5.20}$$

$$\langle \boldsymbol{M} \rangle = \int d^3 r \, \Psi^\dagger \widehat{H}^{-1} \left(\boldsymbol{r} \times \frac{\hbar}{i} \nabla + \hbar \boldsymbol{s} \right) \Psi, \tag{5.21}$$

$$\langle \boldsymbol{N} \rangle = \int d^3 r \, \Psi^\dagger \widehat{H}^{-1} \widehat{H} \boldsymbol{r} \Psi. \tag{5.22}$$

Thus, the operators $\widehat{H}$, $\widehat{\boldsymbol{P}}$, $\widehat{\boldsymbol{J}}$, and $\widehat{\boldsymbol{K}}$ in coordinate representation have the form

$$\widehat{H} = c \left(\boldsymbol{s} \cdot \frac{\hbar}{i} \nabla \right), \tag{5.23}$$

$$\widehat{\boldsymbol{P}} = \frac{\hbar}{i} \nabla, \tag{5.24}$$

$$\widehat{\boldsymbol{J}} = \boldsymbol{r} \times \frac{\hbar}{i} \nabla + \hbar \boldsymbol{s}, \tag{5.25}$$

$$\widehat{\boldsymbol{K}} = \widehat{H} \boldsymbol{r}. \tag{5.26}$$

All these operators preserve the divergence condition (1.6), and are Hermitian with respect to the scalar product (5.17). It is also reassuring to note that the quantum-mechanical operators of momentum and angular momentum in coordinate representation have the same form as in standard quantum mechanics. This can be taken as another indication that $\Psi(\boldsymbol{r},t)$ is a legitimate and useful object.

One may prove directly (without using the Fourier expansions) with the help of the following identities:

$$\widehat{H} \rho_3 \boldsymbol{s}/c \Psi = \frac{\hbar}{i} \nabla \Psi, \quad \widehat{H} \rho_3 \hat{\boldsymbol{r}} \times \boldsymbol{s}/c \Psi = \left(\hat{\boldsymbol{r}} \times \frac{\hbar}{i} \nabla + \hbar \boldsymbol{s} \right) \Psi, \tag{5.27}$$

and with the use of eq. (2.4), that the expectation values (5.19)–(5.21) reduce to the classical expressions (1.8)–(1.11) when the wave function is replaced by the classical electromagnetic field.

The scalar product (5.17) in coordinate representation has been obtained from the scalar product (5.3) in momentum representation. However, the scalar product that contains the division by the Hamiltonian can be derived on more general grounds, and its definition does not depend on the choice of representation. It has been shown (Segal [1963], Ashtekar and Magnon [1975]) that such a scalar product is a general feature of geometric quantization in field theory.

Even though the number of photons is given by a double integral, so that there is no local expression for the photon probability density in coordinate space, the expression for the energy has the form of a single integral over $|\Psi(r)|^2$. Therefore, one may introduce a tentative notion of the "average photon energy in a region of space" and try to associate a probabilistic interpretation of the photon wave function with this quantity (Bialynicki-Birula [1994], Sipe [1995]). More precisely, the quantity $p_E(\Omega)$,

$$p_E(\Omega) = \frac{\int_\Omega d^3 r \, \Psi^\dagger(r) \, \Psi(r)}{\langle E \rangle}, \qquad (5.28)$$

may be interpreted as the probability of finding the energy of the photon localized in the region Ω. In other words, $p_E(\Omega)$ is the fraction of the average total energy of the photon associated with the region Ω. The probability density $\rho_E(r,t)$ of finding the energy of the photon at the point r,

$$\rho_E(r,t) = \frac{\Psi^\dagger(r,t) \, \Psi(r,t)}{\langle E \rangle}, \qquad (5.29)$$

is properly normalized to one and it also satisfies the continuity equation

$$\partial_t \rho_E(r,t) + \nabla \cdot j_E(r,t) = 0, \qquad (5.30)$$

with the normalized average energy flux

$$j_E(r,t) = \frac{\Psi^\dagger(r,t) \rho_3 s \Psi(r,t)}{\langle E \rangle}, \qquad (5.31)$$

as the probability current. The direct connection between the wave function $\Psi(r,t)$ and the average energy density justifies the name "energy wave function" used by Mandel and Wolf [1995]. It is understandable that the localization of photons is associated with their energy because photons do not carry other attributes like charge, fermion number, or rest mass. It is worth noting that for gravitons not only the probability but even the energy cannot be localized (cf., for example, Weinberg and Witten [1980]). The probabilistic interpretation of the energy wave function Ψ is still subject to all the limitations arising from the lack of the photon position operator, as discussed in § 8. In particular, there are no projection operators whose expectation values would give the probabilities $p_E(\Omega)$.

The transition amplitudes, the operators representing important physical quantities, and the expectation values can be expressed with equal ease in momentum representation and in coordinate representation. For photons moving in empty space, both representations are completely equivalent and give the same results. The only relevant issue is whether a particular superposition of wave functions in momentum representation is useful for the description of quantum states of the photon. The distinguished and unique feature of the superposition given by the Fourier integrals (4.20) is that they represent *local* fields. They have local transformation properties (3.10) and they satisfy local boundary conditions. Therefore, for photons moving in an inhomogeneous or bounded medium, it is the coordinate representation that is preferred because only in this representation one may easily take into account the properties of the medium (cf. § 6).

5.3. CONNECTION WITH LANDAU–PEIERLS WAVE FUNCTION

One may easily convert the scalar product (5.7) into a standard form containing a single integration with the use of the following identity:

$$\frac{1}{16\pi} \int d^3 r \, \frac{1}{|r - r'|^{5/2}} \frac{1}{|r - r''|^{5/2}} = \frac{1}{|r' - r''|^2}. \tag{5.32}$$

This enables one to convert the double integral (5.7) into a single integral:

$$\langle \Psi_1 | \Psi_2 \rangle = \int d^3 r \, \Phi_1^\dagger(r) \, \Phi_2(r). \tag{5.33}$$

The new functions Φ are the Landau–Peierls wave functions. They are related to the photon wave functions Ψ through the formula:

$$\Phi(r) = \frac{\pi}{\sqrt{\hbar c}} \int d^3 r' \, \frac{1}{(2\pi |r - r'|)^{5/2}} \, \Psi(r'). \tag{5.34}$$

The form of the scalar product for the Landau–Peierls wave functions is simple, but one must pay for this simplicity with the nonlocality of the wave functions. There is also a simple mathematical argument that shows shortcomings of the Landau–Peierls wave function. While for every integrable wave function Ψ the transformation (5.34) defines the Landau–Peierls wave function Φ, the inverse transformation is singular since it contains a nonintegrable kernel $|r - r'|^{7/2}$ (Amrein [1969], Mandel and Wolf [1995]). This leads to a paradox that for many "reasonable" functions Φ (for example, for every function that becomes zero abruptly at the boundary), the energy density is infinite. Thus, it is much more natural to treat Ψ as the primary and Φ as the derived object.

§ 6. Eigenvalue Problems for the Photon Wave Function

In wave mechanics of photons, as in wave mechanics of massive particles, one may study eigenvalues and eigenfunctions of various interesting observables. The most important observables, of course, are the momentum, angular momentum, energy, and moment of energy – the generators of the Poincaré group. The eigenfunctions of these observables will be given in coordinate representation to underscore the validity and usefulness of the photon wave function in this representation.

6.1. EIGENVALUE PROBLEMS FOR MOMENTUM AND ANGULAR MOMENTUM

The eigenvalue problems for the components of the photon momentum operator have the standard quantum-mechanical form,

$$\widehat{P}_i \, \Psi(r) = \hbar k_i \, \Psi(r), \tag{6.1}$$

and their solutions depend on r through the exponential functions $\exp(i\mathbf{k} \cdot \mathbf{r})$.

The eigenvalue problem for the photon angular momentum also has the standard quantum-mechanical form. It contains, as usual, the eigenvalue problem for the z-component of the total angular momentum,

$$\widehat{J}_z \, \Psi(r) = \hbar M \, \Psi(r), \tag{6.2}$$

and the eigenvalue problem for the square of the total angular momentum

$$\widehat{\mathbf{J}}^2 \, \Psi(r) = \hbar^2 J(J+1) \, \Psi(r). \tag{6.3}$$

The solutions of eqs. (6.2) and (6.3) are well-known vector spherical harmonics (cf., for example, Messiah [1961]). The direct connection between the quantum-mechanical eigenvalue problems and multipole expansion in classical electromagnetism was explored systematically for the first time by Molière [1949].

6.2. EIGENVALUE PROBLEM FOR THE MOMENT OF ENERGY

The solution of the eigenvalue problem for the moment of energy shows the versatility of the calculational methods based on the coordinate representation and sheds some light on the problem of the localizability of the photon that is discussed in § 8. Of course, the same result can be obtained by Fourier transforming the solution of the eigenvalue problem obtained in momentum representation.

The three components of the moment of energy, like the components of angular momentum, do not commute among themselves. Therefore, the eigenvalue problem can be posed only for one component at a time. Choosing, for definiteness, the z-component, one obtains the following eigenvalue equation:

$$-i\rho_3 (\mathbf{s}\cdot\nabla) z\, \Psi = \kappa\, \Psi. \tag{6.4}$$

The solution of this equation becomes unique when one chooses two additional eigenvalue equations to be solved concurrently. For example, eq. (6.4) can be solved together with the eigenvalue problems for the x and y components of momentum, since the three operators $\hat{P}_x$, $\hat{P}_y$, and $\hat{K}_z$ commute. The solutions for the upper and lower components of the wave function differ only in the sign of κ and one can solve them independently. When the wave function in the form $\Psi = \exp(ik_x x + ik_y y)(\psi_x(z), \psi_y(z), \psi_z(z))$ is substituted into eq. (6.4), one obtains a set of ordinary differential equations:

$$ik_y z \psi_z - (z\psi_y)' = \kappa\, \psi_x, \tag{6.5}$$

$$(z\psi_x)' - ik_x z \psi_z = \kappa\, \psi_y, \tag{6.6}$$

$$ik_x z \psi_y - ik_y z \psi_x = \kappa\, \psi_z, \tag{6.7}$$

where the prime denotes the differentiation with respect to z. These equations are solved by the following substitution ($k_\perp^2 = k_x^2 + k_y^2$):

$$\psi_x = \frac{i}{k_\perp^2 z}\left(k_y \kappa + k_x \frac{d}{dz}\right)\psi_z, \tag{6.8}$$

$$\psi_y = \frac{i}{k_\perp^2 z}\left(-k_x \kappa + k_y \frac{d}{dz}\right)\psi_z, \tag{6.9}$$

which results in a Bessel-type equation for ψ_z,

$$z^2 \psi_z'' + z\psi_z' + (\kappa^2 - k_\perp^2 z^2)\psi_z = 0. \tag{6.10}$$

The physically acceptable solution of this equation is given by the Macdonald function of the imaginary index,

$$\psi_z(z) = K_{i\kappa}(k_\perp z) = \int_0^\infty dt\, e^{-k_\perp z \cosh t} \cos(\kappa t). \tag{6.11}$$

The other solution grows exponentially when $z \to \infty$ and must be rejected. The physical solution falls off exponentially for large $|z|$ and represents a photon state that is localized as much as possible in the z-direction. The remaining two components of the eigenfunction are obtained from eqs. (6.8) and (6.9). The photon wave functions which describe eigenstates of $\widehat{K}_z$ are not normalizable, because the spectrum of the eigenvalues is continuous: κ can be any real number.

6.3. PHOTON PROPAGATION ALONG AN OPTICAL FIBER AS A QUANTUM-MECHANICAL BOUND STATE PROBLEM

The eigenvalue problem for the photon energy operator in the absence of a medium is solved by Fourier transformation as described in §4. In the presence of a medium, one can search for eigenstates of the photon Hamiltonian closely following the path traveled in nonrelativistic wave mechanics of massive particles. This procedure usually involves selecting a set of operators commuting with the Hamiltonian and then solving the appropriate set of eigenvalue equations. The photon propagation along an infinite cylindrical optical fiber (cf., for example, Bialynicki-Birula [1994]) is a good illustration of this approach. In order to take care of the boundary conditions at the surface of the fiber, one must work in the coordinate representation.

Consider an infinite, cylindrical optical fiber of diameter a characterized by a dielectric permittivity ϵ. The symmetry of the problem suggests the inclusion in the set of commuting operators, in addition to the Hamiltonian, the projections of the momentum operator and the total angular momentum on the direction of the fiber axis. In cylindrical coordinates the eigenvalue equations for the z-components of momentum and angular momentum and the Hamiltonian have the form:

$$-i\partial_z \Psi = k_z \Psi, \tag{6.12}$$

$$\left(-i\partial_\varphi + (s \cdot e_z)\right) \Psi = M \Psi, \tag{6.13}$$

$$-i\rho_3 \left(s \cdot (e_\rho \partial_\rho + \frac{1}{\rho} e_\varphi \partial_\varphi + e_z \partial_z)\right) \Psi = \frac{\omega}{v} \Psi, \tag{6.14}$$

where e_ρ, e_φ, and e_z are the unit vectors along the coordinate lines and v equals to c outside the fiber. Due to the symmetry of the problem, there is no coupling between the upper and lower components of Ψ and the solution of these eigenvalue equations can be sought in the form of a three-dimensional vector,

$$\Psi = e_\rho \psi_\rho + e_\varphi \psi_\varphi + e_z \psi_z. \tag{6.15}$$

In order to separate the variables and obtain a set of ordinary differential equations, one needs the following differential and algebraic relations:

$$\partial_\varphi e_\rho = e_\varphi, \quad \partial_\varphi e_\varphi = -e_\rho, \tag{6.16}$$

$$(s \cdot e_\rho)e_\varphi = ie_z, \quad (s \cdot e_z)e_\rho = ie_\varphi, \quad (s \cdot e_\varphi)e_z = ie_\rho. \tag{6.17}$$

All unlisted terms of the type (6.16) and (6.17) vanish. The dependence on φ and z of all three components ψ_ρ, ψ_φ, and ψ_z of the photon wave function can be separated out on the basis of eqs. (6.12) and (6.13),

$$\psi = \exp(ik_z z)\exp(iM\varphi)f(\rho). \tag{6.18}$$

The three ρ-dependent components of the wave function satisfy the equations

$$-\frac{M}{\rho}f_z + k_z f_\varphi = \frac{\omega}{v}(if_\rho), \tag{6.19}$$

$$-\partial_\rho f_z + k_z(if_\rho) = \frac{\omega}{v}f_\varphi, \tag{6.20}$$

$$\frac{1}{\rho}\partial_\rho \rho f_\varphi - \frac{M}{\rho}(if_\rho) = \frac{\omega}{v}f_z. \tag{6.21}$$

These equations lead to a Bessel equation for f_z,

$$\left[\partial_\rho^2 + \frac{1}{\rho}\partial_\rho - \frac{m^2}{\rho^2} + k_\perp^2\right]f_z = 0, \tag{6.22}$$

where $k_\perp^2 = \omega^2/v^2 - k_z^2$. The remaining two functions f can be determined in terms of f_z:

$$f_\rho = ik_\perp^{-2}\left(\frac{\omega M}{v\rho} + k_z \partial_\rho\right)f_z, \tag{6.23}$$

$$f_\varphi = k_\perp^{-2}\left(\frac{Mk_z}{\rho} + \frac{\omega}{v}\partial_\rho\right)f_z. \tag{6.24}$$

The photon wave function obeys the Bessel equation inside the fiber with one value of $k_\perp$, and with a different value of $k_\perp$ in the surrounding free space. The behavior of the solution of eq. (6.22) depends on whether $k_\perp$ is real or imaginary. A general solution of this equation is either (for real $k_\perp$) a linear combination of Bessel functions of the first kind $J_M(\rho)$ and the second kind $Y_M(\rho)$, or (for imaginary $k_\perp$) a linear combination of modified Bessel functions $I_M(\rho)$ and $K_M(\rho)$. In full analogy with the problem of a potential well in quantum mechanics, one can search for bound states in the transverse direction by matching a regular oscillatory solution inside (i.e. the $J_M(\rho)$ function) with an exponentially damped solution outside the fiber (i.e. the $K_M(\rho)$ function). The matching conditions, well known from classical electromagnetic theory, are the continuity conditions for the E_z and H_z field components at the surface of the fiber, when $\rho = a$. Bound states occur because the speed of light is greater in the vacuum than inside the fiber. Therefore, it may happen that $k_\perp$ is real inside and imaginary outside the fiber. Since there are two matching conditions and only one ratio of the amplitudes inside and outside the fiber, both conditions can be satisfied only for a set of discrete eigenvalues of the photon energy $\hbar\omega$. It is worth noting that in order to have an imaginary $k_\perp$, one must have a nonvanishing k_z. Thus, a photon may be bound in the plane perpendicular to the fiber, but it is always moving freely along the fiber, as in the quantum-mechanical description of a charged particle moving in a homogeneous magnetic field. This analysis gives an interpretation of electromagnetic evanescent waves as quantum bound states. Of course, true bound states of photons, which are described by a photon wave function decaying exponentially in *all* directions, are not possible.

§ 7. Relativistic Invariance of Photon Wave Mechanics

In a relativistically invariant quantum theory, the Poincaré transformations are represented by unitary operators. The ten Hermitian generators of these transformations must satisfy the commutation relations characteristic of the Poincaré group. The ten generators of the Poincaré group are identified with the operators $\widehat{H}$, $\widehat{P}$, $\widehat{J}$, and $\widehat{K}$. They generate infinitesimal time translation, space translations, rotations, and boosts (special Lorentz transformations), respectively. The structure of the Poincaré group leads to the following commutation relations obeyed by these generators (cf., for example, Bargmann and Wigner [1948], Bialynicki-Birula and Bialynicka-Birula [1975], Itzykson and Zuber [1980], Weinberg [1995] p. 61):

$$\left[\widehat{J}_i, \widehat{P}_j\right] = i\hbar\epsilon_{ijk}\widehat{P}_k, \tag{7.1}$$

$$\left[\widehat{J}_i, \widehat{J}_j\right] = i\hbar\epsilon_{ijk}\widehat{J}_k, \tag{7.2}$$

$$\left[\widehat{J}_i, \widehat{K}_j\right] = i\hbar\epsilon_{ijk}\widehat{K}_k, \tag{7.3}$$

$$\left[\widehat{K}_i, \widehat{P}_j\right] = i\hbar c^{-2}\delta_{ij}\widehat{H}, \tag{7.4}$$

$$\left[\widehat{K}_i, \widehat{H}\right] = i\hbar\widehat{P}_i, \tag{7.5}$$

$$\left[\widehat{K}_i, \widehat{K}_j\right] = -ic^{-2}\hbar\epsilon_{ijk}\widehat{J}_k. \tag{7.6}$$

All the remaining commutators vanish. One may check by a direct calculation that the operators $\widehat{H}$, $\widehat{P}$, $\widehat{J}$, and $\widehat{K}$, given in momentum representation by the formulas (5.11)–(5.14) and in coordinate representation by the formulas (5.23)–(5.26), obey the commutation relation for the generators of the Poincaré group. In the proof of the commutation relations in momentum representation, one needs the condition (4.19). Since all generators of the Poincaré group are represented by operators which are Hermitian with respect to the scalar product (5.3) or (5.7), the Poincaré transformations are represented by unitary operators. Therefore, the scalar product is invariant under these transformations and all transition probabilities are the same for all observers connected by Poincaré transformations. Thus, in both coordinate and momentum representations, wave mechanics of photons is a fully relativistic theory.

§ 8. Localizability of Photons

The problem of localization of relativistic systems was first posed and solved by Newton and Wigner [1949], and later refined by Wightman [1962]. According to this analysis it is possible to define position operators and localized states for massive particles and for massless particles of spin 0, but not for massless particles with spin. Thus, the position operator in the sense of Newton and Wigner does not exist for photons (cf. also a recent tutorial review on that subject by Rosewarne and Sarkar [1992]). As a simple heuristic explanation of why a position operator for the photon does not exist, one may observe (Pryce [1948]) that the multiplication by r cannot be applied to the photon wave function because it destroys the divergence condition (1.6).

A weaker definition of localization that is applicable, even when the position operator does not exist, was proposed by Jauch and Piron [1967] and a very detailed analysis of this problem has been given by Amrein [1969]. The Jauch–Piron localizability allows for noncompatibility of "photon position measurements" in overlapping regions. The main weakness of such an abstract analysis is that an operational definition of the photon position measurement for photons has not been incorporated into it. The existence of position measurements for photons is just taken for granted, regardless of the feasibility of their physical realizations. When a realistic model of the photon detector is brought in, it is the wave function Ψ rather than Φ that appears as the correct probability amplitude for photodetection (Mandel and Wolf [1995]). Thus, in practical applications the energy wave functions Ψ always seem to play a dominant role.

It must, however, be stressed that even for massive particles, the localization is not perfect, because it is not relativistically invariant. Two observers who are in relative motion would not quite agree as to the localization region of a relativistic particle. This follows from the fact that the Newton–Wigner wave function ψ^{NW} is related to the relativistic wave function ψ that transforms locally under Poincaré transformations by a nonlocal transformation (cf. Haag [1993]):

$$\psi^{NW}(r) = \int d^3 r' \, K(r - r') \, \psi(r'), \tag{8.1}$$

where the kernel K can be represented in terms of the Macdonald function,

$$K(r) = \sqrt{\frac{\pi}{2}} \left(\frac{2mc}{r\hbar}\right)^{5/4} K_{5/4}(mcr/\hbar). \tag{8.2}$$

In the limit, when $m \to 0$, $K(r) \to \pi/(2\pi r)^{-5/2}$ and eq. (8.1) becomes the relation (1.2) between the local wave function of the photon and the Landau–Peierls wave function. Thus, the difference between the localizability of massive particles and photons is not that great. In both cases, localization cannot be defined in a relativistic manner. However, for massive particles departures from strict localization are only exponentially small due to the fast decay of the Macdonald function in eq. (8.1). In the nonrelativistic limit, when $c \to \infty$, the exponential tails become infinitely sharp and the localization is restored.

Difficulties with the position operator for relativistic particles have a profound origin connected with the structure of the Poincaré group. In nonrelativistic physics the position operator is the generator (up to a factor of mass) of Galilean transformations (cf., for example, Gottfried [1966], Weinberg [1995] p. 62). In a relativistic theory, the Galilean transformations are replaced by the Lorentz transformations and the position operator (multiplied by the mass) is replaced by the boost generator $\boldsymbol{K}$. The main difference between Galilean and Lorentz transformation affecting the discussion of localizability is that boost generators *do not commute*. Therefore, one may only hope to localize relativistic particles in one direction at a time. The possibility to localize photons in one direction has been discussed in general terms as the "front" description by Acharya and Sudarshan [1960]. The eigenfunctions of the boost operator K_z given in § 6 may serve as an explicit realization of the front description for the photon.

The considerations of photon localizability, while important for the understanding of some fundamental issues, do not much influence the practical applications of the photon wave function. All that should really matter there is that the wave function be defined precisely and that its interpretation be not extended beyond the limits of applicability.

§ 9. Phase-Space Description of a Photon

Distribution functions in phase space are a very convenient tool in the description of statistical properties and the study of the classical limit of wave mechanics. A direct analog of the Wigner function (Wigner [1932]) introduced in wave mechanics may also be introduced for photons with the help of the photon wave function. This is done by Fourier transforming the product of the wave function and its complex conjugate. Fourier transforms of the electromagnetic fields similar to the Wigner function have been introduced in optics, first by Walther [1968] in the two-dimensional context of radiative transfer theory and then by Wolf [1976] and by Sudarshan [1979, 1981a,b] in the three-dimensional case. In these papers, phase-space distribution functions were defined only for the *stationary states* of the electromagnetic field, and they were treated as functions of the frequency. The time-dependent distribution functions can be defined (Bialynicki-Birula [1994]) with the use of the time-dependent wave function. The only formal difference between the standard definition of the Wigner function in nonrelativistic wave mechanics of massive particles and the case of photons is the presence of vector indices. Thus, the photon distribution function in phase space is not a single scalar function but rather a 6×6 Hermitian matrix defined as follows:

$$W_{ab}(\mathbf{r}, \mathbf{k}, t) = \int d^3 s \, e^{-i\mathbf{k} \cdot \mathbf{s}} \, \Psi_a(\mathbf{r} + \tfrac{1}{2}\mathbf{s}, t) \, \Psi_b^*(\mathbf{r} - \tfrac{1}{2}\mathbf{s}, t). \tag{9.1}$$

Similar multi-component distribution functions also arise for a Dirac particle, and one can use some of the techniques developed by Bialynicki-Birula, Górnicki and Rafelski [1991] to deal with such functions.

Every 6×6 Hermitian matrix can be written in the following block form:

$$W_{ab} = \begin{bmatrix} W_{ij}^0 + W_{ij}^3 & W_{ij}^1 - iW_{ij}^2 \\ W_{ij}^1 + iW_{ij}^2 & W_{ij}^0 - W_{ij}^3 \end{bmatrix}, \tag{9.2}$$

where all 3×3 matrices W_{ij}^α are Hermitian. This decomposition can also be expressed in terms of the ρ matrices,

$$W_{ab} = \rho_0 W_{ij}^0 + \rho_1 W_{ij}^1 + \rho_2 W_{ij}^2 + \rho_3 W_{ij}^3. \tag{9.3}$$

The vector indices i and j refer to the components within the upper and lower parts of the wave function and the matrices ρ act on these parts as a whole. The most general photon distribution function, as seen from this analysis, is quite complicated. In general, when the medium induces mixing of the two polarization states, all components of the distribution function are needed. However, when photons propagate in free space, only a subset of these components is sufficient to account for the dynamical properties of photon beams. The simplest case is that of a given helicity. A more interesting case is that of an unpolarized photon beam: a mixture of both helicities with equal weights. This state must described by the distribution function because a mixed state cannot be treated by pure Maxwell's theory. In all these cases, phase-space dynamics can be described by a 3×3 Hermitian matrix; i.e., by just one scalar function and one vector function. To this end, one may introduce the following reduced distribution function:

$$W_{ij}(\mathbf{r},\mathbf{k},t) = \int d^3 s \, e^{-i\mathbf{k}\cdot\mathbf{s}} \, \psi_i(\mathbf{r}+\tfrac{1}{2}\mathbf{s},t) \, \psi_j^*(\mathbf{r}-\tfrac{1}{2}\mathbf{s},t), \tag{9.4}$$

where ψ_i are the upper or the lower components of the original wave function. The Hermitian matrix W_{ij}^a can be decomposed into a real symmetric tensor and a real vector according to the formula

$$W_{ij}^a = w_{ij}^a + \frac{c}{2i} \epsilon_{ijk} u_k^a. \tag{9.5}$$

The tensor corresponds to the symmetric part of W_{ij}^a and the vector corresponds to the antisymmetric part. The factor of c has been separated out in the second term since the vector $\mathbf{u}$ is related to the momentum density, while the trace of w_{ij} is related to the energy density (Bialynicki-Birula [1994]).

The equations satisfied by the components of the photon distribution function in free space can be obtained from Maxwell's equations (1.5) and (1.6) for the vector $\mathbf{F}$,

$$\partial_t W_{ij} = c\left(\mathbf{k}+\frac{i}{2}\nabla\right)_m \epsilon_{mil} W_{lj} - c\left(\mathbf{k}-\frac{i}{2}\nabla\right)_m W_{il} \epsilon_{mlj}, \tag{9.6}$$

$$\left(\mathbf{k}+\frac{i}{2}\nabla\right)_i W_{ij} = 0 = \left(\mathbf{k}-\frac{i}{2}\nabla\right)_j W_{ij}. \tag{9.7}$$

This leads to the following set of coupled evolution equations for the real components w_{ij} and u_i:

$$\partial_t w_{ij} = -c\epsilon_{ilk} k_l w_{kj} - c\epsilon_{jlk} k_l w_{ki} - \frac{c^2}{2}(\nabla_i u_j + \nabla_j u_i - \delta_{ij}\nabla_k u_k), \tag{9.8}$$

$$\partial_t u_i = -c\epsilon_{ijk} k_j u_k - \tfrac{1}{2}(\nabla_j w_{ij} - \nabla_i w_{jj}), \tag{9.9}$$

and to the subsidiary conditions

$$c\epsilon_{ijk}k_j u_k = \nabla_j w_{ij}, \qquad (9.10)$$

$$c\epsilon_{ijk}\nabla_j u_k = 4k_j w_{ij}. \qquad (9.11)$$

The k-dependent terms in the evolution equations describe a uniform rotation of the vector u_i and of the tensor w_{ij} around the wave vector k so that these terms can be eliminated by "going to a rotating coordinate frame".

With the help of the subsidiary conditions (9.10, 9.11) one can eliminate the remaining components and obtain from the evolution equations (9.8, 9.9) the equations for $w = \sum w_{ii}$ and u,

$$\partial_t w = -c^2 \nabla_i u, \qquad (9.12)$$

$$\partial_t u_i = -2c\epsilon_{ijk}k_j u_k - \nabla_i w. \qquad (9.13)$$

These evolution equations do form a simple, self-contained set. However, as is always the case with the phase-space distribution functions in wave mechanics, not all solutions of these equations are admissible. Only those distribution functions are allowed which can be represented in the form (9.4) at the initial time (with u and w satisfying the subsidiary conditions (9.10) and (9.11)).

§ 10. Hydrodynamic Formulation

It was shown by Madelung [1926] that the Schrödinger wave equation can be cast into a hydrodynamic form. In this form, the complex wave function is replaced by real variables: the probability density ρ and the velocity u of the probability flow. The wave equation is replaced by the hydrodynamic evolution equations for the variables ρ and u. In order to reduce the number of functions from four to the original two, one must impose auxiliary conditions – the quantization condition – on the velocity field. Later, other wave equations in quantum mechanics (Pauli, Dirac, Weyl) were also presented in the hydrodynamic form. The wave equation for the photon wave function is not an exception in this respect. It can also be written (Bialynicki-Birula [1996b]) as a set of equations for real hydrodynamic-like variables. Since the Riemann–Silberstein vector F carries all the information about the photon wave function, one may use $\boldsymbol{F}$ to define these variables. They comprise the energy density ρ and the velocity of the energy flow v,

$$\rho(\boldsymbol{r},t) = \boldsymbol{F}^*(\boldsymbol{r},t) \cdot \boldsymbol{F}(\boldsymbol{r},t), \qquad \rho(\boldsymbol{r},t)v(\boldsymbol{r},t) = \frac{c}{2i}\boldsymbol{F}^*(\boldsymbol{r},t) \times \boldsymbol{F}(\boldsymbol{r},t), \qquad (10.1)$$

the components t_{ij} of the following tensor,

$$t_{ij}(\boldsymbol{r},t) = F_i^*(\boldsymbol{r},t)F_j(\boldsymbol{r},t) + F_j^*(\boldsymbol{r},t)F_i(\boldsymbol{r},t), \qquad (10.2)$$

and another vector $\boldsymbol{u}$,

$$\rho(\boldsymbol{r},t)\boldsymbol{u}(\boldsymbol{r},t) = \frac{c}{2i}\left(\boldsymbol{F}^*(\boldsymbol{r},t)\nabla\boldsymbol{F}(\boldsymbol{r},t) - (\nabla\boldsymbol{F}^*(\boldsymbol{r},t))\boldsymbol{F}(\boldsymbol{r},t)\right). \qquad (10.3)$$

Owing to the existence of the following identities satisfied by the hydrodynamic variables,

$$t_{ii} = 2c, \qquad v_i t_{ik} = 0, \qquad t_{ij}t_{ij} = 4c^2 - 2\vec{v}^2, \qquad (10.4)$$

only one component of t_{ij} is arbitrary, but the hydrodynamic equations look more symmetric when all the components are treated on equal footing. The number of algebraically independent hydrodynamic variables is reduced from eight to six (the number of degrees of freedom described by $\boldsymbol{F}$) by the following quantization condition

$$\int d\vec{S} \cdot \left[\nabla \times \vec{u} - \frac{1}{8c^3}\varepsilon_{ijk}(v_i\nabla v_j \times \nabla v_k + v_i\nabla t_{jl} \times \nabla t_{kl} - 2t_{il}\nabla t_{jl} \times \nabla v_k)\right] = 2\pi n, \qquad (10.5)$$

where n is a natural number. This condition must hold for every choice of the integration surface and it states, in essence, that the phase of the wave function is defined uniquely (up to an overall constant phase).

The evolution equations for the hydrodynamic variables are

$$\partial_t \rho + (\vec{v}\cdot\nabla)\rho = -\rho(\nabla\cdot\vec{v}), \qquad (10.6)$$

$$\partial_t v_i + (\vec{v}\cdot\nabla)v_i = \frac{1}{\rho}\partial_j(-c^2\rho\delta_{ij} + \rho v_i v_j + \rho t_{ij}), \qquad (10.7)$$

$$\begin{aligned}\partial_t t_{ij} + (\vec{v}\cdot\nabla)t_{ij} = &\frac{1}{\rho}(t_{ij}v_k\partial_k\rho - c\delta_{ij}v_k\partial_k\rho + \frac{c}{2}(v_i\partial_j + v_j\partial_i)\rho) \\ &+ \delta_{ij}v_k\partial_l t_{kl} + 2v_k\partial_k t_{ij} + c\varepsilon_{ikl}u_k t_{lj} + c\varepsilon_{jkl}u_k t_{li} \\ &+ (v_k\partial_k t_{ij} - t_{ij}\partial_k v_k) + \tfrac{1}{2}(t_{ik}\partial_k v_j + t_{jk}\partial_k v_i) \\ &- \tfrac{1}{2}(v_i\partial_k t_{kj} + v_j\partial_k t_{ki} + v_k\partial_i t_{kj} + v_k\partial_j t_{ki}),\end{aligned} \qquad (10.8)$$

$$\partial_t u_i + (\vec{v}\cdot\nabla)u_i = \frac{1}{4c\rho}\partial_j\left[\rho\varepsilon_{jkl}(t_{km}\partial_i t_{ml} + v_k\partial_i v_l)\right]. \tag{10.9}$$

They must be supplemented by the equations which express the divergence condition (1.6):

$$\frac{1}{2}\partial_k(\rho t_{ik}) + \rho\varepsilon_{ijk}v_j u_k$$
$$+ \frac{\rho}{4c}(t_{jk}\partial_k t_{ij} - t_{ij}\partial_k t_{jk} + v_k\partial_k v_i - v_i\partial_k v_k) = 0, \tag{10.10}$$
$$\frac{1}{2}\partial_k(\rho\varepsilon_{ikl}v_l) + \rho t_{ik}u_k$$
$$+ \frac{\rho}{4c}\left[\varepsilon_{jkl}(t_{il}\partial_k v_j - v_j\partial_k t_{il}) + \varepsilon_{ijl}(t_{kl}\partial_k v_j - v_j\partial_k t_{kl})\right] = 0. \tag{10.11}$$

Hydrodynamic description of the photon dynamics is not simple, but its existence again underscores the unification of the quantum theory of the photon with the rest of quantum mechanics. Everything that can be done with other particles can also be done for photons.

§ 11. Photon Wave Function in Non-Cartesian Coordinate Systems and in Curved Space

It was shown by Skrotskiĭ [1957] and Plebanski [1960] (for a pedagogical review, see Schleich and Scully [1984]) that the propagation of the electromagnetic field in arbitrary coordinate systems, including the case of curved space–time, may be described by Maxwell's equations, with all the information about the space–time geometry contained in the relations connecting the field vectors E, B and D, H. This discovery can be enhanced further by the observation (Bialynicki-Birula [1994]) that in contradistinction to the case of an inhomogeneous medium, in the gravitational field the two photon helicities do not mix. This follows from the fact that for arbitrary metric $g_{\mu\nu}$ the constitutive relations can be written as a single equation connecting two complex vectors: the vector $F(r,t)$ defined by eq. (1.7) and a new vector $G(r,t)$ defined as

$$G(r,t) = \frac{1}{\sqrt{2}}\left(\frac{E(r,t)}{\sqrt{\mu_0}} + i\frac{H(r,t)}{\sqrt{\epsilon_0}}\right). \tag{11.1}$$

In curved space (or in curvilinear coordinates), the constitutive relations for two complex vectors $F(r,t)$ and $G(r,t)$ have the form

$$F^i = -\frac{1}{g_{00}}\left(\sqrt{-g}\,g^{ij} + ig_{0k}\varepsilon^{ikj}\right)G_j, \tag{11.2}$$

$$G_i = -\frac{1}{g^{00}}\left(g_{ij}/\sqrt{-g} - ig^{0k}\varepsilon_{ikj}\right)F^j, \tag{11.3}$$

where $g_{\mu\nu}$ is the metric tensor, $g^{\mu\nu}$ is its inverse, and g is the determinant of $g_{\mu\nu}$. The Maxwell equations expressed in terms of vectors G and F in curved space–time are the same as in flat space:

$$i\partial_t F(r,t) = \nabla \times G(r,t), \tag{11.4}$$

$$\nabla \cdot F(r,t) = 0. \tag{11.5}$$

In these equations, all derivatives are *ordinary* (not covariant) derivatives as in flat space. The whole difference is in the form of the constitutive relations (11.2) and (11.3). These relations contain all information about the gravitational field or the curvilinear coordinate system. Since the relations between G and F are linear, one may again write two separate wave equations for the two helicity states as in flat space. By combining these two equations, one obtains the following wave equation for the six-component photon wave function $\mathcal{F}(r,t)$:

$$i\partial_t \mathcal{F}(r,t) = \rho_3 \nabla \times \mathcal{G}(r,t), \tag{11.6}$$

where

$$\mathcal{G}_i = -\frac{1}{g^{00}}\left(g_{ij}/\sqrt{-g} - i\rho_3 g^{0k}\varepsilon_{ikj}\right)\mathcal{F}^j. \tag{11.7}$$

These equations contain only the matrix ρ_3 that does not mix the helicity states.

The true photon wave function $\Psi(r,t)$ may be introduced only in the time-independent case, when the separation of $\mathcal{F}(r,t)$ into positive and negative energy parts is well defined.

§ 12. Photon Wave Function as a Spinor Field

Soon after the formulation of spinor calculus by van der Waerden in the context of the Dirac equation, it was discovered by Laporte and Uhlenbeck [1931] that the Maxwell equations can also be cast into a spinorial form. The spinor representation of the electromagnetic field and the Riemann–Silberstein vector are closely connected. The components of the vector F are related to the components of a second-rank symmetric spinor ϕ_{AB},

$$\phi_{00} = -F_x + iF_y, \tag{12.1}$$

$$\phi_{01} = F_z, \tag{12.2}$$

$$\phi_{11} = F_x + iF_y, \tag{12.3}$$

and the components of the complex conjugate vector $\boldsymbol{F}^*$ are related to the components of a second-rank symmetric primed spinor $\phi^{A'B'}$,

$$\phi^{0'0'} = -F_x^* - iF_y^*, \tag{12.4}$$

$$\phi^{0'1'} = -F_z^*, \tag{12.5}$$

$$\phi^{1'1'} = F_x^* - iF_y^*. \tag{12.6}$$

The property that even in curved space both helicities propagate without mixing is, in the spinorial formalism, a simple consequence of the fact that both spinors ϕ_{AB} and $\phi^{A'B'}$ satisfy separate wave equations (Laporte and Uhlenbeck [1931], Penrose and Rindler [1984])

$$\sigma^{\mu C'A} \partial_\mu \phi_{AB}(\boldsymbol{r}, t) = 0, \tag{12.7}$$

$$\sigma^{\mu}{}_{CA'} \partial_\mu \phi^{A'B'}(\boldsymbol{r}, t) = 0, \tag{12.8}$$

where the matrices $\sigma^{\mu A'B}$ and $\sigma^{\mu}{}_{A'B}$ are built from the unit matrix and the Pauli matrices,

$$(\sigma^{\mu A'B}) = (I, \boldsymbol{\sigma})^{A'B}, \tag{12.9}$$

$$(\sigma^{\mu}{}_{AB'}) = (I, -\boldsymbol{\sigma})_{AB'}. \tag{12.10}$$

Under a Lorentz transformation the second-rank spinor changes according to the formula

$$\phi'_{AB}(\boldsymbol{r}', t') = S_A{}^C S_B{}^D \phi_{CD}(\boldsymbol{r}, t), \tag{12.11}$$

where $S_A{}^C$ is a 2×2 matrix of the fundamental (spinor) representation of the Lorentz group. Thus, from the point of view of the representation theory of the Lorentz and Poincaré groups (Streater and Wightman [1978], Weinberg [1995] p. 231), the photon wave functions for a given helicity are just the three-component fields which transform as irreducible representations $(1, 0)$ or $(0, 1)$ of the proper Lorentz group (without reflections). In order to accommodate reflections, one must combine both representations and introduce the six-dimensional objects $\mathcal{F}$ or Ψ.

Equations (12.7) [and similarly eqs. (12.8)] represent a set of four equations obeyed by four components $(\phi_{11}, \phi_{12}, \phi_{21}, \phi_{22})$ of the second order spinor. All four equations can be written in the form of the Dirac equation (Ohmura [1956], Moses [1959]):

$$i\hbar\partial_t \phi(\mathbf{r},t) = \boldsymbol{\alpha} \cdot \frac{\hbar}{i} \nabla \phi(\mathbf{r},t), \tag{12.12}$$

where the matrices $\boldsymbol{\alpha}$ are

$$\alpha_x = \begin{bmatrix} 0 & 0 & 1 & 0 \\ 0 & 0 & 0 & 1 \\ 1 & 0 & 0 & 0 \\ 0 & 1 & 0 & 0 \end{bmatrix}, \quad \alpha_y = \begin{bmatrix} 0 & 0 & -i & 0 \\ 0 & 0 & 0 & -i \\ i & 0 & 0 & 0 \\ 0 & i & 0 & 0 \end{bmatrix}, \quad \alpha_z = \begin{bmatrix} 1 & 0 & 0 & 0 \\ 0 & 1 & 0 & 0 \\ 0 & 0 & -1 & 0 \\ 0 & 0 & 0 & -1 \end{bmatrix}. \tag{12.13}$$

One may check that in this formulation the divergence condition takes on the following simple algebraic form: $\phi_{12} = \phi_{21}$.

The Maxwell equations expressed in spinor notation and the Weyl equation provide just the simplest examples from a hierarchy of wave equations for massless fields described by symmetric spinors $\phi_{B_1 B_2 \cdots B_n}$ or $\phi^{B'_1 B'_2 \cdots B'_n}$. All these equations have the form (Penrose and Rindler [1984])

$$\sigma^{\mu C'A} \nabla_\mu \phi_{AB_1 B_2 \cdots B_{n-1}}(\mathbf{r},t) = 0, \tag{12.14}$$

$$\sigma^\mu{}_{CA'} \nabla_\mu \phi^{A' B'_1 B'_2 \cdots B'_{n-1}}(\mathbf{r},t) = 0. \tag{12.15}$$

This universality of massless wave equations for all spins gives an additional argument for treating the Riemann–Silberstein vector as the photon wave function.

§ 13. Photon Wave Functions and Mode Expansion of the Electromagnetic Field

The concept of the photon wave function is also useful in the process of quantization of the electromagnetic field. One may simply apply the procedure of second quantization to the photon wave function in the same manner as one does for other field operators. In order to see this analogy, one may recall that the field operator $\hat{\psi}(\mathbf{r})$ for, say the electron field, is built from a complete set of wave functions for the electrons $\psi_n^+(\mathbf{r})$ and from a complete set of wave functions for positrons $\psi_n^-(\mathbf{r})$ according to the following rule (cf., for example, Schweber [1961], Bialynicki-Birula and Bialynicka-Birula [1975], Weinberg [1995]):

$$\hat{\psi}(\mathbf{r},t) = \sum_n (\psi_n^+(\mathbf{r},t)\hat{a}_n + \psi_n^-(\mathbf{r},t)\hat{b}_n^\dagger), \tag{13.1}$$

where $\hat{a}_n$ and $\hat{b}_n$ are the annihilation operators for electrons and positrons, respectively. The second part of the field operator is related to the first by the operation of charge conjugation performed on the wave functions and on the operators. The analog of charge conjugation for photon wave functions is given by eq. (3.2). Following this procedure, one may construct the field operator of the electromagnetic field in the form

$$\widehat{\mathcal{F}}(r,t) = \sum_n (\Psi_n(r,t)\,\hat{c}_n + \rho_1\,\Psi_n^*(r,t)\,\hat{c}_n^\dagger), \tag{13.2}$$

where the identity of particles and antiparticles for photons has been taken into account by using only one set of creation and annihilation operators. The field operator (13.2) is non-Hermitian, but it satisfies the particle–antiparticle conjugation condition (3.2). Therefore, it has only six Hermitian components. These Hermitian operators are identified as the field operators $\widehat{D}(r,t)$ and $\widehat{B}(r,t)$, and they are obtained from $\widehat{\mathcal{F}}$ through the formula

$$\begin{bmatrix}\widehat{D}\\\widehat{B}\end{bmatrix} = \frac{1}{2}\begin{bmatrix}1 & 1\\-i & i\end{bmatrix}\widehat{\mathcal{F}}. \tag{13.3}$$

As a direct consequence of the formula (13.2), one may identify the photon wave functions in the second-quantized theory with the matrix elements of the electromagnetic field operators $\widehat{\mathcal{F}}$ or $\widehat{D}$ and $\widehat{B}$ taken between one-particle states and the vacuum:

$$\Psi_n(r,t) = \langle 0|\widehat{\mathcal{F}}(r,t)\hat{c}_n^\dagger|0\rangle. \tag{13.4}$$

In the simplest case of the free field, when the complete set of photon wave functions may be labeled by the wave vector k and helicity λ, the formula (13.2) takes on the form

$$\widehat{\mathcal{F}}(r,t) = \sqrt{\hbar c}\int\frac{d^3k}{(2\pi)^3}\begin{bmatrix}e(k,1)(\hat{c}(k,1)e^{-i\omega t+ik\cdot r}+\hat{c}^\dagger(k,-1)e^{i\omega t-ik\cdot r})\\ e(k,-1)(\hat{c}(k,-1)e^{-i\omega t+ik\cdot r}+\hat{c}^\dagger(k,1)e^{i\omega t-ik\cdot r})\end{bmatrix}. \tag{13.5}$$

In the presence of a medium, the expansion (13.2) of the electromagnetic field operator requires the knowledge of a complete set of wave functions Ψ_n which satisfy the photon wave equation in the medium. These functions are usually called the mode functions of the electromagnetic field (cf., for example, Louisell [1973] p. 240 and Mandel and Wolf [1995] p. 905) but the term photon wave functions is perhaps more appropriate (Moses [1973], Bialynicki-Birula and Bialynicka-Birula [1975]). The advantage of using the terminology of wave functions is that it brings in all the associations with wave mechanics and makes the classification of the modes more transparent. In particular, one may use the quantum-mechanical notion of eigenfunctions and eigenvalues to classify the functions used in the mode expansion (Moses [1973]) and also borrow from quantum mechanics the methods of proving their completeness (Bialynicki-Birula and Brojan [1972]).

This discussion shows that the photon wave function is not restricted to the *wave mechanics* of photons. The same wave functions also appear as the mode functions in the expansion of the electromagnetic field operators.

§ 14. Summary

The aim of this review was to collect and explain all basic properties of a certain well-defined mathematical object – a six-component function of space–time variables – that describes the quantum state of the photon. Whether or not one decides to call this object the photon wave function in coordinate representation is a matter of opinion, since some properties known from wave mechanics of massive particles are missing. The most essential property that does not hold for the photon wave function is that the argument r of the wave function cannot be directly associated with the position operator of the photon. The position operator for the photon simply *does not exist*. However, one should remember that for massive particles also, the true position operator exists only in the nonrelativistic approximation. The concept of localization associated with the Newton–Wigner position operator is *not relativistically invariant*. Since photons cannot be described in a nonrelativistic manner, there is no approximate position operator.

The strongest argument that can be made for the photon wave function in coordinate representation is based on the most fundamental property of quantum states – on the *principle of superposition*. According to the superposition principle, wave functions form a linear space. By adding wave functions, one again obtains a legitimate wave function. Once this principle is accepted, the existence of photon wave functions in coordinate representation follows from the existence of the photon wave functions in momentum representation, and these functions are genuine by all standards; their existence follows simply from relativistic quantum kinematics (or more precisely, from the representation theory of the Poincaré group). The Fourier integral (4.20) represents a special combination of momentum space wave functions with different momenta and as a matter of principle, such linear combinations are certainly allowed. One may only argue which superpositions to take as more natural or useful, but rejecting totally the very concept of the photon wave function in coordinate representation is tantamount to rejecting altogether the superposition principle.

There is not much advantage in using the photon wave function in coordinate representation to perform calculations for photons moving in *free space*. The relation of this wave function to momentum wave function is so straightforward that one may as well stick to momentum representation. It is only in the presence of a medium, especially in an inhomogeneous medium, that the photon wave function in coordinate representation becomes useful and even essential. Only in the coordinate representation may one hope to solve the eigenvalue problems and to take into account the boundary conditions.

The introduction of the wave function for the photon has many significant benefits. The photon wave function enables one to formulate a consistent wave mechanics of photons that could be often used as a convenient tool in the quantum description of electromagnetic fields, *independent* of the formalism of second quantization. In other words, in constructing quantum theories of photons one may proceed, as in quantum theory of all other particles, through two stages. At the first stage, one introduces wave functions and a wave equation obeyed by these wave functions. At the second stage, one upgrades the wave functions to the level of field operators in order to deal more effectively with the states involving many indistinguishable particles and to allow for processes in which the number of particles is not conserved. Many methods which have proven very useful in the study of particles described by the Schrödinger wave functions can also be implemented for photons, leading to some new insights. These methods include relationships between symmetries and operators, the definitions of various sets of modes for the electromagnetic field and their completeness relations, eigenvalue problems for various observables (§ 6), phase-space representation (§ 9), and the hydrodynamic formulation (§ 10). Finally, there are important logical and pedagogical advantages coming from the use of the photon wave function. The quantum-mechanical description of *all* particles, including photons, becomes uniform.

Acknowledgments

Four lectures based on this review were presented while I was a Senior Visiting Fellow at the Rochester Theory Center for Optical Science and Engineering in February 1996. I would also like to acknowledge discussions with Z. Bialynicka-Birula, M. Czachor, J.H. Eberly, A. Orłowski, W. Schleich, M.O. Scully, and E. Wolf.

References

Acharya, R., and E.C.G. Sudarshan, 1960, "Front" description in relativistic quantum mechanics, J. Math. Phys. **1**, 532.

Akhiezer, A.I., and V.B. Berestetskii, 1965, Quantum Electrodynamics (Interscience, New York) ch. 1.

Amrein, W.O., 1969, Localizability for particles of zero mass, Helv. Phys. Acta **42**, 149.

Ashtekar, A., and A. Magnon, 1975, Quantum fields in curved space–times, Proc. R. Soc. (London) A **346**, 375.

Bargmann, V., and E.P. Wigner, 1948, Group theoretical discussion of relativistic wave equations, Proc. Natl. Acad. Sci. USA **34**, 211. Reprinted: 1966, in: Symmetry Groups in Nuclear and Particle Physics, ed. F.J. Dyson (Benjamin, New York) p. 205.

Bateman, H., 1915, The Mathematical Analysis of Electrical and Optical Wave Motion on the Basis of Maxwell's Equations (Cambridge). Reprinted: 1955 (Dover, New York).

Baym, G., 1969, Lectures on Quantum Mechanics (Benjamin, Reading, MA) ch. 1.

Bialynicki-Birula, I., 1963, Quantum electrodynamics without electromagnetic field, Phys. Rev. **130**, 465.

Bialynicki-Birula, I., 1994, On the wave function of the photon, Acta Phys. Polon. A **86**, 97.

Bialynicki-Birula, I., 1996a, The photon wave function, in: Coherence and Quantum Optics VII, eds J.H. Eberly, L. Mandel, and E. Wolf (Plenum Press, New York).

Bialynicki-Birula, I., 1996b, Hydrodynamics of relativistic probability flows, in: Nonlinear Dynamics, Chaotic and Complex Systems, eds E. Infeld, R. Żelazny and A. Gałkowski (Cambridge University, Cambridge).

Bialynicki-Birula, I., and Z. Bialynicka-Birula, 1975, Quantum Electrodynamics (Pergamon Press, Oxford).

Bialynicki-Birula, I., and Z. Bialynicka-Birula, 1976, Quantum electrodynamics of intense photon beams. New approximation method, Phys. Rev. A **14**, 1101.

Bialynicki-Birula, I., and Z. Bialynicka-Birula, 1987, Berry's phase in the relativistic theory of spinning particles, Phys. Rev. D **35**, 2383.

Bialynicki-Birula, I., and J. Brojan, 1972, Completeness of evanescent waves, Phys. Rev. D5, 485.

Bialynicki-Birula, I., P. Górnicki and J. Rafelski, 1991, Phase-space structure of the Dirac vacuum, Phys. Rev. D **44**, 1825.

Bjorken, J., 1963, A dynamical origin of the electromagnetic field, Ann. Phys. (New York) **24**, 174.

Bohm, D., 1954, Quantum Theory (Constable, London) p. 91.

Born, M., and E. Wolf, 1980, Principles of Optics (Pergamon, Oxford) p. 494.

Cohen-Tannoudji, C., B. Diu and F. Laloë, 1977, Quantum Mechanics, Vol. I (Wiley, New York).

Cohen-Tannoudji, C., J. Dupont-Roc and G. Grynberg, 1989, Photons and Atoms, Introduction to Quantum Electrodynamics (Wiley, New York).

Cook, R.J., 1982a, Photon dynamics, Phys. Rev. A **25**, 2164.

Cook, R.J., 1982b, Lorentz covariance of photon dynamics, Phys. Rev. A **26**, 2754.

Dirac, P.A.M., 1958, The Principles of Quantum Theory, 4th Ed. (Clarendon Press, Oxford) p. 9.

Good Jr, R.H., 1957, Particle aspect of the electromagnetic field equations, Phys. Rev. **105**, 1914.

Gottfried, K., 1966, Quantum Mechanics (Benjamin, New York) p. 252.

Gross, L., 1964, Norm invariance of mass-zero equations under the conformal group, J. Math. Phys. **5**, 687.

Haag, R., 1993, Local Quantum Physics (Springer, Berlin) p. 33.

Inagaki, T., 1994, Quantum-mechanical approach to a free photon, Phys. Rev. A **49**, 2839.

Itzykson, C., and J.-B. Zuber, 1980, Quantum Field Theory (McGraw-Hill, New York) p. 50.

Jackson, J.D., 1975, Classical Electrodynamics (Wiley, New York) p. 552.

Jauch, J.M., and C. Piron, 1967, Generalized localizability, Helv. Phys. Acta **40**, 559.
Kaiser, G., 1992, Wavelet electrodynamics, Phys. Lett. A **168**, 28.
Kidd, R., J. Ardini and A. Anton, 1989, Evolution of the modern photon, Am. J. Phys. **57**, 27.
Kramers, H.A., 1938, Quantentheorie des Elektrons und der Strahlung, in: Hand- und Jahrbuch der Chemischen Physik (Eucken-Wolf, Leipzig) [English translation: 1957, Quantum Mechanics (North-Holland, Amsterdam)].
Landau, L.D., and R. Peierls, 1930, Quantenelektrodynamik im Konfigurationsraum, Z. Phys. **62**, 188.
Laporte, O., and G.E. Uhlenbeck, 1931, Application of spinor analysis to the Maxwell and Dirac equations, Phys. Rev. **37**, 1380.
Lipkin, H.J., 1973, Quantum Mechanics (North-Holland, Amsterdam) ch. 1.
Lomont, J.S., and H.E. Moses, 1962, Simple realizations of the infinitesimal generators of the proper orthochronous inhomogeneous Lorentz group for zero mass, J. Math. Phys. **3**, 405.
London, F., 1927, Winkelvariable und kanonische Transformationen in der Undulationsmechanik, Z. Phys. **40**, 193.
Louisell, W.H., 1973, Quantum Statistical Properties of Radiation (Wiley, New York).
Madelung, E., 1926, Quantentheorie in Hydrodynamischer Form, Z. Phys. **40**, 322.
Mandel, L., and E. Wolf, 1995, Optical coherence and quantum optics (Cambridge University Press, Cambridge).
Messiah, A., 1961, Quantum Mechanics, Vol. II (North-Holland, Amsterdam) p. 1034.
Mignani, R., E. Recami and M. Baldo, 1974, About a Dirac-like equation for the photon according to Ettore Majorana, Lett. Nuovo Cimento **11**, 568.
Misner, C.W., and J.A. Wheeler, 1957, Classical physics as geometry: Gravitation, electromagnetism, unquantized charge, and mass as properties of curved empty space, Ann. Phys. **2**, 525.
Molière, G., 1949, Laufende elektromagnetische Multipolwellen und eine neue Methode der Feld-Quantisierung, Ann. Phys. **6**, 146.
Moses, H.E., 1959, Solution of Maxwell's equations in terms of a spinor notation: the direct and inverse problem, Phys. Rev. **113**, 1670.
Moses, H.E., 1973, Photon wave functions and the exact electromagnetic matrix elements for hydrogenic atoms, Phys. Rev. A **8**, 1710.
Newton, T.D., and E.P. Wigner, 1949, Localized states for elementary systems, Rev. Mod. Phys. **21**, 400.
Ohmura, T., 1956, A new formulation on the electromagnetic field, Prog. Theor. Phys. **16**, 684.
Oppenheimer, J.R., 1931, Note on light quanta and the electromagnetic field, Phys. Rev. **38**, 725.
Pauli, W., 1933, Prinzipien der Quantentheorie, in: Handbuch der Physik, Vol. 24 (Springer, Berlin) [English translation: 1980, General Principles of Quantum Mechanics (Springer, Berlin)] p. 189.
Pegg, D.T., and S.M. Barnett, 1988, Unitary phase operator in quantum mechanics, Europhys. Lett. **6**, 483.
Penrose, R., and W. Rindler, 1984, Spinors and Space-Time, Vol. I (Cambridge University Press, Cambridge) ch. 5.
Plebanski, J., 1960, Electromagnetic waves in gravitational fields, Phys. Rev. **118**, 1396.
Power, E.A., 1964, Introductory Quantum Electrodynamics (Longmans, London) p. 61.
Pryce, M.H.L., 1948, The mass-centre in the restricted theory of relativity and its connection with the quantum theory of elementary particles, Proc. R. Soc. (London) A **195**, 62.
Rosewarne, D., and S. Sarkar, 1992, Rigorous theory of photon localizability, Quantum Opt. **4**, 405.
Schiff, L.I., 1968, Quantum Mechanics (McGraw-Hill, New York) p. 203.
Schleich, W., and M.O. Scully, 1984, General relativity and modern optics in: New Trends in Atomic Physics, Les Houches, Session XXXVIII, eds G. Grynberg and R. Stora (Elsevier, Amsterdam) p. 995.

Schweber, S.S., 1961, An Introduction to Relativistic Quantum Field Theory (Row, Peterson, and Co., Evanston, IL).
Segal, I.E., 1963, Mathematical Problems of Relativistic Physics (American Mathematical Society, Providence, RI).
Silberstein, L., 1907a, Elektromagnetische Grundgleichungen in bivectorieller Behandlung, Ann. Phys. **22**, 579.
Silberstein, L., 1907b, Nachtrag zur Abhandlung über "Elektromagnetische Grundgleichungen in bivectorieller Behandlung", Ann. Phys. **24**, 783.
Silberstein, L., 1914, The Theory of Relativity (MacMillan, London). 2nd enlarged Ed.: 1924.
Sipe, J.F., 1995, Photon wave functions, Phys. Rev. A **52**, 1875.
Skrotskiĭ, G.B., 1957, The influence of gravitation on the propagation of light, Sov. Phys. Dokl. **2**, 226.
Staruszkiewicz, A., 1973, On affine properties of the light cone and their application in quantum electrodynamics, Acta Phys. Polon. B **4**, 57.
Streater, R.F., and A.S. Wightman, 1978, PCT, Spin and Statistics, and All That (Benjamin, Reading, MA).
Sudarshan, E.C.G., 1979, Pencils of rays in wave optics, Phys. Lett. A **73**, 269.
Sudarshan, E.C.G., 1981a, Quantum electrodynamics and light rays, Physica A **96**, 315.
Sudarshan, E.C.G., 1981b, Quantum theory of radiative transfer, Phys. Rev. A **23**, 2803.
Walther, A., 1968, Radiometry and coherence, J. Opt. Soc. Am. **68**, 1256.
Weber, H., 1901, Die partiellen Differential-Gleichungen der mathematischen Physik nach Riemann's Vorlesungen (Friedrich Vieweg und Sohn, Braunschweig) p. 348.
Weinberg, S., 1995, The quantum theory of fields, Vol. I (Cambridge University, Cambridge) p. 61.
Weinberg, S., and E. Witten, 1980, Limits on massless particles, Phys. Lett. B **96**, 59.
Weyl, H., 1929, Elektron und Gravitation, Z. Phys. **56**, 330.
Wightman, A.S., 1962, On the localizability of quantum mechanical systems, Rev. Mod. Phys. **34**, 845.
Wigner, E.P., 1932, On the quantum correction for thermodynamic equilibrium, Phys. Rev. **40**, 749.
Wigner, E.P., 1939, On unitary representations of the inhomogeneous Lorentz group, Ann. Math. **40**, 39. Reprinted: 1966, Symmetry Groups in Nuclear and Particle Physics, ed. F.J. Dyson (Benjamin, New York) p. 120.
Wolf, E., 1976, New theory of radiative energy transfer in free electromagnetic field, Phys. Rev. D **13**, 869.
Zeldovich, Ya.B., 1965, Number of quanta as an invariant of the classical electromagnetic field, Dokl. Acad. Sci. USSR, **163**, 1359. In Russian.

AUTHOR INDEX FOR VOLUME XXXVI

A

Abbe, E. 131, 135
Abbiss, J.B. 160
Abrahams, E. 219
Acharya, R. 280
Ackley, D.E. 13–17
Acquista, C. 206
Adam, A. 60
Agarwal, G.S. 73, 113
Akhiezer, A.I. 250, 263
Akkerman, E. 216
Alford, W.P. 79
Alley, C.O. 92
Altshuler, B.L. 215, 234
Amrein, W.O. 267, 273, 279
Anderlain, R. 26
Anderson, J.L. 206
Anderson, P.W. 217, 219
Andreev, A.A. 26
Andrews, H.C. 139
Andriesh, A.M. 3, 4, 17, 19, 20, 23–25, 28, 33, 37–41
Antaramian, A. 74
Anton, A. 249
Antonetti, A. 17
Apai, P. 6, 8, 9
Apresyan, L.A. 185, 187, 189, 205, 206, 211, 212, 216, 219
Ardini, J. 249
Arecchi, F.T. 61
Aronov, A.G. 215
Arsenin, V.Y. 147
Ash, E.A. 172
Ashtekar, A. 271
Asobe, M. 4, 37
Aspect, A. 58, 60, 61, 66, 92, 108

B

Babinets, Iu.Iu. 12
Baldo, M. 252
Baldzuhn, J. 110
Balescu, R. 193
Balkanski, M. 3, 6
Ballik, E.A. 70
Baltes, H.P. 189, 198, 206
Barabanenkov, Yu.N. 185, 205–207, 211, 215, 216
Barbosa, G.A. 104, 105
Barends, P. 162, 164
Bargmann, V. 254, 278
Barnett, S.M. 117, 250
Bartelt, H.O. 193
Bateman, H. 252
Batho, H.F. 65
Baumgartner, R. 27
Baym, G. 248
Beck, M. 118
Beichman, C.A. 235
Bekefi, G. 182
Belenkii, M.S. 216
Bell, D.A. 145
Bell, J.S. 53, 91
Berestetskii, V.B. 250, 263
Bergman, C. 219
Bergman, D.J. 204
Berkovits, R. 216, 229, 231, 232, 234, 236
Berné, A. 61
Berntsen, S. 172
Berry, M.V. 52, 73, 74
Bertero, M. 140, 147, 150, 152–154, 156, 161, 162, 164–168, 170–172
Bertolotti, M. 4, 17, 19, 24, 37–41
Besieris, I.M. 204
Beswick, J.A. 58
Betzig, E. 172

295

Bhandari, R. 74
Bialynicka-Birula, Z. 250, 265, 266, 278, 288, 289
Bialynicki-Birula, I. 249, 250, 253, 258, 260, 265, 266, 272, 276, 278, 281–283, 285, 288–290
Biemond, J. 154, 161
Billing, H. 120
Bissonnette, D. 217
Bjorken, J. 249
Blom, P. 162, 164
Boccacci, P. 154, 164, 166–168
Boekee, D.E. 161
Bogdan, O.I. 20, 24, 28, 33
Bohm, D. 91, 249
Bohr, N. 110
Bonch-Bruevichi, V.L. 26
Bond, W.L. 70
Bondurant, R.S. 120
Born, M. 134, 135, 139, 182, 261
Borovoi, A.G. 205, 222
Bosch, M.A. 12–17
Bostan, G. 4
Bouger, P. 201
Bozhevolnaya, E. 172
Bozhevolnyi, S. 172
Braginsky, V.B. 108
Brakenhoff, G.J. 162–164, 166–168
Branning, D. 99, 100
Brendel, J. 62, 87, 89, 90, 97
Brenner, K.N. 193
Brianzi, P. 164
Brojan, J. 290
Brown, N. 70–73
Broyer, M. 58
Brune, M. 109
Bruynseraede, Y. 219
Buchler, U. 117
Bugnolo, D. 205
Burlamacchi, P. 61
Burnett, K. 120
Burnham, D.C. 60
Büttiker, M. 113
Byer, R.L. 60
Bykovskii, Y.A. 3

C

Cai, S.F. 6, 10
Campos, R.A. 62, 86, 87
Carruthers, P. 117
Carter, W.H. 185, 194, 205, 206
Casabella, P.A. 72
Caves, C.M. 62–64, 119–121
Cerbari, P.G. 3, 4, 20, 24
Cernii, M.R. 3
Cesini, G. 145, 160
Chandrasekhar, S. 182, 201
Chauhan, G. 36
Chiao, R.Y. 74, 76–78, 86, 97, 98, 109–115
Choi, W.K. 5, 6, 10
Choudhury, A. 162
Chu, Q.-J. 219
Chuang, S.L. 229
Chumash, V.N. 3, 4, 17, 19, 20, 23–25, 28, 33, 37–41
Chyba, T.H. 74
Clauser, J.F. 59, 61, 92, 96, 98
Cohen, E.G.D. 204
Cohen, L. 191, 193
Cohen, M.H. 204
Cohen-Tannoudji, C. 248, 250
Cojocaru, I.A. 3, 4, 17, 19, 20, 23, 24, 28, 33, 37, 38
Commins, E.D. 59
Cook, R.J. 251
Cooper, J. 118
Courjon, D. 172
Courty, J. 109
Cox, I.J. 164
Croca, J.R. 90, 91
Cwilich, G. 234

D

Dainty, J.C. 235
Dalibard, J. 92
Davies, A.R. 147
Davies, R.E. 164–166
Davoust, J. 163
Dawar, A.L. 36
de Broglie, L. 73, 90
De Mol, C. 140, 150, 152, 153, 160, 162, 164, 166, 170–172
De Neufville, J.P. 35
De Santis, P. 142, 149, 150, 152, 159, 160
de Voe, R.G. 64
De Wolf, D.A. 216
Defrise, M. 164, 166
Dempster, A.J. 65
Dhadwal, H.S. 160
Diedrich, F. 61

Dirac, P.A.M.   51, 117, 122, 248
Diu, B.   248
Dolghinov, A.Z.   206
Dolginov, A.Z.   205
Dolin, L.S.   184, 187, 209, 221
Dombrovskii, L.A.   228
Donoho, D.L.   154, 161
Dorfman, J.R.   219
Druhl, K.   110
Dudarev, S.L.   216
Dultz, W.   89, 90
Dupont-Roc, J.   250

E
Eberhard, P.   98
Eberhard, P.H.   98
Eberly, J.H.   193
Edelstein, W.A.   119
Edrei, I.   216
Einstein, A.   91, 100
Elitzur, A.C.   115
Eliyahu, D.   231
Elliot, S.R.   4
Enaki, N.A.   3, 4, 20, 23, 24, 28, 33
Engelhardt, M.   27
England, A.W.   226
Englert, B.-G.   110
Erdös, P.   219
Espinosa, G.P.   6, 10
Esser, B.   26
Etkin, V.S.   229
Ewen, P.J.S.   43

F
Fauchet, P.M.   16, 17
Fazekas, P.   6, 9, 11
Fazio, E.   4, 17, 19, 24, 37–41
Fearn, H.   62, 81, 85
Fekeshgazi, I.V.   12
Felderhof, B.U.   204
Feng, S.   229, 233, 234
Ferguson, J.B.   70
Fernandez Guasti, M.   3, 6
Ferrari, A.   4, 19
Fertig, H.A.   113
Feynman, R.P.   57, 67
Filonovich, S.R.   228
Finkelberg, V.M.   185, 204–206, 211
Firth, A.   6, 10, 11
Ford, G.W.   204

Fork, R.L.   12–17
Forrester, A.T.   70
Forward, R.L.   119
Fougères, A.   118
Franklin, J.N.   145
Franson, J.D.   97, 113
Freedman, S.J.   59
Freund, I.   216, 229, 234, 236
Friberg, A.T.   185, 189
Friberg, S.   60
Friedburg, H.   117
Frieden, B.R.   139, 157, 161
Frölich, D.   117
Fujima, I.   71, 72
Fukuyama, H.   219
Fung, A.K.   228
Fustoss-Wegner, M.   6, 9

G
Gabor, D.   136
Galinas, R.J.   205
Ganga, K.M.   74
Gans, R.   65
Garuccio, A.   90, 91
Gazso, J.   6, 8
Genchev, J.D.   228
Gerchberg, R.W.   159
Gerhardt, H.   117
Gertsenshtein, M.E.   204
Ghielmetti, F.   73
Ghose, P.   113
Ghosh, R.   79–82
Gibbs, H.M.   19, 21
Glass, A.M.   12–17
Glauber, R.J.   52, 54–56
Glogower, J.   117
Gnedin, Yu.N.   205, 206
Goetze, W.   219
Gold, A.   79
Golden, K.   204
Golubentsev, A.A.   216
Gonsiorowski, T.   72
Good Jr, R.H.   253, 269, 270
Goodman, J.W.   135
Gori, F.   142, 144, 149, 150, 159, 160, 164, 170, 171
Górnicki, P.   281
Gorodnichev, E.E.   216
Gottfried, K.   280
Grangier, P.   58, 60, 61, 66, 108, 109, 120

Grayson, T.P. 104, 105
Gribkovskii, V.P. 25, 30
Griffiths, J.E. 6, 10
Grochmalicki, J. 166
Groetsch, C.W. 147, 150
Gross, L. 269
Grozescu, A. 23, 33
Grynberg, G. 250
Guattari, G. 142, 144, 145, 149, 160, 164
Gudmundsen, R.A. 70
Gurvich, A.S. 216, 227
Gustafson, T.K. 109
Gzara, K. 17

H
Haag, R. 280
Habashy, T. 161
Hagan, D.J. 38
Hajto, J. 3–11, 36, 43
Hall, J.L. 65
Hanbury Brown, R. 61, 79
Handa, Y. 3
Hardy, L. 99
Hariharan, P. 70–76
Haro-Poniatowski, M. 3, 6
Haroche, S. 109
Harris, J.L. 140
Harris, R.W. 140
Harris, S.E. 60
Hashin, Z. 204
Hauge, E.H. 113
Haus, H.A. 109
Heitler, W. 108
Hellmuth, T. 109
Heo, J. 4
Herndon, R.C. 219
Herzog, T. 101, 102, 116
Hoch, J.C. 154, 161
Hoenders, B.J. 172
Holland, M.J. 120
Holland, P.R. 91
Hollberg, L.W. 64
Holt, R.A. 92, 98
Home, D. 113
Hong, C.K. 60, 62, 79–82, 84, 86, 89, 106, 113
Hori, H. 204
Horne, M.A. 92, 94, 95, 98
Hough, J. 119
Howe, M.S. 205

Hulin, D. 16, 17, 19, 24, 37, 38
Hunt, B.R. 139

I
Igo, T. 35
Imai, Y. 36
Imoto, N. 108, 109
Inagaki, T. 251
Irisov, V.G. 229
Ishimaru, A. 206, 208, 216, 226
Itzykson, C. 278

J
Jackson, J.D. 261
Jagadish, C. 36
Jakeman, E. 86, 87
Janossy, I. 3, 5–8, 10, 11
Janossy, L. 60, 65
Jauch, J.M. 279
Javan, A. 70
Jerrard, H.G. 75
Jerri, A.J. 136
Jiao, H. 74
Johnson, P.O. 70
Johnstone, I.M. 154, 161
Jordan, T.F. 73, 99
Joshi, J.C. 36

K
Kaiber, R. 26
Kaiser, G. 269
Kalinin, V.I. 227
Kalmikova, N.P. 37
Kamiya, K. 4
Kanamori, T. 4, 37
Kane, C. 229, 233, 234
Kaplan, R. 163
Kar, A.K. 43
Kasevich, M.A. 116
Kashkarov, S.S. 216
Kastner, M.A. 4
Kaveh, M. 216, 229, 231, 232, 234, 236
Khmelnitskii, D.E. 215
Khvolson, O.D. 181, 201
Kidd, R. 249
Kikuchi, M. 36
Kimble, H.J. 65, 74
Kirkpatrick, T.R. 219
Kitagawa, M. 109
Klauder, J.R. 120, 121

Kleeorin, N.I.   222, 226
Klyshko, D.N.   60
Kobayashi, M.   4
Kocher, C.A.   59
Kohler, W.   204
Kohler, W.E.   204
Kolomeiko, E.P.   3
Kolomiec, B.T.   11
Kolomiets, B.T.   26
Kong, J.A.   226, 229
Kosa, T.   43
Kosa Somogyi, I.   5, 6, 8, 11
Koyama, J.   3
Kozlov, V.P.   145
Kramers, B.   219
Kramers, H.A.   253, 262
Kravtsov, V.E.   234
Kravtsov, Yu.A.   185, 187, 189, 193, 203, 205–208, 211, 212, 216, 221, 222, 226–229
Krulikovskii, B.K.   12
Kubodera, K.   4, 37
Kuga, Y.   216
Kumar, P.   64
Kwiat, P.G.   76–78, 86, 97, 98, 110–116

L

Lagendijk, A.   216
Lagendijk, R.L.   154, 161
Laloë, F.   248
Lambert, J.H.   201
Lampard, D.G.   193
Landau, L.D.   251
Landauer, R.   113, 115, 204, 205
Landweber, L.   160
Lange, R.   62
LaPorta, A.   120
Laporte, O.   253, 286, 287
Larchuk, T.   86
Larchuk, T.S.   87
Larkin, A.I.   215
Lax, M.   204
Lee, A.P.   219
Lee, B.   109
Lee, I.K.   17
Lee, J.M.   40
Lee, P.A.   229, 233, 234
Leighton, R.B.   57, 67
Lent, A.   160, 161
Lepore, V.L.   90, 91
Lerner, I.V.   234

Levenson, M.D.   64
Levin, M.L.   190
Liakhou, G.   37, 41
Licciardello, D.C.   219
Lin, S.L.   229
Lipkin, H.J.   248
Lisitsa, M.P.   12
Litfin, G.   117
Lohmann, A.W.   193
Lomont, J.S.   254
London, F.   250
Loudon, R.   62, 81, 85
Louisell, W.H.   117, 289
Lucarini, G.   145, 160
Lucy, L.   161

M

Ma, S.-K.   218
Machida, S.   108
Mackenzie, J.D.   4
Madelung, E.   283
Maeda, M.W.   64
Maggio, F.   154
Magnon, A.   271
Magyar, G.   67
Maischberger, K.   120
Makovkin, A.V.   3
Malakhov, A.N.   230
Malfanti, F.   166–168
Malinovskii, V.K.   27
Malkevich, M.S.   145
Mandel, L.   59, 60, 62, 67–69, 73, 74, 79–82, 84–94, 99–106, 110, 111, 113, 118, 123, 261, 272, 273, 279, 289
Manning, R.M.   223, 225
Manukian, A.L.   26
Mao, X.   19
Marc, W.D.   193
Maret, G.   216
Marsman, H.J.B.   163
Martienssen, W.   62, 87, 89, 90, 97, 110
Martin, W.   119
Mathur, P.C.   36
Matsuda, A.   36
Matveev, D.H.   227
Maynard, J.D.   172
Maynard, R.   216
Mazets, T.F.   26, 37
McCall, S.L.   120, 121
McCaughan, L.   37

McCutchen, C.W. 140
Meinel, E.S. 154
Melekhin, V.N. 69
Mendez, E.P. 3, 6
Mereminskii, A.E. 222, 226
Mersereau, R.M. 154, 161
Mertz, J.C. 64
Messiah, A. 274
Michelotti, F. 17, 19, 24, 37–41
Middleton, D. 138
Mignani, R. 252
Miguez, P. 65
Migus, A. 12–17
Milburn, G.J. 64, 108, 109, 121
Miller, D.A.B. 25, 30
Milton, G.W. 204
Mironov, A.G. 26
Mironov, V.L. 216
Mirovskaya, E.A. 229
Mirovskii, V.G. 222, 226
Mishin, S.A. 69
Misner, C.W. 253
Mitsa, V.N. 12
Mizobuchi, Y. 113
Mizuno, H. 36
Mohler, E. 87, 97, 110
Molière, G. 253, 254, 274
Monken, C.H. 99
Moreira, R.N. 90, 91
Mori, H. 193
Morris, R.H. 70
Moses, H.E. 254, 255, 288–290
Moss, S.C. 35
Mott, N.F. 218
Mourchid, A. 16, 17

N

Nagaoka, Y. 219
Nakamura, M. 4
Nanninga, N. 163
Naray, Z. 65
Nasu, H. 4
Nasyrov, U. 12
Nathel, H. 74, 86
Natterer, F. 173
Nesterihin, Iu.E. 27
Newton, T.D. 85, 279
Nicholls, G. 172
Nieto, M.M. 117
Nieto-Vesperinas, M. 168

Nighan Jr, W.L. 16
Nishihara, H. 3
Nityananda, R. 74
Noh, J.W. 118
Novikov, V.N. 27

O

O'Byrne, J.W. 74–76
Odajima, A. 36
Odelevskii, V.I. 205
Ohmachi, Y. 35
Ohmura, T. 255, 288
Ohtake, Y. 113
Ohtsuka, Y. 35, 36
Oppenheim, I. 193
Oppenheimer, J.R. 253–255
Orbach, R. 27
Orlov, Yu.I. 206
Oshman, M.K. 60
Ostafeichuk, N.D. 3, 20, 24
Ott, R.L. 205
Ou, Z.Y. 62, 79–82, 84–89, 92–94, 100, 101, 106, 111, 113
Ovchinnikov, G.I. 185, 205
Ovshinsky, S.R. 35
Owen, A.E. 6, 10, 43
Ozrin, V.D. 216

P

Paesler, M.A. 40
Page, C.H. 193
Paley, R.E.A.C. 149, 155
Palma, C. 142, 145, 149, 150, 152, 160
Pancharatnam, S. 74
Pang, D. 17
Paolucci, S. 150
Papanicolaou, G.C. 204
Papoulis, A. 135, 155, 159
Paul, H. 67, 117
Paul, W. 13–17
Pauli, W. 251
Pavičić, M. 106–108
Pavlov, S.K. 26, 37
Peacher, J.I. 205
Pegg, D.T. 117, 250
Peierls, R. 251
Penrose, R. 287, 288
Perlmutter, S.H. 64
Petersen, D.P. 138
Petroff, M. 98

Pfleegor, R.L.   67–69
Philips, J.C.   6, 10
Pike, E.R.   150, 153, 156, 161, 162, 164–166
Pinzenik, V.P.   12
Pipkin, F.M.   65
Piron, C.   279
Plebanski, J.   285
Pnini, R.   234
Podolsky, B.   91, 100
Podolyak, E.R.   69
Pohl, D.W.   172
Polder, D.   204
Pollack, H.O.   156
Popescu, M.   23, 33
Popov, A.E.   229
Power, E.A.   249
Prise, M.E.   25, 30
Pryce, M.H.L.   255, 279
Pugh, J.R.   119

R
Rafelski, J.   281
Raimond, J.M.   109
Raizer, V.Yu.   228
Ramakrishnan, T.V.   219
Ramaseshan, S.   74
Ramshaw, J.D.   204
Rangel-Rojo, R.   43
Rarity, J.G.   82, 86, 87, 96, 97, 101, 102
Rayleigh, J.W.S.   131, 134
Raymer, M.G.   118
Recami, E.   252
Remeika, J.P.   6, 10
Renk, F.F.   27
Reynaud, S.   109
Richards, B.   167
Richards, M.A.   161
Richardson, W.H.   161
Ridgway, S.   235
Rindler, W.   287, 288
Robinson, P.A.   74–76
Roch, J.F.   109
Rochon, P.   217
Roger, G.   58, 60, 61, 66, 92, 108, 109
Rogozkin, D.B.   216
Ronchi, L.   150, 170, 171
Rosen, N.   91, 100
Rosenbluh, M.   229, 234
Rosewarne, D.   279
Ross, J.   193

Roth, L.M.   204
Roy, M.   74–76
Rozenberg, G.V.   186, 201
Rubin, M.H.   97
Rüdiger, A.   120
Rushforth, C.K.   140
Rytov, S.M.   184, 190, 193, 203, 206, 208, 221, 227, 228
Ryzhov, Y.A.   204

S
Saichev, A.I.   216, 230
Said, Ali A.   38
Saito, M.   36
Saleh, B.E.A.   62, 86, 87
Samuel, J.   74
Sanders, B.C.   70–73, 109, 121
Sands, M.   57, 67
Sanghera, J.S.   4
Santos, E.   98
Sarkar, S.   279
Scala, C.   204
Schafer, R.W.   161
Schiff, E.   17
Schiff, L.I.   254
Schilling, R.   120
Schleich, W.   109, 285
Schnupp, L.   120
Schumaker, B.L.   64
Schuster, A.   181, 201
Schütrumpf, S.   62
Schweber, S.S.   250, 288
Scully, M.O.   62, 110, 285
Seaton, C.T.   25, 30
Segal, I.E.   271
Sen, P.N.   204
Senesi, F.   37–41
Serber, R.   117
Sergienko, A.V.   97
Shah, I.   12–17
Shank, C.V.   12–17
Shannon, C.E.   136
Shapiro, B.   234
Shapiro, J.H.   64, 120
Sharkov, E.A.   228
Sheik-Bahae, M.   38
Shelby, R.M.   64
Sheng, P.   219
Shepp, L.A.   161
Sheppard, C.J.R.   162, 164, 166

Sherman, G.C. 168
Shewell, J.R. 168, 170
Shifrin, E.I. 37
Shih, Y.H. 92, 97
Shimony, A. 92, 94–96, 98
Shishodia, P.K. 36
Shmal'ko, A.V. 3
Shtrikman, S. 204
Sibilia, C. 4, 19
Silant'ev, N.A. 206
Silberstein, L. 252, 253, 256
Silo, V.P. 11
Simon, R. 74
Sipe, J.F. 253, 272
Skrotskiĭ, G.B. 285
Slepian, D. 156, 160
Slusher, R.E. 64, 120
Smirnov, V.L. 3
Smith, S.D. 25, 30
Smithey, D.T. 118
Smorgonskaya, E.A. 37
Sobolev, V.V. 182
Sokolov, A.P. 27
Spivak, B.Z. 234
Staruszkiewicz, A. 266
Steinberg, A.M. 97, 98, 110–115
Stelzer, E.H.K. 163
Stephen, M.J. 234
Stern, A.S. 154, 161
Stogryn, A. 228
Stone, A.D. 229, 233, 234
Stott, P.E. 205
Stourac, L. 11
Støvneng, J.A. 113
Strand, O.N. 145
Streater, R.F. 287
Stricker, R. 163
Stroud, D. 205
Sudarshan, E.C.G. 52, 56, 74, 280, 281
Suhara, T. 3
Summhammer, J. 106
Susskind, L. 117
Suzuki, K. 4
Svechnikov, G.S. 12
Synge, E.H. 172

T

Takayama, T. 36
Tamoikin, V.V. 204
Tanaka, K. 4, 24, 35, 36

Tanguy, C. 17
Tapster, P.R. 82, 86, 87, 96, 97
Tatarskii, V.I. 185, 193, 203–208, 216, 221, 227, 228
Tauc, J. 13–17
Taylor, G.I. 65
Teich, M.C. 62, 86, 87
Tikhonov, A.N. 147
Tiwari, S.C. 78
Tomita, A. 74
Toraldo di Francia, G. 136, 140, 141, 157
Torgerson, J.R. 99, 100, 104
Toth, L. 4–7
Townes, C.H. 117
Trautman, J.K. 172
Troitskii, I.A. 229
Trokhimovskii, Yu.G. 229
True, E.M. 37
Tsang, L. 216, 226
Tsolakis, A.I. 204
Turchin, V.F. 145
Tuy, H. 160, 161
Twersky, V. 204
Twiss, R.Q. 61, 79

U

Uhlenbeck, G.E. 253, 286, 287

V

Vaidman, L. 115
Vain'shtein, L.A. 69
Valley, J.F. 64
Van Albada, M.P. 216
van Cittert, P.H. 161
van der Voort, H.T.M. 163, 167, 168
van Santen, J.H. 204
van Spronsen, E.A. 163
Van Stryland, E.W. 38
Vanderhaden, R. 16
Vardanyan, R.C. 223, 225
Vardeny, Z. 15–17, 25
Vardi, Y. 161
Vareka, W.A. 86
Varga, P. 60
Viano, G.A. 140, 152, 155, 172
Vigier, J.P. 91
Vigué, J. 58
Vinogradov, A.G. 185, 205, 207, 208, 216
Vlasenko, Iu.V. 12
Vollhardt, D. 219
von Weiszäcker, C.F. 109

W

Walker, J.G.   150, 162, 164–166
Walls, D.F.   62, 64, 65, 73, 108
Walther, A.   185, 192, 198, 205, 281
Walther, H.   61, 109, 110
Wang, L.   233, 234
Wang, L.J.   74, 86–88, 90, 91, 100–104, 110, 111
Warniak, J.S.   70
Watson, K.M.   186, 205, 214, 215
Webb, H.   160
Weber, H.   252
Wei, T.   38
Weinberg, D.L.   60
Weinberg, S.   267, 272, 278, 280, 287, 288
Weinfurter, H.   101, 102, 116
Welling, H.   117
Wen, B.   226
Wentz, F.J.   228
Westwater, E.R.   145
Weyl, H.   255
Wheeler, J.A.   109, 253
Wherrett, B.S.   43
Whittaker, E.T.   136
Wiener, N.   149, 155
Wightman, A.S.   279, 287
Wigner, E.P.   85, 193, 250, 254, 278, 279, 281
Wijnaendts-van-Resandt, R.W.   163
Wilkinson, S.R.   74
Williams, E.G.   172
Wilson, T.   162, 164
Winebrenner, D.R.   226
Winkler, W.   120
Witten, E.   272
Wodkiewicz, K.   193
Wolf, E.   52, 54, 59, 134, 135, 139, 161, 167, 168, 170, 182, 185, 189, 194, 198–200, 205, 206, 261, 272, 273, 279, 281, 289

Wolf, P.E.   216
Wölfle, P.   219
Wolter, H.   140, 155
Wootters, W.K.   108
Wraback, M.   17
Wu, H.   65
Wu, L.   65
Wu, S.T.   228

X

Xie, C.X.   6, 10
Xu, L.B.   6, 10
Xue, W.   219

Y

Yablonovitch, E.   114
Yamamoto, Y.   108, 109
Yang, P.   19
Yin, E.   109
Youla, D.C.   160
Young, M.R.   165, 166
Yuen, H.P.   62
Yurke, B.   64, 120, 121

Z

Zajonc, A.   109
Zajonc, A.G.   110, 111
Zeeman, P.   65
Zeilinger, A.   94, 95, 101, 102, 116
Zeldovich, Ya.B.   269
Zentai, G.   4–8, 11
Zhang, Z.-Q.   219
Zheleznyakov, V.V.   182
Zou, X.Y.   86–88, 90, 91, 100–105, 110, 111
Zuber, J.-B.   278
Zurek, W.H.   108
Zuzin, A.Yu.   234
Zveaghin, I.P.   26

SUBJECT INDEX FOR VOLUME XXXVI

A

Aharonov–Bohm effect 109
Airy pattern 134
Anderson localization 183, 215, 217, 219
average field approximation 204

B

Beer law 20
Bell's inequality 53, 92–99, 122
– –, interferometric tests 94–98
– –, loophole-free test 98
– –, two-photon tests 106–108
– theorem 92
Berry's phase 77, 78, 266
Bessel equation 277
Bethe–Salpeter equation 186, 203–207, 219
Born approximation 161
Bourret approximation 223

C

chalcogenide glass 35
Clausius–Mossotti formula 205
coherence, fourth-order 53, 55, 79
–, second-order 52, 53
coherent potential approximation 204
– state 57, 58, 117
complementarity 108–116
confocal microscopy 162–168
correlation, second-order 55
correspondence principle 78

D

Debye approximation 167
delayed-choice experiments 109, 110
Dirac equation 253, 259
Dyson equation 203–207, 221, 223

E

effective medium approximation 204
evanescent wave 168
Ewald sphere 161

F

Fabry–Perot interferometer 104
– – resonator 20
Feynman diagrams 215
Fock state 59, 81
– –, two-photon 81
Fourier slice theorem 173

G

geometric phase 73–78, 89, 90, 104
geometrical optics 196, 197
Glauber–Sudarshan P-representation 52, 53, 55, 56, 66

H

hidden variables 91
homodyne detection 117

I

idler photon 60
impulse response 135, 142–144
intensity fluctuations 79
interaction-free measurements 115, 116
interference, fourth-order 78–91
–, second-order 65–73, 89, 102
–, two-photon 100–106
– in the time domain 70–72
– with independent sources 67–69
– – single-photon states 66, 67
interferometry, quantum limits 117–122
–, two-photon 91–108
Ioffe–Regel condition 183
iterated dilute approximation 204

J
Jauch–Piron localizability 279

K
Kerr effect 109
Kirchhoff law 226, 227
Kramers–Kronig relations 35

L
Landau–Peierls wave function 251, 252, 254, 273
light-induced transmission oscillations 8–10
Lorentz transformation 251, 261
Lorenz–Lorentz formula 205

M
Macdonald function 280
Mach–Zehnder interferometer 52, 67, 74, 88, 109, 117
Maxwell equations 253–256, 264, 282
Maxwell–Garnett formula 205
Maxwell–Odelevskii formula 205
Michelson interferometer 52, 76, 89, 120

N
n-photon state 55, 56
Newton–Wigner wave function 280
nonclassical effects 102
– light 52, 55, 56, 59
Nyquist distance 137, 139, 140

O
optical bistability 3, 11, 19

P
Paley–Wiener space 137
Pancharatnam phase 74, 75, 89, 104
parabolic equation 208, 209
parametric down-conversion 60, 64
Parseval's relation 147
phase operator 117
– problem 132
photometry 201
photon, localizability 279, 280
–, localization 249
–, phase-space description of 281–283
– bunching 59
photon-counting statistics 119
– position operator 250
– wave function 245–291
– – –, as a spinor field 286–288

– – –, eigenvalue problems for 273–278
– – –, in coordinate representation 259–263
– – –, – momentum representation 263–267
– – –, – non-cartesian coordinate systems 285, 286
Poincaré group 250, 254, 274
– sphere 74
– transformation 267
point spread function (PSF) 136
projection theorem 173

Q
quantum eraser 53, 110–112
quantum-nondemolition measurements 108, 109
quasi-crystalline approximation 204

R
radiance 181, 189
–, generalized 198–200
radiation pressure 119
radiative transfer equation (RTE) 179–237
– – – – in scattering media 200–203
– – in free space 188–200
– – theory 179–237
– transport theory 201
radiometry, diffraction 208, 209
–, nonclassical 186, 187, 199
Radon transform 173
Raman scattering 183
Rayleigh limit 133, 141
– resolution distance 140
– – limit 134, 135
resonance fluorescence 58
Riemann–Silberstein vector 253, 254, 256, 260, 261, 267, 283, 288
– – wave function 252–254

S
Schwarz's inequality 67
semiconductors, amorphous 3, 4, 20, 27, 28
–, nonlinear optical effects 3
Shannon number 133, 141, 157, 158
signal photon 60
space–bandwidth product 141
spectral density 189–191
squeezed light 62–65
standard quantum limit (SQL) 54, 62, 119, 120
superposition states 72, 73

T

transfer function (TF) 136, 148, 166
Twersky's multiple scattering theory 204

U

uncertainty relation, number–phase 117–119

V

vacuum fluctuations 62

W

Watson's channels 214, 215
wave equation for photons 254–259
Weyl equation 254

which-path measurement 52
Wiener filter 133, 145, 147
Wiener–Khinchin formula 191
Wigner function 118, 193, 281
Wolf effect 200

Y

Young's double-slit experiment 109

Z

Z-Scan signal 38–41
– spectroscopy 4
– technique 37, 38, 44

CONTENTS OF PREVIOUS VOLUMES

VOLUME I (1961)

I	The Modern Development of Hamiltonian Optics, R.J. PEGIS	1– 29
II	Wave Optics and Geometrical Optics in Optical Design, K. MIYAMOTO	31– 66
III	The Intensity Distribution and Total Illumination of Aberration-Free Diffraction Images, R. BARAKAT	67–108
IV	Light and Information, D. GABOR	109–153
V	On Basic Analogies and Principal Differences between Optical and Electronic Information, H. WOLTER	155–210
VI	Interference Color, H. KUBOTA	211–251
VII	Dynamic Characteristics of Visual Processes, A. FIORENTINI	253–288
VIII	Modern Alignment Devices, A.C.S. VAN HEEL	289–329

VOLUME II (1963)

I	Ruling, Testing and Use of Optical Gratings for High-Resolution Spectroscopy, G.W. STROKE	1– 72
II	The Metrological Applications of Diffraction Gratings, J.M. BURCH	73–108
III	Diffusion Through Non-Uniform Media, R.G. GIOVANELLI	109–129
IV	Correction of Optical Images by Compensation of Aberrations and by Spatial Frequency Filtering, J. TSUJIUCHI	131–180
V	Fluctuations of Light Beams, L. MANDEL	181–248
VI	Methods for Determining Optical Parameters of Thin Films, F. ABELÈS	249–288

VOLUME III (1964)

I	The Elements of Radiative Transfer, F. KOTTLER	1– 28
II	Apodisation, P. JACQUINOT, B. ROIZEN-DOSSIER	29–186
III	Matrix Treatment of Partial Coherence, H. GAMO	187–332

VOLUME IV (1965)

I	Higher Order Aberration Theory, J. FOCKE	1– 36
II	Applications of Shearing Interferometry, O. BRYNGDAHL	37– 83
III	Surface Deterioration of Optical Glasses, K. KINOSITA	85–143
IV	Optical Constants of Thin Films, P. ROUARD, P. BOUSQUET	145–197
V	The Miyamoto–Wolf Diffraction Wave, A. RUBINOWICZ	199–240
VI	Aberration Theory of Gratings and Grating Mountings, W.T. WELFORD	241–280
VII	Diffraction at a Black Screen, Part I: Kirchhoff's Theory, F. KOTTLER	281–314

VOLUME V (1966)

I	Optical Pumping, C. COHEN-TANNOUDJI, A. KASTLER	1– 81
II	Non-Linear Optics, P.S. PERSHAN	83–144
III	Two-Beam Interferometry, W.H. STEEL	145–197
IV	Instruments for the Measuring of Optical Transfer Functions, K. MURATA	199–245
V	Light Reflection from Films of Continuously Varying Refractive Index, R. JACOBSSON	247–286
VI	X-Ray Crystal-Structure Determination as a Branch of Physical Optics, H. LIPSON, C.A. TAYLOR	287–350
VII	The Wave of a Moving Classical Electron, J. PICHT	351–370

VOLUME VI (1967)

I	Recent Advances in Holography, E.N. LEITH, J. UPATNIEKS	1– 52
II	Scattering of Light by Rough Surfaces, P. BECKMANN	53– 69
III	Measurement of the Second Order Degree of Coherence, M. FRANÇON, S. MALLICK	71–104
IV	Design of Zoom Lenses, K. YAMAJI	105–170
V	Some Applications of Lasers to Interferometry, D.R. HERRIOT	171–209
VI	Experimental Studies of Intensity Fluctuations in Lasers, J.A. ARMSTRONG, A.W. SMITH	211–257
VII	Fourier Spectroscopy, G.A. VANASSE, H. SAKAI	259–330
VIII	Diffraction at a Black Screen, Part II: Electromagnetic Theory, F. KOTTLER	331–377

VOLUME VII (1969)

I	Multiple-Beam Interference and Natural Modes in Open Resonators, G. KOPPELMAN	1– 66
II	Methods of Synthesis for Dielectric Multilayer Filters, E. DELANO, R.J. PEGIS	67–137
III	Echoes at Optical Frequencies, I.D. ABELLA	139–168
IV	Image Formation with Partially Coherent Light, B.J. THOMPSON	169–230
V	Quasi-Classical Theory of Laser Radiation, A.L. MIKAELIAN, M.L. TER-MIKAELIAN	231–297
VI	The Photographic Image, S. OOUE	299–358
VII	Interaction of Very Intense Light with Free Electrons, J.H. EBERLY	359–415

VOLUME VIII (1970)

I	Synthetic-Aperture Optics, J.W. GOODMAN	1– 50
II	The Optical Performance of the Human Eye, G.A. FRY	51–131
III	Light Beating Spectroscopy, H.Z. CUMMINS, H.L. SWINNEY	133–200
IV	Multilayer Antireflection Coatings, A. MUSSET, A. THELEN	201–237
V	Statistical Properties of Laser Light, H. RISKEN	239–294
VI	Coherence Theory of Source-Size Compensation in Interference Microscopy, T. YAMAMOTO	295–341
VII	Vision in Communication, L. LEVI	343–372
VIII	Theory of Photoelectron Counting, C.L. MEHTA	373–440

VOLUME IX (1971)

I	Gas Lasers and their Application to Precise Length Measurements, A.L. BLOOM	1– 30
II	Picosecond Laser Pulses, A.J. DEMARIA	31– 71
III	Optical Propagation Through the Turbulent Atmosphere, J.W. STROHBEHN	73–122
IV	Synthesis of Optical Birefringent Networks, E.O. AMMANN	123–177

V	Mode Locking in Gas Lasers, L. ALLEN, D.G.C. JONES	179–234
VI	Crystal Optics with Spatial Dispersion, V.M. AGRANOVICH, V.L. GINZBURG	235–280
VII	Applications of Optical Methods in the Diffraction Theory of Elastic Waves, K. GNIADEK, J. PETYKIEWICZ	281–310
VIII	Evaluation, Design and Extrapolation Methods for Optical Signals, Based on Use of the Prolate Functions, B.R. FRIEDEN	311–407

VOLUME X (1972)

I	Bandwidth Compression of Optical Images, T.S. HUANG	1– 44
II	The Use of Image Tubes as Shutters, R.W. SMITH	45– 87
III	Tools of Theoretical Quantum Optics, M.O. SCULLY, K.G. WHITNEY	89–135
IV	Field Correctors for Astronomical Telescopes, C.G. WYNNE	137–164
V	Optical Absorption Strength of Defects in Insulators, D.Y. SMITH, D.L. DEXTER	165–228
VI	Elastooptic Light Modulation and Deflection, E.K. SITTIG	229–288
VII	Quantum Detection Theory, C.W. HELSTROM	289–369

VOLUME XI (1973)

I	Master Equation Methods in Quantum Optics, G.S. AGARWAL	1– 76
II	Recent Developments in Far Infrared Spectroscopic Techniques, H. YOSHINAGA	77–122
III	Interaction of Light and Acoustic Surface Waves, E.G. LEAN	123–166
IV	Evanescent Waves in Optical Imaging, O. BRYNGDAHL	167–221
V	Production of Electron Probes Using a Field Emission Source, A.V. CREWE	223–246
VI	Hamiltonian Theory of Beam Mode Propagation, J.A. ARNAUD	247–304
VII	Gradient Index Lenses, E.W. MARCHAND	305–337

VOLUME XII (1974)

I	Self-Focusing, Self-Trapping, and Self-Phase Modulation of Laser Beams, O. SVELTO	1– 51
II	Self-Induced Transparency, R.E. SLUSHER	53–100
III	Modulation Techniques in Spectrometry, M. HARWIT, J.A. DECKER JR	101–162
IV	Interaction of Light with Monomolecular Dye Layers, K.H. DREXHAGE	163–232
V	The Phase Transition Concept and Coherence in Atomic Emission, R. GRAHAM	233–286
VI	Beam-Foil Spectroscopy, S. BASHKIN	287–344

VOLUME XIII (1976)

I	On the Validity of Kirchhoff's Law of Heat Radiation for a Body in a Nonequilibrium Environment, H.P. BALTES	1– 25
II	The Case For and Against Semiclassical Radiation Theory, L. MANDEL	27– 68
III	Objective and Subjective Spherical Aberration Measurements of the Human Eye, W.M. ROSENBLUM, J.L. CHRISTENSEN	69– 91
IV	Interferometric Testing of Smooth Surfaces, G. SCHULZ, J. SCHWIDER	93–167
V	Self-Focusing of Laser Beams in Plasmas and Semiconductors, M.S. SODHA, A.K. GHATAK, V.K. TRIPATHI	169–265
VI	Aplanatism and Isoplanatism, W.T. WELFORD	267–292

VOLUME XIV (1976)

I	The Statistics of Speckle Patterns, J.C. DAINTY	1– 46
II	High-Resolution Techniques in Optical Astronomy, A. LABEYRIE	47– 87
III	Relaxation Phenomena in Rare-Earth Luminescence, L.A. RISEBERG, M.J. WEBER	89–159
IV	The Ultrafast Optical Kerr Shutter, M.A. DUGUAY	161–193
V	Holographic Diffraction Gratings, G. SCHMAHL, D. RUDOLPH	195–244
VI	Photoemission, P.J. VERNIER	245–325
VII	Optical Fibre Waveguides – A Review, P.J.B. CLARRICOATS	327–402

VOLUME XV (1977)

I	Theory of Optical Parametric Amplification and Oscillation, W. BRUNNER, H. PAUL	1– 75
II	Optical Properties of Thin Metal Films, P. ROUARD, A. MEESSEN	77–137
III	Projection-Type Holography, T. OKOSHI	139–185
IV	Quasi-Optical Techniques of Radio Astronomy, T.W. COLE	187–244
V	Foundations of the Macroscopic Electromagnetic Theory of Dielectric Media, J. VAN KRANENDONK, J.E. SIPE	245–350

VOLUME XVI (1978)

I	Laser Selective Photophysics and Photochemistry, V.S. LETOKHOV	1– 69
II	Recent Advances in Phase Profiles Generation, J.J. CLAIR, C.I. ABITBOL	71–117
III	Computer-Generated Holograms: Techniques and Applications, W.-H. LEE	119–232
IV	Speckle Interferometry, A.E. ENNOS	233–288
V	Deformation Invariant, Space-Variant Optical Pattern Recognition, D. CASASENT, D. PSALTIS	289–356
VI	Light Emission From High-Current Surface-Spark Discharges, R.E. BEVERLY III	357–411
VII	Semiclassical Radiation Theory Within a Quantum-Mechanical Framework, I.R. SENITZKY	413–448

VOLUME XVII (1980)

I	Heterodyne Holographic Interferometry, R. DÄNDLIKER	1– 84
II	Doppler-Free Multiphoton Spectroscopy, E. GIACOBINO, B. CAGNAC	85–161
III	The Mutual Dependence Between Coherence Properties of Light and Nonlinear Optical Processes, M. SCHUBERT, B. WILHELMI	163–238
IV	Michelson Stellar Interferometry, W.J. TANGO, R.Q. TWISS	239–277
V	Self-Focusing Media with Variable Index of Refraction, A.L. MIKAELIAN	279–345

VOLUME XVIII (1980)

I	Graded Index Optical Waveguides: A Review, A. GHATAK, K. THYAGARAJAN	1–126
II	Photocount Statistics of Radiation Propagating Through Random and Nonlinear Media, J. PEŘINA	127–203
III	Strong Fluctuations in Light Propagation in a Randomly Inhomogeneous Medium, V.I. TATARSKII, V.U. ZAVOROTNYI	204–256
IV	Catastrophe Optics: Morphologies of Caustics and their Diffraction Patterns, M.V. BERRY, C. UPSTILL	257–346

VOLUME XIX (1981)

I	Theory of Intensity Dependent Resonance Light Scattering and Resonance Fluorescence, B.R. MOLLOW	1– 43
II	Surface and Size Effects on the Light Scattering Spectra of Solids, D.L. MILLS, K.R. SUBBASWAMY	45–137
III	Light Scattering Spectroscopy of Surface Electromagnetic Waves in Solids, S. USHIODA	139–210
IV	Principles of Optical Data-Processing, H.J. BUTTERWECK	211–280
V	The Effects of Atmospheric Turbulence in Optical Astronomy, F. RODDIER	281–376

VOLUME XX (1983)

I	Some New Optical Designs for Ultra-Violet Bidimensional Detection of Astronomical Objects, G. COURTÈS, P. CRUVELLIER, M. DETAILLE, M. SAÏSSE	1– 61
II	Shaping and Analysis of Picosecond Light Pulses, C. FROEHLY, B. COLOMBEAU, M. VAMPOUILLE	63–153
III	Multi-Photon Scattering Molecular Spectroscopy, S. KIELICH	155–261
IV	Colour Holography, P. HARIHARAN	263–324
V	Generation of Tunable Coherent Vacuum-Ultraviolet Radiation, W. JAMROZ, B.P. STOICHEFF	325–380

VOLUME XXI (1984)

I	Rigorous Vector Theories of Diffraction Gratings, D. MAYSTRE	1– 67
II	Theory of Optical Bistability, L.A. LUGIATO	69–216
III	The Radon Transform and its Applications, H.H. BARRETT	217–286
IV	Zone Plate Coded Imaging: Theory and Applications, N.M. CEGLIO, D.W. SWEENEY	287–354
V	Fluctuations, Instabilities and Chaos in the Laser-Driven Nonlinear Ring Cavity, J.C. ENGLUND, R.R. SNAPP, W.C. SCHIEVE	355–428

VOLUME XXII (1985)

I	Optical and Electronic Processing of Medical Images, D. MALACARA	1– 76
II	Quantum Fluctuations in Vision, M.A. BOUMAN, W.A. VAN DE GRIND, P. ZUIDEMA	77–144
III	Spectral and Temporal Fluctuations of Broad-Band Laser Radiation, A.V. MASALOV	145–196
IV	Holographic Methods of Plasma Diagnostics, G.V. OSTROVSKAYA, YU.I. OSTROVSKY	197–270
V	Fringe Formations in Deformation and Vibration Measurements using Laser Light, I. YAMAGUCHI	271–340
VI	Wave Propagation in Random Media: A Systems Approach, R.L. FANTE	341–398

VOLUME XXIII (1986)

I	Analytical Techniques for Multiple Scattering from Rough Surfaces, J.A. DESANTO, G.S. BROWN	1– 62
II	Paraxial Theory in Optical Design in Terms of Gaussian Brackets, K. TANAKA	63–111
III	Optical Films Produced by Ion-Based Techniques, P.J. MARTIN, R.P. NETTERFIELD	113–182
IV	Electron Holography, A. TONOMURA	183–220
V	Principles of Optical Processing with Partially Coherent Light, F.T.S. YU	221–275

VOLUME XXIV (1987)

I	Micro Fresnel Lenses, H. Nishihara, T. Suhara	1– 37
II	Dephasing-Induced Coherent Phenomena, L. Rothberg	39–101
III	Interferometry with Lasers, P. Hariharan	103–164
IV	Unstable Resonator Modes, K.E. Oughstun	165–387
V	Information Processing with Spatially Incoherent Light, I. Glaser	389–509

VOLUME XXV (1988)

I	Dynamical Instabilities and Pulsations in Lasers, N.B. Abraham, P. Mandel, L.M. Narducci	1–190
II	Coherence in Semiconductor Lasers, M. Ohtsu, T. Tako	191–278
III	Principles and Design of Optical Arrays, Wang Shaomin, L. Ronchi	279–348
IV	Aspheric Surfaces, G. Schulz	349–415

VOLUME XXVI (1988)

I	Photon Bunching and Antibunching, M.C. Teich, B.E.A. Saleh	1–104
II	Nonlinear Optics of Liquid Crystals, I.C. Khoo	105–161
III	Single-Longitudinal-Mode Semiconductor Lasers, G.P. Agrawal	163–225
IV	Rays and Caustics as Physical Objects, Yu.A. Kravtsov	227–348
V	Phase-Measurement Interferometry Techniques, K. Creath	349–393

VOLUME XXVII (1989)

I	The Self-Imaging Phenomenon and Its Applications, K. Patorski	1–108
II	Axicons and Meso-Optical Imaging Devices, L.M. Soroko	109–160
III	Nonimaging Optics for Flux Concentration, I.M. Bassett, W.T. Welford, R. Winston	161–226
IV	Nonlinear Wave Propagation in Planar Structures, D. Mihalache, M. Bertolotti, C. Sibilia	227–313
V	Generalized Holography with Application to Inverse Scattering and Inverse Source Problems, R.P. Porter	315–397

VOLUME XXVIII (1990)

I	Digital Holography – Computer-Generated Holograms, O. Bryngdahl, F. Wyrowski	1– 86
II	Quantum Mechanical Limit in Optical Precision Measurement and Communication, Y. Yamamoto, S. Machida, S. Saito, N. Imoto, T. Yanagawa, M. Kitagawa, G. Björk	87–179
III	The Quantum Coherence Properties of Stimulated Raman Scattering, M.G. Raymer, I.A. Walmsley	181–270
IV	Advanced Evaluation Techniques in Interferometry, J. Schwider	271–359
V	Quantum Jumps, R.J. Cook	361–416

VOLUME XXIX (1991)

I	Optical Waveguide Diffraction Gratings: Coupling between Guided Modes, D.G. HALL	1– 63
II	Enhanced Backscattering in Optics, YU.N. BARABANENKOV, YU.A. KRAVTSOV, V.D. OZRIN, A.I. SAICHEV	65–197
III	Generation and Propagation of Ultrashort Optical Pulses, I.P. CHRISTOV	199–291
IV	Triple-Correlation Imaging in Optical Astronomy, G. WEIGELT	293–319
V	Nonlinear Optics in Composite Materials. 1. Semiconductor and Metal Crystallites in Dielectrics, C. FLYTZANIS, F. HACHE, M.C. KLEIN, D. RICARD, PH. ROUSSIGNOL	321–411

VOLUME XXX (1992)

I	Quantum Fluctuations in Optical Systems, S. REYNAUD, A. HEIDMANN, E. GIACOBINO, C. FABRE	1– 85
II	Correlation Holographic and Speckle Interferometry, YU.I. OSTROVSKY, V.P. SHCHEPINOV	87–135
III	Localization of Waves in Media with One-Dimensional Disorder, V.D. FREILIKHER, S.A. GREDESKUL	137–203
IV	Theoretical Foundation of Optical-Soliton Concept in Fibers, Y. KODAMA, A. HASEGAWA	205–259
V	Cavity Quantum Optics and the Quantum Measurement Process., P. MEYSTRE	261–355

VOLUME XXXI (1993)

I	Atoms in Strong Fields: Photoionization and Chaos, P.W. MILONNI, B. SUNDARAM	1–137
II	Light Diffraction by Relief Gratings: A Macroscopic and Microscopic View, E. POPOV	139–187
III	Optical Amplifiers, N.K. DUTTA, J.R. SIMPSON	189–226
IV	Adaptive Multilayer Optical Networks, D. PSALTIS, Y. QIAO	227–261
V	Optical Atoms, R.J.C. SPREEUW, J.P. WOERDMAN	263–319
VI	Theory of Compton Free Electron Lasers, G. DATTOLI, L. GIANNESSI, A. RENIERI, A. TORRE	321–412

VOLUME XXXII (1993)

I	Guided-Wave Optics on Silicon: Physics, Technology and Status, B.P. PAL	1– 59
II	Optical Neural Networks: Architecture, Design and Models, F.T.S. YU	61–144
III	The Theory of Optimal Methods for Localization of Objects in Pictures, L.P. YAROSLAVSKY	145–201
IV	Wave Propagation Theories in Random Media Based on the Path-Integral Approach, M.I. CHARNOTSKII, J. GOZANI, V.I. TATARSKII, V.U. ZAVOROTNY	203–266
V	Radiation by Uniformly Moving Sources. Vavilov–Cherenkov effect, Doppler effect in a medium, transition radiation and associated phenomena, V.L. GINZBURG	267–312
VI	Nonlinear Processes in Atoms and in Weakly Relativistic Plasmas, G. MAINFRAY, C. MANUS	313–361

VOLUME XXXIII (1994)

I	The Imbedding Method in Statistical Boundary-Value Wave Problems, V.I. KLYATSKIN	1–127
II	Quantum Statistics of Dissipative Nonlinear Oscillators, V. PEŘINOVÁ, A. LUKŠ	129–202
III	Gap Solitons, C.M. DE STERKE, J.E. SIPE	203–260
IV	Direct Spatial Reconstruction of Optical Phase from Phase-Modulated Images, V.I. VLAD, D. MALACARA	261–317
V	Imaging through Turbulence in the Atmosphere, M.J. BERAN, J. OZ-VOGT	319–388
VI	Digital Halftoning: Synthesis of Binary Images, O. BRYNGDAHL, T. SCHEERMESSER, F. WYROWSKI	389–463

VOLUME XXXIV (1995)

I	Quantum Interference, Superposition States of Light, and Nonclassical Effects, V. BUŽEK, P.L. KNIGHT	1–158
II	Wave Propagation in Inhomogeneous Media: Phase-Shift Approach, L.P. PRESNYAKOV	159–181
III	The Statistics of Dynamic Speckles, T. OKAMOTO, T. ASAKURA	183–248
IV	Scattering of Light from Multilayer Systems with Rough Boundaries, I. OHLÍDAL, K. NAVRÁTIL, M. OHLÍDAL	249–331
V	Random Walk and Diffusion-Like Models of Photon Migration in Turbid Media, A.H. GANDJBAKHCHE, G.H. WEISS	333–402

VOLUME XXXV (1996)

I	Transverse Patterns in Wide-Aperture Nonlinear Optical Systems, N.N. ROSANOV	1–60
II	Optical Spectroscopy of Single Molecules in Solids, M. ORRIT, J. BERNARD, R. BROWN, B. LOUNIS	61–144
III	Interferometric Multispectral Imaging, K. ITOH	145–196
IV	Interferometric Methods for Artwork Diagnostics, D. PAOLETTI, G. SCHIRRIPA SPAGNOLO	197–255
V	Coherent Population Trapping in Laser Spectroscopy, E. ARIMONDO	257–354
VI	Quantum Phase Properties of Nonlinear Optical Phenomena, R. TANAŚ, A. MIRANOWICZ, TS. GANTSOG	355–446

CUMULATIVE INDEX – VOLUMES I–XXXVI

ABELÈS, F., Methods for Determining Optical Parameters of Thin Films	II, 249
ABELLA, I.D., Echoes at Optical Frequencies	VII, 139
ABITBOL, C.I., *see* Clair, J.J.	XVI, 71
ABRAHAM, N.B., P. MANDEL, L.M. NARDUCCI, Dynamical Instabilities and Pulsations in Lasers	XXV, 1
AGARWAL, G.S., Master Equation Methods in Quantum Optics	XI, 1
AGRANOVICH, V.M., V.L. GINZBURG, Crystal Optics with Spatial Dispersion	IX, 235
AGRAWAL, G.P., Single-Longitudinal-Mode Semiconductor Lasers	XXVI, 163
ALLEN, L., D.G.C. JONES, Mode Locking in Gas Lasers	IX, 179
AMMANN, E.O., Synthesis of Optical Birefringent Networks	IX, 123
APRESYAN, L.A., *see* Kravtsov, Yu.A.	XXXVI, 179
ARIMONDO, E., Coherent Population Trapping in Laser Spectroscopy	XXXV, 257
ARMSTRONG, J.A., A.W. SMITH, Experimental Studies of Intensity Fluctuations in Lasers	VI, 211
ARNAUD, J.A., Hamiltonian Theory of Beam Mode Propagation	XI, 247
ASAKURA, T., *see* Okamoto, T.	XXXIV, 183
BALTES, H.P., On the Validity of Kirchhoff's Law of Heat Radiation for a Body in a Nonequilibrium Environment	XIII, 1
BARABANENKOV, YU.N., YU.A. KRAVTSOV, V.D. OZRIN, A.I. SAICHEV, Enhanced Backscattering in Optics	XXIX, 65
BARAKAT, R., The Intensity Distribution and Total Illumination of Aberration-Free Diffraction Images	I, 67
BARRETT, H.H., The Radon Transform and its Applications	XXI, 217
BASHKIN, S., Beam-Foil Spectroscopy	XII, 287
BASSETT, I.M., W.T. WELFORD, R. WINSTON, Nonimaging Optics for Flux Concentration	XXVII, 161
BECKMANN, P., Scattering of Light by Rough Surfaces	VI, 53
BERAN, M.J., J. OZ-VOGT, Imaging through Turbulence in the Atmosphere	XXXIII, 319
BERNARD, J., *see* Orrit, M.	XXXV, 61
BERRY, M.V., C. UPSTILL, Catastrophe Optics: Morphologies of Caustics and their Diffraction Patterns	XVIII, 257
BERTERO, M., C. DE MOL, Super-resolution by Data Inversion	XXXVI, 129
BERTOLOTTI, M., *see* Mihalache, D.	XXVII, 227
BERTOLOTTI, M., *see* Chumash, V.	XXXVI, 1
BEVERLY III, R.E., Light Emission From High-Current Surface-Spark Discharges	XVI, 357
BIALYNICKI-BIRULA, I., Photon Wave Function	XXXVI, 245
BJÖRK, G., *see* Yamamoto, Y.	XXVIII, 87
BLOOM, A.L., Gas Lasers and their Application to Precise Length Measurements	IX, 1
BOUMAN, M.A., W.A. VAN DE GRIND, P. ZUIDEMA, Quantum Fluctuations in Vision	XXII, 77

BOUSQUET, P., see Rouard, P. IV, 145
BROWN, G.S., see DeSanto, J.A. XXIII, 1
BROWN, R., see Orrit, M. XXXV, 61
BRUNNER, W., H. PAUL, Theory of Optical Parametric Amplification and Oscillation XV, 1
BRYNGDAHL, O., Applications of Shearing Interferometry IV, 37
BRYNGDAHL, O., Evanescent Waves in Optical Imaging XI, 167
BRYNGDAHL, O., F. WYROWSKI, Digital Holography – Computer-Generated Holograms XXVIII, 1
BRYNGDAHL, O., T. SCHEERMESSER, F. WYROWSKI, Digital Halftoning: Synthesis of Binary Images XXXIII, 389
BURCH, J.M., The Metrological Applications of Diffraction Gratings II, 73
BUTTERWECK, H.J., Principles of Optical Data-Processing XIX, 211
BUŽEK, V., P.L. KNIGHT, Quantum Interference, Superposition States of Light, and Nonclassical Effects XXXIV, 1

CAGNAC, B., see Giacobino, E. XVII, 85
CASASENT, D., D. PSALTIS, Deformation Invariant, Space-Variant Optical Pattern Recognition XVI, 289
CEGLIO, N.M., D.W. SWEENEY, Zone Plate Coded Imaging: Theory and Applications XXI, 287
CHARNOTSKII, M.I., J. GOZANI, V.I. TATARSKII, V.U. ZAVOROTNY, Wave Propagation Theories in Random Media Based on the Path-Integral Approach XXXII, 203
CHRISTENSEN, J.L., see Rosenblum, W.M. XIII, 69
CHRISTOV, I.P., Generation and Propagation of Ultrashort Optical Pulses XXIX, 199
CHUMASH, V., I. COJOCARU, E. FAZIO, F. MICHELOTTI, M. BERTOLOTTI, Nonlinear Propagation of Strong Laser Pulses in Chalcogenide Glass Films XXXVI, 1
CLAIR, J.J., C.I. ABITBOL, Recent Advances in Phase Profiles Generation XVI, 71
CLARRICOATS, P.J.B., Optical Fibre Waveguides – A Review XIV, 327
COHEN-TANNOUDJI, C., A. KASTLER, Optical Pumping V, 1
COJOCARU, I., see Chumash, V. XXXVI, 1
COLE, T.W., Quasi-Optical Techniques of Radio Astronomy XV, 187
COLOMBEAU, B., see Froehly, C. XX, 63
COOK, R.J., Quantum Jumps XXVIII, 361
COURTÈS, G., P. CRUVELLIER, M. DETAILLE, M. SAÏSSE, Some New Optical Designs for Ultra-Violet Bidimensional Detection of Astronomical Objects XX, 1
CREATH, K., Phase-Measurement Interferometry Techniques XXVI, 349
CREWE, A.V., Production of Electron Probes Using a Field Emission Source XI, 223
CRUVELLIER, P., see Courtès, G. XX, 1
CUMMINS, H.Z., H.L. SWINNEY, Light Beating Spectroscopy VIII, 133

DAINTY, J.C., The Statistics of Speckle Patterns XIV, 1
DÄNDLIKER, R., Heterodyne Holographic Interferometry XVII, 1
DATTOLI, G., L. GIANNESSI, A. RENIERI, A. TORRE, Theory of Compton Free Electron Lasers XXXI, 321
DE MOL, C., see Bertero, M. XXXVI, 129
DE STERKE, C.M., J.E. SIPE, Gap Solitons XXXIII, 203
DECKER JR, J.A., see Harwit, M. XII, 101
DELANO, E., R.J. PEGIS, Methods of Synthesis for Dielectric Multilayer Filters VII, 67
DEMARIA, A.J., Picosecond Laser Pulses IX, 31
DESANTO, J.A., G.S. BROWN, Analytical Techniques for Multiple Scattering from Rough Surfaces XXIII, 1
DETAILLE, M., see Courtès, G. XX, 1
DEXTER, D.L., see Smith, D.Y. X, 165

DREXHAGE, K.H., Interaction of Light with Monomolecular Dye Layers	XII, 163
DUGUAY, M.A., The Ultrafast Optical Kerr Shutter	XIV, 161
DUTTA, N.K., J.R. SIMPSON, Optical Amplifiers	XXXI, 189
EBERLY, J.H., Interaction of Very Intense Light with Free Electrons	VII, 359
ENGLUND, J.C., R.R. SNAPP, W.C. SCHIEVE, Fluctuations, Instabilities and Chaos in the Laser-Driven Nonlinear Ring Cavity	XXI, 355
ENNOS, A.E., Speckle Interferometry	XVI, 233
FABRE, C., see Reynaud, S.	XXX, 1
FANTE, R.L., Wave Propagation in Random Media: A Systems Approach	XXII, 341
FAZIO, E., see Chumash, V.	XXXVI, 1
FIORENTINI, A., Dynamic Characteristics of Visual Processes	I, 253
FLYTZANIS, C., F. HACHE, M.C. KLEIN, D. RICARD, PH. ROUSSIGNOL, Nonlinear Optics in Composite Materials. 1. Semiconductor and Metal Crystallites in Dielectrics	XXIX, 321
FOCKE, J., Higher Order Aberration Theory	IV, 1
FRANÇON, M., S. MALLICK, Measurement of the Second Order Degree of Coherence	VI, 71
FREILIKHER, V.D., S.A. GREDESKUL, Localization of Waves in Media with One-Dimensional Disorder	XXX, 137
FRIEDEN, B.R., Evaluation, Design and Extrapolation Methods for Optical Signals, Based on Use of the Prolate Functions	IX, 311
FROEHLY, C., B. COLOMBEAU, M. VAMPOUILLE, Shaping and Analysis of Picosecond Light Pulses	XX, 63
FRY, G.A., The Optical Performance of the Human Eye	VIII, 51
GABOR, D., Light and Information	I, 109
GAMO, H., Matrix Treatment of Partial Coherence	III, 187
GANDJBAKHCHE, A.H., G.H. WEISS, Random Walk and Diffusion-Like Models of Photon Migration in Turbid Media	XXXIV, 333
GANTSOG, TS., see Tanaś, R.	XXXV, 355
GHATAK, A., K. THYAGARAJAN, Graded Index Optical Waveguides: A Review	XVIII, 1
GHATAK, A.K., see Sodha, M.S.	XIII, 169
GIACOBINO, E., B. CAGNAC, Doppler-Free Multiphoton Spectroscopy	XVII, 85
GIACOBINO, E., see Reynaud, S.	XXX, 1
GIANNESSI, L., see Dattoli, G.	XXXI, 321
GINZBURG, V.L., see Agranovich, V.M.	IX, 235
GINZBURG, V.L., Radiation by Uniformly Moving Sources. Vavilov–Cherenkov effect, Doppler effect in a medium, transition radiation and associated phenomena	XXXII, 267
GIOVANELLI, R.G., Diffusion Through Non-Uniform Media	II, 109
GLASER, I., Information Processing with Spatially Incoherent Light	XXIV, 389
GNIADEK, K., J. PETYKIEWICZ, Applications of Optical Methods in the Diffraction Theory of Elastic Waves	IX, 281
GOODMAN, J.W., Synthetic-Aperture Optics	VIII, 1
GOZANI, J., see Charnotskii, M.I.	XXXII, 203
GRAHAM, R., The Phase Transition Concept and Coherence in Atomic Emission	XII, 233
GREDESKUL, S.A., see Freilikher, V.D.	XXX, 137
HACHE, F., see Flytzanis, C.	XXIX, 321
HALL, D.G., Optical Waveguide Diffraction Gratings: Coupling between Guided Modes	XXIX, 1
HARIHARAN, P., Colour Holography	XX, 263

HARIHARAN, P., Interferometry with Lasers — XXIV, 103
HARIHARAN, P., B.C. SANDERS, Quantum Phenomena in Optical Interferometry — XXXVI, 49
HARWIT, M., J.A. DECKER JR, Modulation Techniques in Spectrometry — XII, 101
HASEGAWA, A., see Kodama, Y. — XXX, 205
HEIDMANN, A., see Reynaud, S. — XXX, 1
HELSTROM, C.W., Quantum Detection Theory — X, 289
HERRIOT, D.R., Some Applications of Lasers to Interferometry — VI, 171
HUANG, T.S., Bandwidth Compression of Optical Images — X, 1

IMOTO, N., see Yamamoto, Y. — XXVIII, 87
ITOH, K., Interferometric Multispectral Imaging — XXXV, 145

JACOBSSON, R., Light Reflection from Films of Continuously Varying Refractive Index — V, 247
JACQUINOT, P., B. ROIZEN-DOSSIER, Apodisation — III, 29
JAMROZ, W., B.P. STOICHEFF, Generation of Tunable Coherent Vacuum-Ultraviolet Radiation — XX, 325
JONES, D.G.C., see Allen, L. — IX, 179

KASTLER, A., see Cohen-Tannoudji, C. — V, 1
KHOO, I.C., Nonlinear Optics of Liquid Crystals — XXVI, 105
KIELICH, S., Multi-Photon Scattering Molecular Spectroscopy — XX, 155
KINOSITA, K., Surface Deterioration of Optical Glasses — IV, 85
KITAGAWA, M., see Yamamoto, Y. — XXVIII, 87
KLEIN, M.C., see Flytzanis, C. — XXIX, 321
KLYATSKIN, V.I., The Imbedding Method in Statistical Boundary-Value Wave Problems — XXXIII, 1
KNIGHT, P.L., see Bužek, V. — XXXIV, 1
KODAMA, Y., A. HASEGAWA, Theoretical Foundation of Optical-Soliton Concept in Fibers — XXX, 205
KOPPELMAN, G., Multiple-Beam Interference and Natural Modes in Open Resonators — VII, 1
KOTTLER, F., The Elements of Radiative Transfer — III, 1
KOTTLER, F., Diffraction at a Black Screen, Part I: Kirchhoff's Theory — IV, 281
KOTTLER, F., Diffraction at a Black Screen, Part II: Electromagnetic Theory — VI, 331
KRAVTSOV, YU.A., Rays and Caustics as Physical Objects — XXVI, 227
KRAVTSOV, YU.A., see Barabanenkov, Yu.N. — XXIX, 65
KRAVTSOV, YU.A., L.A. APRESYAN, Radiative Transfer: New Aspects of the Old Theory — XXXVI, 179
KUBOTA, H., Interference Color — I, 211

LABEYRIE, A., High-Resolution Techniques in Optical Astronomy — XIV, 47
LEAN, E.G., Interaction of Light and Acoustic Surface Waves — XI, 123
LEE, W.-H., Computer-Generated Holograms: Techniques and Applications — XVI, 119
LEITH, E.N., J. UPATNIEKS, Recent Advances in Holography — VI, 1
LETOKHOV, V.S., Laser Selective Photophysics and Photochemistry — XVI, 1
LEVI, L., Vision in Communication — VIII, 343
LIPSON, H., C.A. TAYLOR, X-Ray Crystal-Structure Determination as a Branch of Physical Optics — V, 287
LOUNIS, B., see Orrit, M. — XXXV, 61
LUGIATO, L.A., Theory of Optical Bistability — XXI, 69
LUKŠ, A., see Peřinová, V. — XXXIII, 129

MACHIDA, S., see Yamamoto, Y. — XXVIII, 87
MAINFRAY, G., C. MANUS, Nonlinear Processes in Atoms and in Weakly Relativistic Plasmas — XXXII, 313

MALACARA, D., Optical and Electronic Processing of Medical Images	XXII, 1
MALACARA, D., see Vlad, V.I.	XXXIII, 261
MALLICK, S., see Françon, M.	VI, 71
MANDEL, L., Fluctuations of Light Beams	II, 181
MANDEL, L., The Case For and Against Semiclassical Radiation Theory	XIII, 27
MANDEL, P., see Abraham, N.B.	XXV, 1
MANUS, C., see Mainfray, G.	XXXII, 313
MARCHAND, E.W., Gradient Index Lenses	XI, 305
MARTIN, P.J., R.P. NETTERFIELD, Optical Films Produced by Ion-Based Techniques	XXIII, 113
MASALOV, A.V., Spectral and Temporal Fluctuations of Broad-Band Laser Radiation	XXII, 145
MAYSTRE, D., Rigorous Vector Theories of Diffraction Gratings	XXI, 1
MEESSEN, A., see Rouard, P.	XV, 77
MEHTA, C.L., Theory of Photoelectron Counting	VIII, 373
MEYSTRE, P., Cavity Quantum Optics and the Quantum Measurement Process.	XXX, 261
MICHELOTTI, F., see Chumash, V.	XXXVI, 1
MIHALACHE, D., M. BERTOLOTTI, C. SIBILIA, Nonlinear Wave Propagation in Planar Structures	XXVII, 227
MIKAELIAN, A.L., M.L. TER-MIKAELIAN, Quasi-Classical Theory of Laser Radiation	VII, 231
MIKAELIAN, A.L., Self-Focusing Media with Variable Index of Refraction	XVII, 279
MILLS, D.L., K.R. SUBBASWAMY, Surface and Size Effects on the Light Scattering Spectra of Solids	XIX, 45
MILONNI, P.W., B. SUNDARAM, Atoms in Strong Fields: Photoionization and Chaos	XXXI, 1
MIRANOWICZ, A., see Tanaś, R.	XXXV, 355
MIYAMOTO, K., Wave Optics and Geometrical Optics in Optical Design	I, 31
MOLLOW, B.R., Theory of Intensity Dependent Resonance Light Scattering and Resonance Fluorescence	XIX, 1
MURATA, K., Instruments for the Measuring of Optical Transfer Functions	V, 199
MUSSET, A., A. THELEN, Multilayer Antireflection Coatings	VIII, 201
NARDUCCI, L.M., see Abraham, N.B.	XXV, 1
NAVRÁTIL, K., see Ohlídal, I.	XXXIV, 249
NETTERFIELD, R.P., see Martin, P.J.	XXIII, 113
NISHIHARA, H., T. SUHARA, Micro Fresnel Lenses	XXIV, 1
OHLÍDAL, I., K. NAVRÁTIL, M. OHLÍDAL, Scattering of Light from Multilayer Systems with Rough Boundaries	XXXIV, 249
OHLÍDAL, M., see Ohlídal, I.	XXXIV, 249
OHTSU, M., T. TAKO, Coherence in Semiconductor Lasers	XXV, 191
OKAMOTO, T., T. ASAKURA, The Statistics of Dynamic Speckles	XXXIV, 183
OKOSHI, T., Projection-Type Holography	XV, 139
OOUE, S., The Photographic Image	VII, 299
ORRIT, M., J. BERNARD, R. BROWN, B. LOUNIS, Optical Spectroscopy of Single Molecules in Solids	XXXV, 61
OSTROVSKAYA, G.V., YU.I. OSTROVSKY, Holographic Methods of Plasma Diagnostics	XXII, 197
OSTROVSKY, YU.I., see Ostrovskaya, G.V.	XXII, 197
OSTROVSKY, YU.I., V.P. SHCHEPINOV, Correlation Holographic and Speckle Interferometry	XXX, 87
OUGHSTUN, K.E., Unstable Resonator Modes	XXIV, 165
OZ-VOGT, J., see Beran, M.J.	XXXIII, 319
OZRIN, V.D., see Barabanenkov, Yu.N.	XXIX, 65

PAL, B.P., Guided-Wave Optics on Silicon: Physics, Technology and Status — XXXII, 1
PAOLETTI, D., G. SCHIRRIPA SPAGNOLO, Interferometric Methods for Artwork Diagnostics — XXXV, 197
PATORSKI, K., The Self-Imaging Phenomenon and Its Applications — XXVII, 1
PAUL, H., see Brunner, W. — XV, 1
PEGIS, R.J., The Modern Development of Hamiltonian Optics — I, 1
PEGIS, R.J., see Delano, E. — VII, 67
PEŘINA, J., Photocount Statistics of Radiation Propagating Through Random and Nonlinear Media — XVIII, 127
PEŘINOVÁ, V., A. LUKŠ, Quantum Statistics of Dissipative Nonlinear Oscillators — XXXIII, 129
PERSHAN, P.S., Non-Linear Optics — V, 83
PETYKIEWICZ, J., see Gniadek, K. — IX, 281
PICHT, J., The Wave of a Moving Classical Electron — V, 351
POPOV, E., Light Diffraction by Relief Gratings: A Macroscopic and Microscopic View — XXXI, 139
PORTER, R.P., Generalized Holography with Application to Inverse Scattering and Inverse Source Problems — XXVII, 315
PRESNYAKOV, L.P., Wave Propagation in Inhomogeneous Media: Phase-Shift Approach — XXXIV, 159
PSALTIS, D., see Casasent, D. — XVI, 289
PSALTIS, D., Y. QIAO, Adaptive Multilayer Optical Networks — XXXI, 227

QIAO, Y., see Psaltis, D. — XXXI, 227

RAYMER, M.G., I.A. WALMSLEY, The Quantum Coherence Properties of Stimulated Raman Scattering — XXVIII, 181
RENIERI, A., see Dattoli, G. — XXXI, 321
REYNAUD, S., A. HEIDMANN, E. GIACOBINO, C. FABRE, Quantum Fluctuations in Optical Systems — XXX, 1
RICARD, D., see Flytzanis, C. — XXIX, 321
RISEBERG, L.A., M.J. WEBER, Relaxation Phenomena in Rare-Earth Luminescence — XIV, 89
RISKEN, H., Statistical Properties of Laser Light — VIII, 239
RODDIER, F., The Effects of Atmospheric Turbulence in Optical Astronomy — XIX, 281
ROIZEN-DOSSIER, B., see Jacquinot, P. — III, 29
RONCHI, L., see Wang Shaomin — XXV, 279
ROSANOV, N.N., Transverse Patterns in Wide-Aperture Nonlinear Optical Systems — XXXV, 1
ROSENBLUM, W.M., J.L. CHRISTENSEN, Objective and Subjective Spherical Aberration Measurements of the Human Eye — XIII, 69
ROTHBERG, L., Dephasing-Induced Coherent Phenomena — XXIV, 39
ROUARD, P., P. BOUSQUET, Optical Constants of Thin Films — IV, 145
ROUARD, P., A. MEESSEN, Optical Properties of Thin Metal Films — XV, 77
ROUSSIGNOL, PH., see Flytzanis, C. — XXIX, 321
RUBINOWICZ, A., The Miyamoto–Wolf Diffraction Wave — IV, 199
RUDOLPH, D., see Schmahl, G. — XIV, 195

SAICHEV, A.I., see Barabanenkov, Yu.N. — XXIX, 65
SAÏSSE, M., see Courtès, G. — XX, 1
SAITO, S., see Yamamoto, Y. — XXVIII, 87
SAKAI, H., see Vanasse, G.A. — VI, 259
SALEH, B.E.A., see Teich, M.C. — XXVI, 1
SANDERS, B.C., see Hariharan, P. — XXXVI, 49
SCHEERMESSER, T., see Bryngdahl, O. — XXXIII, 389

SCHIEVE, W.C., see Englund, J.C.	XXI, 355
SCHIRRIPA SPAGNOLO, G., see Paoletti, D.	XXXV, 197
SCHMAHL, G., D. RUDOLPH, Holographic Diffraction Gratings	XIV, 195
SCHUBERT, M., B. WILHELMI, The Mutual Dependence Between Coherence Properties of Light and Nonlinear Optical Processes	XVII, 163
SCHULZ, G., J. SCHWIDER, Interferometric Testing of Smooth Surfaces	XIII, 93
SCHULZ, G., Aspheric Surfaces	XXV, 349
SCHWIDER, J., see Schulz, G.	XIII, 93
SCHWIDER, J., Advanced Evaluation Techniques in Interferometry	XXVIII, 271
SCULLY, M.O., K.G. WHITNEY, Tools of Theoretical Quantum Optics	X, 89
SENITZKY, I.R., Semiclassical Radiation Theory Within a Quantum-Mechanical Framework	XVI, 413
SHCHEPINOV, V.P., see Ostrovsky, Yu.I.	XXX, 87
SIBILIA, C., see Mihalache, D.	XXVII, 227
SIMPSON, J.R., see Dutta, N.K.	XXXI, 189
SIPE, J.E., see Van Kranendonk, J.	XV, 245
SIPE, J.E., see De Sterke, C.M.	XXXIII, 203
SITTIG, E.K., Elastooptic Light Modulation and Deflection	X, 229
SLUSHER, R.E., Self-Induced Transparency	XII, 53
SMITH, A.W., see Armstrong, J.A.	VI, 211
SMITH, D.Y., D.L. DEXTER, Optical Absorption Strength of Defects in Insulators	X, 165
SMITH, R.W., The Use of Image Tubes as Shutters	X, 45
SNAPP, R.R., see Englund, J.C.	XXI, 355
SODHA, M.S., A.K. GHATAK, V.K. TRIPATHI, Self-Focusing of Laser Beams in Plasmas and Semiconductors	XIII, 169
SOROKO, L.M., Axicons and Meso-Optical Imaging Devices	XXVII, 109
SPREEUW, R.J.C., J.P. WOERDMAN, Optical Atoms	XXXI, 263
STEEL, W.H., Two-Beam Interferometry	V, 145
STOICHEFF, B.P., see Jamroz, W.	XX, 325
STROHBEHN, J.W., Optical Propagation Through the Turbulent Atmosphere	IX, 73
STROKE, G.W., Ruling, Testing and Use of Optical Gratings for High-Resolution Spectroscopy	II, 1
SUBBASWAMY, K.R., see Mills, D.L.	XIX, 45
SUHARA, T., see Nishihara, H.	XXIV, 1
SUNDARAM, B., see Milonni, P.W.	XXXI, 1
SVELTO, O., Self-Focusing, Self-Trapping, and Self-Phase Modulation of Laser Beams	XII, 1
SWEENEY, D.W., see Ceglio, N.M.	XXI, 287
SWINNEY, H.L., see Cummins, H.Z.	VIII, 133
TAKO, T., see Ohtsu, M.	XXV, 191
TANAKA, K., Paraxial Theory in Optical Design in Terms of Gaussian Brackets	XXIII, 63
TANAŚ, R., A. MIRANOWICZ, TS. GANTSOG, Quantum Phase Properties of Nonlinear Optical Phenomena	XXXV, 355
TANGO, W.J., R.Q. TWISS, Michelson Stellar Interferometry	XVII, 239
TATARSKII, V.I., V.U. ZAVOROTNYI, Strong Fluctuations in Light Propagation in a Randomly Inhomogeneous Medium	XVIII, 204
TATARSKII, V.I., see Charnotskii, M.I.	XXXII, 203
TAYLOR, C.A., see Lipson, H.	V, 287
TEICH, M.C., B.E.A. SALEH, Photon Bunching and Antibunching	XXVI, 1
TER-MIKAELIAN, M.L., see Mikaelian, A.L.	VII, 231
THELEN, A., see Musset, A.	VIII, 201

THOMPSON, B.J., Image Formation with Partially Coherent Light — VII, 169
THYAGARAJAN, K., see Ghatak, A. — XVIII, 1
TONOMURA, A., Electron Holography — XXIII, 183
TORRE, A., see Dattoli, G. — XXXI, 321
TRIPATHI, V.K., see Sodha, M.S. — XIII, 169
TSUJIUCHI, J., Correction of Optical Images by Compensation of Aberrations and by Spatial Frequency Filtering — II, 131
TWISS, R.Q., see Tango, W.J. — XVII, 239

UPATNIEKS, J., see Leith, E.N. — VI, 1
UPSTILL, C., see Berry, M.V. — XVIII, 257
USHIODA, S., Light Scattering Spectroscopy of Surface Electromagnetic Waves in Solids — XIX, 139

VAMPOUILLE, M., see Froehly, C. — XX, 63
VAN DE GRIND, W.A., see Bouman, M.A. — XXII, 77
VAN HEEL, A.C.S., Modern Alignment Devices — I, 289
VAN KRANENDONK, J., J.E. SIPE, Foundations of the Macroscopic Electromagnetic Theory of Dielectric Media — XV, 245
VANASSE, G.A., H. SAKAI, Fourier Spectroscopy — VI, 259
VERNIER, P.J., Photoemission — XIV, 245
VLAD, V.I., D. MALACARA, Direct Spatial Reconstruction of Optical Phase from Phase-Modulated Images — XXXIII, 261

WALMSLEY, I.A., see Raymer, M.G. — XXVIII, 181
WANG SHAOMIN, L. RONCHI, Principles and Design of Optical Arrays — XXV, 279
WEBER, M.J., see Riseberg, L.A. — XIV, 89
WEIGELT, G., Triple-Correlation Imaging in Optical Astronomy — XXIX, 293
WEISS, G.H., see Gandjbakhche, A.H. — XXXIV, 333
WELFORD, W.T., Aberration Theory of Gratings and Grating Mountings — IV, 241
WELFORD, W.T., Aplanatism and Isoplanatism — XIII, 267
WELFORD, W.T., see Bassett, I.M. — XXVII, 161
WHITNEY, K.G., see Scully, M.O. — X, 89
WILHELMI, B., see Schubert, M. — XVII, 163
WINSTON, R., see Bassett, I.M. — XXVII, 161
WOERDMAN, J.P., see Spreeuw, R.J.C. — XXXI, 263
WOLTER, H., On Basic Analogies and Principal Differences between Optical and Electronic Information — I, 155
WYNNE, C.G., Field Correctors for Astronomical Telescopes — X, 137
WYROWSKI, F., see Bryngdahl, O. — XXVIII, 1
WYROWSKI, F., see Bryngdahl, O. — XXXIII, 389

YAMAGUCHI, I., Fringe Formations in Deformation and Vibration Measurements using Laser Light — XXII, 271
YAMAJI, K., Design of Zoom Lenses — VI, 105
YAMAMOTO, T., Coherence Theory of Source-Size Compensation in Interference Microscopy — VIII, 295
YAMAMOTO, Y., S. MACHIDA, S. SAITO, N. IMOTO, T. YANAGAWA, M. KITAGAWA, G. BJÖRK, Quantum Mechanical Limit in Optical Precision Measurement and Communication — XXVIII, 87
YANAGAWA, T., see Yamamoto, Y. — XXVIII, 87

YAROSLAVSKY, L.P., The Theory of Optimal Methods for Localization of Objects in
   Pictures   XXXII, 145
YOSHINAGA, H., Recent Developments in Far Infrared Spectroscopic Techniques   XI, 77
YU, F.T.S., Principles of Optical Processing with Partially Coherent Light   XXIII, 221
YU, F.T.S., Optical Neural Networks: Architecture, Design and Models   XXXII, 61

ZAVOROTNY, V.U., *see* Charnotskii, M.I.   XXXII, 203
ZAVOROTNYI, V.U., *see* Tatarskii, V.I.   XVIII, 204
ZUIDEMA, P., *see* Bouman, M.A.   XXII, 77